T0136870

21st Century Prometheus

Maurizio Martellini • Ralf Trapp
Editors

21st Century Prometheus

Managing CBRN Safety and Security Affected by Cutting-Edge Technologies

Editors
Maurizio Martellini
Università degli Studi dell'Insubria
Como, Italy

Ralf Trapp
Independent Disarmament Consultant
Chessanaz, France

ISBN 978-3-030-28287-5 ISBN 978-3-030-28285-1 (eBook)
https://doi.org/10.1007/978-3-030-28285-1

This Springer imprint is published by the registered company Springer Nature Switzerland AG
The registered company address is: Gewerbestrasse 11, 6330 Cham, Switzerland

Contents

Part II Evolving Risk Mitigation Strategies and Technologies

Acronyms and Abbreviations

ABEO	Advisory Board on Education and Outreach of the OPCW
AChE	Acetylcholinesterase
AI	Artificial Intelligence
AM	Additive Manufacturing (also “3D Printing”)
AMDIS	Automated Mass Spectral Deconvolution and Identification System
APS	Action Protection System
ASAT	Antisatellite (missile, system, etc.)
ATR	Automatic/Automated Target Recognition
AUV	Autonomous Underwater Vehicle
AWS	Autonomous Weapon Systems
BRAIN	Brain Research through Advancing Innovative Neurotechnologies (initiative of the US National Institutes of Health (NIH))
BuChE/HuBChE	Butyrylcholine Esterase/Human Butyrylcholine Esterase
BWC, BTWC	Biological Weapons Convention (also Biological and Toxin Weapons Convention)
C4ISR	Command, Control, Communications, Computers, Intelligence, Surveillance and Reconnaissance
CANDU	Canada Deuterium Uranium pressurized (a Canadian pressurized heavy-water reactor design)
CAS	Chemical Attribution Signature
CBRNe	Chemical, Biological, Radiological, Nuclear and Explosive (weapons, material, technologies)
CCW	Convention on Prohibitions or Restrictions on the Use of Certain Conventional Weapons
CGC	Cyber Grand Challenge (a cybersecurity competition hosted by DARPA)
CIWS	Close-In Weapon System
CNS	Central Nervous System

CoE	Centres of Excellence (CoE Initiative: CoE initiative of the EU to strengthen CBRN risk mitigation capacities in EU partner countries)
COTS	Commercial Off-The-Shelf (in the context of unmanned aerial vehicles – UAV)
CRISPR-Cas	Clustered Regularly Interspaced Short Palindromic Repeats (These are DNA sequences found in the genomes of prokaryotic organisms such as bacteria; they are derived from DNA related to previous viral infections and form part of the prokaryote's immune defence. The Cas proteins (CRISPR-associated proteins) are enzymes that use the CRISPR sequence to recognise and cleave specific DNA strands. CRISPR-Cas systems are being used as efficient genome editing tools – one of the first systems used to this end was CRSIPR-Cas9.)
CRS	Cyber Reasoning System
CTA	Chemical Threat Agent
CTITF	Counterterrorism Implementation Task Force (of the United Nations)
CWA	Chemical Warfare Agent
CWC	Chemical Weapons Convention
Daesh	See ISIS
DARPA	Defense Advanced Research Projects Agency (a US DOD agency)
DNA	Deoxyribonucleic acid
DOD	Department of Defense (of the USA)
ELF	Extremely Low Frequency (in communications)
EMP	Electromagnetic Pulse
EU	European Union
FAO	Food and Agriculture Organisation of the United Nations
FBI	Federal Bureau of Investigation
FFM	Fact-Finding Mission
GA	Tabun (nerve agent)
GB	Sarin (nerve agent)
GC	Gas Chromatography
GC-MS	Gas Chromatography Coupled with Mass Spectrometry
GD	Soman (nerve agent)
GF	Cyclosarin (nerve agent)
GGE	Group of Governmental Experts
GNEP	Global Nuclear Energy Partnership
GoG	Group of Governmental Experts
GPS	Global Positioning System
HBP	Human Brain Project
HCA	Hierarchical Cluster Analysis
HCM	Hypersonic Cruise Missile
HEU	Highly Enriched Uranium
HGV	Hypersonic Boost-Glide Vehicle

IADM	Improvised Air-Delivered Munition(s)
IAEA	International Atomic Energy Agency
ICA	Incapacitating Chemical Agent (also CNS-acting chemical) – a chemical agent that targets the CNS to cause “selective malfunctions of the human machine” (such as impairments of vision, cognition, state of alertness, etc.)
ICBM	Intercontinental Ballistic Missile
ICT	Information and Communications Technology
IDLH	Immediately Dangerous to Life or Health
IED	Improvised Explosive Device(s)
IEEE	Institute of Electrical and Electronics Engineers
IFNEC	International Framework for Nuclear Energy Cooperation
IHL	International Humanitarian Law
IIT	Investigation and Identification team (of the OPCW, established pursuant to the decision of the fourth Special Session of the Conference of the States Parties of the CWC to investigate and gather information to identify the individuals or parties responsible for CW uses in Syria; the decision also authorizes the OPCW to support national investigations of responsibility for CW uses)
IMS	Ion Mobility Spectrometry
IR	Infrared (Spectroscopy)
ISD	Instructional Systems Design
ISIS, ISIL, IS	Islamic State, also referred to as the Islamic State of Iraq and the Levant (ISIL), the Islamic State of Iraq and Syria (ISIS), Daesh in Arabic
ISTAR	Intelligence, Surveillance, Target Acquisition, and Reconnaissance
ISU	Implementation Support Unit (of the Biological Weapons Convention, part of the UNODA)
JCPOA	Joint Comprehensive Plan of Action (also known as Iran nuclear deal) between Iran, the P5 + 1 (the five permanent members of the UNSC – China, France, Russia, UK and the USA – plus Germany) and the European Union
JIM	Joint Investigative Mechanism (of the OPCW and the United Nations, established by the UNSC to identify those responsible for cases of CW use confirmed by the OPCW FFM in Syria)
LAWS	Lethal Autonomous Weapon System
LC	Liquid Chromatography
LCt_{50}	Median Lethal Dose by Inhalation – the dose (expressed as air concentration multiplied by exposure time) of a toxic, pathogenic or radioactive agent required to kill 50% of a test population after a specified time after acute exposure
LD_{50}	Median Lethal Dose – the dose of a toxic, pathogenic or radioactive agent required to kill 50% of a test population after a specified time after acute exposure

LEO	Low Earth Orbit
LEU	Low-Enriched Uranium
LRASM	Long Range Anti-ship Missile
LWR	Light-Water Reactor
MaRV	Manoeuvring Re-entry Vehicle
ML	Machine Learning
MS, MS/MS	Mass Spectrometry/Tandem Mass Spectrometry
MSP	(annual) Meeting of the States Party to the BWC
MTA	Material Threat Assessment
MX	Meeting of Experts (of the States party to the BWC)
NAS	National Academy of Sciences
NATO	North Atlantic Treaty Organization
NGS	Next-Generation Sequencing
NMR	Nuclear Magnetic Resonance (spectrometry)
NPT	Nuclear Non-proliferation Treaty
NSG	Nuclear Suppliers Group
NTI	Nuclear Threat Initiative
OCAD	OPCW Central Analytical Database – the OPCW's reference library of analytical data (mostly GC-MS, IR and NMR data of CWC-relevant chemicals)
OIE	World Organisation for Animal Health (its original name was Office International des Épizooties, still reflected in the organisation's abbreviation)
OODA	Observe Orient Decide Act
OPCW	Organisation for the Prohibition of Chemical Weapons
PCA	Principal Component Analysis
PCR	Polymerase Chain Reaction (a method to generate copies of specific DNA segments)
PPE	Personal Protective Equipment
R&D	Research and Development
RCA	Riot Control Agent
RCS	Reaction Control System (a spacecraft system using thrusters to provide attitude control and sometimes translation)
RDD	Radiological Dispersion Device
RNA	Ribonucleic Acid
RRI	Responsible Research and Innovation (a concept of the EU used in its framework programmes to take into account effects and potential impacts on the environment and society)
RV	Re-entry Vehicle
S&T	Science and Technology
SAB	Scientific Advisory Board (a statutory organ of the OPCW)
SAP	Source Attribution Profile
SLBM	Submarine Launched Ballistic Missile
SOP	Standard Operating Procedure
TIC	Toxic Industrial Chemical

TOF-MS	Time-of-Flight Mass Spectrometer
TRA	Terrorism Risk Assessment
TWG	Temporary Working Group (of the OPCW SAB)
UAS	Unmanned Aerial System
UAV	Unmanned Aerial Vehicle
UCLASS	Unmanned Carrier-Launched Airborne Surveillance and Strike
UGS	Unmanned Ground System
UGV	Unmanned Ground Vehicle
UN	United Nations
UNITAR	United Nations Institute for Training and Research
UNOCHA	United Nations Office for the Coordination of Humanitarian Affairs
UNODA	United Nations Office of Disarmament Affairs
UNODC	United Nations Office on Drugs and Crimes
UNOSAT	UNITAR's Operational Satellite Applications Programme; it delivers imagery analysis and satellite solutions to relief and development organisations within and outside the UN system
UNSC	United Nations Security Council
UNSGM	United Nations Secretary-General's Mechanism (to investigate reports of alleged uses of chemical, biological or toxin weapons)
UUV	Unmanned Underwater Vehicle
WHO	World Health Organization
WMD	Weapons of Mass Destruction

Introduction

Maurizio Martellini and Ralf Trapp

This book deals with the changing landscape of risks posed by chemical, biological, radiological, nuclear and explosive (CBRNe) materials and technologies. It looks, in particular, at how advances in science and technology are changing and reshaping this risk landscape. These risks range from natural events such as major disease outbreaks to industrial, transportation and other accidents leading to the release of CBRNe materials, and to their use for hostile purposes as weapons of terror or war.

At the same time, the book looks at new opportunities that these scientific and technological advances are expected to bring about for mitigating these risks, and it looks more generally at governance strategies for the twenty-first century to manage the evolving CBRNe risk landscape.

Chemical, biological, radiological, and nuclear weapons and materials remain among the most recognised security threats of today. Associated with explosive materials and devices (including improvised ones) that are used for their employment, the threat spectrum associated with these materials and technologies is often set within a broader CBRNe framework.

This CBRNe threat spectrum is typically associated with certain types of materials ("agents") – toxic chemicals and their precursors, pathogenic agents, radiological materials, nuclear isotopes suitable for the construction of nuclear devices, and explosives. However, their threat level is not merely associated with the properties of the materials themselves, but it also critically depends on their means of delivery/employment and the context and mode within which they are used. Furthermore, their acquisition requires technologies and equipment for their manufacturing or extraction from natural sources. A discussion of CBRNe threats therefore cannot be

M. Martellini (✉)
Università degli Studi dell'Insubria, Como, Italy
e-mail: maurizio.martellini@fondazionealessandrovolta.it

R. Trapp
Independent Disarmament Consultant, Chessenaz, France

M. Martellini, R. Trapp (eds.), *21st Century Prometheus*,
https://doi.org/10.1007/978-3-030-28285-1_1

limited to the agents of concern but must also include the technologies associated with their manufacturing, weaponisation and employment.

On the other hand, the threats associated with these weapons and materials will be moderated by the effectiveness of protections against their effects, as well was the degree to which such protection can be provided to troops as well as populations. In addition, science and technology can augment other capabilities that affect the threat landscape, such as the ability to detect and identify agents after release, technical means of verifying regulatory and treaty compliance, medical and other countermeasures to mitigate their effects, forensic methods to help identifying those responsible for any deliberate release/use of such material and weapons, means of recovery from their effects, and technologies and methods used to shape the information environment within which such weapons and agents are being acquired and deployed.

CBRNe threats and their mitigation can be addressed from a number of different perspectives:

- The actors associated with the development, acquisition and use of such weapons and materials (States, terrorist groups, criminal organisations, individuals), and their capabilities and intents
- The types of materials of concern (toxic chemicals including traditional chemical warfare agents and toxic industrial chemicals, natural toxins and other biomolecule, explosives, propellants, precursors of such chemicals, microorganisms, radionuclides)
- Their origin (naturally occurring such as in natural disease outbreaks, synthetic materials or synthetically modified natural materials)
- The level of sophistication, technological maturity and financial power needed to acquire and deploy them in relevant amounts and qualities
- The level of sophistication needed to use them to achieve a desired effect (effects can range from mass casualties or large-scale physical destruction to the terrorisation of groups of people or an entire population, to assassinations of individuals).

In addition, threats posed by CBRNe materials need to be assessed in conjunction with the technologies that are enablers for their development, manufacturing, storage, or employment.

Furthermore, facilities where CBRNe materials are present (manufacturing plants, storage facilities, transportation infrastructure, waste treatment facilities and the like) can suffer accidents as a result of natural catastrophes, or technical malfunctioning, or they can be damaged as a result of sabotage, or become targets of physical, cyber or other forms of attack. Such incidents can lead to the release of agents into the environment, resulting in immediate effects on people and the environment or causing long-term contamination.

Finally, CBRNe threats are embedded in broader contexts (for example an armed conflict with its different belligerents and their capabilities and intentions, an industrial location with its equipment, infrastructure and nearby populations, or a geographical area with its endemic disease patterns and public health capabilities). Risk mitigation strategies will need to take account of these contextual factors.

As a result, assessments of CBRNe risks are complex and highly scenario-dependent. This poses serious challenges to traditional approaches to assessing the risks involved. Traditional risk assessment and management strategies build on probabilistic approaches: they identify the relevant hazards, analyse the factors that determine the corresponding impact of incidents, and assess their likelihood (including with regard to the intentions and capabilities of the different actors concerned). Such assessments highlight vulnerabilities in prevention, response and recovery systems, and allow a prioritization of necessary measures to strengthen prevention, response and recovery. On that basis, effective risk mitigation strategies can be designed for a variety of relevant scenarios.

An example of this approach is the CBRNe response planning developed by the US Department of Homeland Security (DHS). DHS is responsible for conducting risk assessments for CBRN agents under the 2004 BioShield Act and various Homeland Security Presidential Directives. To implement these requirements, DHS develops CBRN Terrorist Risk Assessments – TRAs – and Material Threat Assessments – MTAs (GAO 2011). From time to time, these TRAs and MTAs are combined into integrated CBRN Terrorist Risk Assessments (ITRAs). By 2012, the DHS had conducted three TRAs covering biological risk factors, two chemical TRAs, and two Integrated CBRN Terrorism Risk Assessments (GAO 2012).

The methodology used in such probabilistic risk assessments defines a set of specific scenarios on the basis of event trees to capture a relevant and realistic range of risks of concern. It then estimates the probability and consequences for each scenario and from these data calculates risks. The complexity of this approach becomes apparent when one looks at the number of variables and, consequently, scenarios that need to be systematically worked through. For a BTRA, the scope defined by DHS is based on three terrorist organisation categories, 37 different biological hazards, 2 exposure routes, 20 different targets, and 7 possible modes of dissemination – resulting in more than 5 billion (theoretical) scenarios of which more than 600,000 result in consequences (White 2016).

In such an approach that is based on specific threat agents, uncertainty and complexity are the major challenges. Uncertainties exist, amongst others, in the understanding of the agent(s) concerned and their behaviour, in the modelling of how scenarios would play out, and in the input data used in the models. Furthermore, uncertainties are accumulative. This and the complexity of the matter may well be a reason why, according to a 2012 GAO audit, only 2 out of 12 CBRN-specific response plans developed by DHS at the time were directly informed by CBRN risk assessments; another seven were indirectly informed by the existing risk assessments (GAO 2012). Risk assessment, management and communication approaches and their tools remain important for implementing an effective response to a CBRNe incident: together with other considerations and data, they inform the decision making processes during an incident response, from strategic to operational and tactical levels. But for broader CBRNe preparedness planning and prevention, more generic approaches are important, aiming at developing generic response capabilities that can reach back to specialised competencies and resources as and when needed. Such generic preparedness aims at increasing resilience in society to CBRNe incidents.

Experience in Europe underlines that the CBRNe threat spectrum is highly fragmented and dynamic. Basic CBRNe materials as well as information about how to employ them are widely available in open sources including the Internet, and risk assessments for such scenarios are extremely complicated if not impossible given the large number of variables, the variances in those variables, and situational unpredictabilities – hence the need for a refocus on strengthening generic consequence management capabilities and enhanced resilience, combined with reach-back to specialised expertise as and when required (Herzog 2019).

At the heart of any CBRNe risk mitigation strategy, however, must be a sound understanding of the sciences and technologies involved, of the trends and drivers that steer their development and application, and of the manner in which these developments may alter existing or create new risk potential as well as opportunities to manage evolving risks. The role of science and technology in shaping and managing the evolving CBRNe threat landscape, then, is the focus of this book. The different chapters will look at a number of technologies and scientific disciplines that shape this landscape. But this cannot be an all-encompassing, comprehensive review. Other areas of science and technology that may be equally relevant are only touched upon briefly. It was therefore desirable to open the book with a more general reflection on how advances in science and technology may affect the overall CBRNe security landscape.

A first observation is that science and technology are evolving at a fast pace. In recent decades, the time it takes for scientific discoveries to move from the laboratory bench to practical application in industrial development and manufacturing has shrunk considerably. In 2011, an international study commissioned by the US National Academies concluded that "there has been particularly rapid progress in the power of, and access to, enabling technologies, especially those depending upon increased computing power. These include high throughput laboratory technologies and computational and communication resources" (NRC 2011, p. 8). The study highlighted as possible consequences the increase of collaborations at a global scale including the emergence of "virtual laboratories"; increased access to and diffusion of sophisticated reagents, kits and services; the facilitation of the transfer of tacit knowledge across the web; and a reduction of barriers to the spread of S&T knowledge for responsible, but also for malevolent purposes (Ibid.).

It is important, therefore, to recognise the potential of scientific advances and emerging technologies at an early stage, to be able to respond to new safety and security challenges in time. Often called "horizon screening", formal assessments of the possible impact of advances in science and technology are recognised as a means of ensuring that regulatory systems (including arms control agreements, export control systems, national laws and regulations concerning safety and security) as well as administrative measures and industry standards are adapted *before* science and technology have moved from the laboratory to practical use, and *before* these applications have spread widely. In practice, however, regulatory adaptations tend to be slow.

At the same time, "horizon screening" must be aware of certain pitfalls. Evaluations at an early stage of technology development (at the initial up-swing of the so-called "hype curve", when the real capabilities and limitations of a technology are yet to be fully understood) carry the risk of exaggeration or may be altogether misleading; the selection of technologies that appear to be of concern can be somewhat arbitrary when the technologies considered are still fast evolving; and new technologies do not evolve into an empty space but, most of the time, must compete with well established and mature technologies that have established markets and proven technological and economic viability. In many application fields, tacit knowledge remains important for the effectiveness of technological solutions, and/or for the economy of their exploitation. All this sets limits to how quickly regulatory systems can and should respond to scientific and technological progress, or it may lead to regulatory adaptations that subsequently turn out to be ill-advised "knee-jerk" reactions.

Secondly, advances in science and technology can happen in a non-linear fashion – leaps in understanding as well as chance discoveries (serendipity) that remove obstacles in the way of faster or more widely distributed application of certain techniques, or that result in new understandings or approaches that open up new avenues for scientific inquiry. Such qualitative leaps are more likely when different scientific and engineering disciplines are deployed together, in particular if they have not hitherto been combined. An example is the convergence in the life sciences – the combined deployment of a range of different scientific methods and technologies at the intersection of chemistry and biology. This convergence brings together theoretical concepts and investigative methods from different science and engineering disciplines; from mathematics, modelling and simulation of complex systems; and from manufacturing technologies such as additive manufacturing and DNA origami. It makes use of miniaturization, automation and robotics in research, development and manufacturing. And it exploits new approaches to analysing very large sets of data, including deep learning and artificial intelligence. Any one of these approaches has the potential of creating significant change in science, technology and industrial application – their combination is expected by many to revolutionise the life sciences and their application in society.

Thirdly, one development is causing particular concern today: the shift in risk potential from materials and equipment to information. The nexus of disruptive technologies and CBRNe security will require more and more to take a holistic and bold approach towards CBRNe risk mitigation. The "dematerialization" of the CBRNe threat spectrum is the result, amongst others, of the growths of the "internet of things" in our digital age. Indeed, the theoretical possibility provided by AI-driven storage and analyses of mega-data can render the role of the material substrate on which traditional CBRNe threats have traditionally been based, less essential.

For example: if State actors, or even terrorist or criminal groups, were to deploy Stuxnet-like worms to attack national critical infrastructure, even when they are "air-gapped", traditional protections and threat reduction measures would become irrelevant. In such a scenario, traditional protections (detecting and fending off an attack, securing the perimeter and preventing access by the attacker to critical assets) are bound to fail – what is needed instead are means that enable critical systems to continue functioning and to recover even if they are "infected" by threats.

Such a paradigm shift from CBRNe protection to CBRNe resilience needs to be promoted and advanced in the near future, but it will call for a rethinking of some of our current risk mitigation approaches.

A fourth issue is S&T diffusion: one possible effect of scientific and technological progress is the lowering of thresholds for the conduct of certain scientific, technical or economic activities in the CBRN domain. This may include a spread of capabilities for conducting certain types of experiments, manufacturing certain types of products, or gaining access to certain types of materials that in the past have stood in the way of their broader application. CRISPR-Cas9 based genome editing, additive manufacturing ("3D printing"), the emergence of an Internet-based supply chain for complex biomolecules, and cloud manufacturing to customer specification of chemicals and biologicals are recent examples of such "democratization" trends. Whilst such new tools don't do away with the importance of theoretical understanding and tacit knowledge, they certainly contribute to the fast and global diffusion of new methods, technologies, equipment and materials that may pose concerns with regard to the proliferation of CBRNe capabilities.

At the same time, science and technology evolve in a particular economic, institutional and societal context. New scientific discoveries or methods don't automatically result in new types of weapons, or lead to activities that may carry risks to society. The directions which scientific inquiry takes towards technological development and industrial application are not unconstrained, but influenced by external as well as internal factors, amongst them (but not limited to):

- Laws and regulations that create barriers that institutionalised science and technology organisations respect as constraints for their projects
- Expectations and demands from society that create a context within which scientists and engineers define their objectives, and societal recognition of achievements reinforces this "self-image" and channels the utilisation of science and technology in certain directions
- Funding priorities that reflect both internal interests and objectives of the S&T enterprise and expectations and demands of funders in governments, societal actors, and private investors
- The scientific method itself (systematic observation, measurement, and experimentation; the formulation, testing, and modification of hypotheses; rigid documentation of experimental methods and result and their publication to allow results to be reproduced by others), which sets standards for what is generally accepted as proper scientific conduct, which in turn guides the ethics of science and its own perceptions of proper conduct.

From a scientific perspective, or perhaps more accurately from the perspective of the science and technology enterprise, CBRNe risk mitigation therefore is in essence a matter of building a culture of responsible behaviour, and of shaping the space available for scientific inquiry and its practical application in ways that minimises CBRNe risks. Like in other domains of influencing behaviour (such as laboratory safety, or the prevention of falsification of research results), developing such a culture of responsibility does not replace other means of risk mitigation (such as laws and regulations and their enforcement including sanctions), but without it, compliance with accepted norms would remain fragile and norms could easily be undermined by new developments.

A crucial issue in this context is the different time scale between developments in science and technology, and the normatization process which includes law framing and the enactment and application of regulations by the States bodies. It has been argued in several contexts that there exists a time gap between these two realities and that State as well as international institutions are not able to catch up with the pace of scientific and technological change. Due to this ineluctable reality, the only practical solution to this conundrum is to boost the ethical responsibility of scientists and engineers, and to train young scientists to become drivers in finding solutions to risk mitigation rather than becoming part of the problem.

Until now, many of the approaches to CBRNe risk mitigation have their origin in prevention and protection concepts developed against weapons of mass destruction during the Cold War. These were concepts developed with State-level weapons programmes in mind. They involved strategic deterrence and force protection, technology transfer controls and embargos to deny certain countries access to critical technologies, as well as arms control and disarmament measures aimed at limiting certain types of weaponry or their delivery systems, in terms of qualitative characteristics as well as overall numbers, and where possible reducing their numbers or altogether eliminating them. International humanitarian law, too, remains an important constraint on the manner in which such weapons may be used, or must not be used, in armed conflict.

Arms control measures developed during the Cold War included data exchanges and verification measures such as the acceptance of the use of certain national technical means, on-site inspections conducted bilaterally or by international inspectors, transparency measures to build confidence and create a degree of predictability, and political and institutional compliance management tools to clarify compliance concerns and resolve problems and disputes. Technical measures to verify compliance with arms control measures were constructed in ways that would ensure the detection of militarily significant violations in a timely manner to deny a violator any significant advantage from cheating.

Whilst this system of international, regional and bilateral treaties and arrangements in the arms control domain remains important until today, the security environment within which risk mitigation strategies need to function has changed profoundly from that of the Cold War and the immediate post-Cold-War period. In a review of the priorities of the OPCW conducted by a high-level panel set up by the OPCW Director-General in 2011 under the leadership of Swedish diplomat Rolf Ekeus, this evolving security environment was characterised thus (OPCW 2011):

- Conflict is no longer framed in the context of opposing military alliances in a bipolar world. The number of inter-State conflicts has declined yet the level of violence has not. The borderline between war, civil war, large-scale violations of human rights, revolutions and uprising, insurgencies and terrorism as well as organised crime are blurred.

- In addition to traditional military forces, more non-State actors have appeared on the battlefield, i.e. paramilitary groups, warlords and their militias and volunteers, mercenaries and private military companies, terrorists and criminal groups. As a consequence, contemporary threat perceptions are also driven by attacks on populations and critical infrastructure, in addition to more traditional state-based threats.
- Furthermore, there are worries, in such types of conflict and with such actors, that the rules of international law applicable in armed conflict, and in particular the principles and rules of international humanitarian law, may be undermined.

As the science and technology CBRNe landscape is changing rapidly, the advancement of disruptive technologies such as artificial intelligence, the exploitation of cyberspace, synthetic biology, omics and big data, raises more concerns due to the absence of a unified arrangement to mitigate these CBRNe challenges and of the related technologies.

An early indication that, funding and time permitting, non-State actors are *capable* of deploying a WMD capability was the Sarin attack of the Aum Shinrikyo in the Tokyo subway in 1995. The attacks of September 11 (2001) and the subsequent Anthrax mail attacks in the USA confirmed that non-State actors were capable of inflicting mass casualties, or causing terror and economic damage, at a significant scale. They and subsequent developments and the use of improvised explosive devices and chemical weapons in Iraq and Syria also signalled a growing *intent* of certain kinds of non-State actors to resort to unconventional attack schemes.

These incidents, furthermore, demonstrated a level of sophistication that had not been seen previously with regard to non-State actors. The Aum Shinrikyo sect was able to set up a pilot-plant-scale production line for the nerve agent Sarin (it failed, however, to develop effective dissemination techniques); al Q'eda turned commercial aircraft into weapons to huge destructive effect and reportedly developed and successfully tested a blueprint for a device to disseminate toxic chemicals; and Bruce Irwin, the presumed instigator of the Anthrax letter attacks, managed to formulate anthrax spores in ways that some observers described as "military-grade".

These and other developments raised concerns about the possibility that terrorists and criminal organisations might be able to harvest new technological advances and scientific discoveries to acquire even more effective means of causing large-scale terror and destruction.

Subsequent developments in particular in the Middle East, on the other hand, underscored that CBRNe threats are not only associated with scientific and technological advances exploited for novel types of weapons and war fighting, but can equally involve the employment of old technologies and long-established types of weapons and materials, albeit in ways and circumstances that may differ from traditional military scenarios. A striking example for the association of CBRN threats with advances in science and technology was the assessment by the US Director of National Intelligence in 2016 that genome editing constituted a global security threat alongside other WMD threats (Clapper 2016). A striking example for the use of old and widely distributed technology for mass killing and the terrorisation of populations was the use of chlorine gas in the Syrian armed conflict.

The Syrian conflict is also an example for the increasing association of CBRNe attack scenarios with propaganda warfare. This is not *per se* a new phenomenon – chemical as well as biological methods of warfare have long been associated with misinformation and propaganda. Partly this was the result of certain technical complexities of these methods of warfare that made it more difficult to establish precisely what had happened in an attack; partly it emanated from the secrecy that surrounds WMD programmes – some attack scenarios were in fact clandestine operations involving concealed agent releases; and partly it reflected the perfidy that is often associated with the use of these types of weapons – the use of poison and disease as weapons is considered a dishonourable means of war fighting in many cultures (and of course some are prohibited under international law). What has changed is the ease with which misinformation (from half-truth to outright lies) can be spread through informal yet highly influential and widely distributed channels – social media and Internet platforms – and the intensity and sophistication with which States as well as non-State actors are using such platforms as well as traditional media outlets to create narratives that serve their objectives.

These complexities – the association of CBRNe threats with a multitude of State, State-sponsored and non-State actors, the wide range of scenarios, the impact of enabling technologies on their effectiveness, and combination of CBRNe weapons uses with information warfare – are increasingly being mirrored by national security strategies. In its 2018 National Strategy for Countering Weapons of Mass Destruction (WMD) Terrorism, the United States emphasized the need for continuous pressure against WMD-capable terrorist groups, enhanced security for dangerous materials throughout the world, and increased burden sharing among the US' foreign partners (USA 2018). The strategy defines a set of US policy objectives, which include putting agents, precursors and materials to acquire WMD beyond the reach of terrorists and other malicious actors, deterring States and individuals from providing support to would-be WMD terrorists, establishing an effective architecture to detect and defeat terrorist WMD networks, strengthening defences against and preparedness for mitigating WMD threats at all levels, and identifying and responding to technological trends that may enable terrorists to develop, acquire and use WMD.

The EU, too, in the context of its Action Plan to enhance preparedness to mitigate CBRN security risks, points to a range of terrorist and other violent threats, from both networked groups and lone actors (EC 2017). It observes that terrorist organisations have not used CBRN agents in Europe, but points to credible indications suggesting that they might have the intention of acquiring CBRN materials or weapons and are developing the knowledge and capacity to use them. The action plan emphasises the need for significant investments by EU Member States to reinforce resilience against CBRN threats in terms of prevention, preparedness and response, and calls for a more focused and better coordinated, all-hazards approach (including with regard to mitigating large scale CBRN hazards unconnected to terrorism). To a certain degree, its objectives mirror those of the US National Strategy: reducing the accessibility of CBRN materials, ensuring a more robust preparedness for and response to CBRN incidents, building stronger links both within the EU and with external partners, and enhancing the knowledge of CBRN risks.

As a kind of a "spin-off" of the EU's Strategy against Weapons of Mass Destruction and its CBRN Action Plan, and aligned to its evolving Common Foreign and Security Policy, the EU has been implementing its CBRN Centre of Excellence Initiative since 2010. The initiative is currently funded under an external financial instrument know as the Instrument contributing to Stability and Peace (IcSP)[1] – see also at the end of this introduction. This EU mechanism, as well as the programmes implemented by the United States to strengthen security and cooperation in CBRN risk mitigation to develop global capacity, are cooperative, bottom-up, voluntary and all-stakeholders approaches based on the idea that promoting and ensuring the security of CBRNe materials and technologies worldwide is a common objective not restricted to zero-sum geopolitical calculations or selected geographic domains.

The Canadian Security Intelligence Service in its 2018 security outlook study (Canada 2016), too, noted that development and use of WMDs would be a continuing concern. While weapons technology will not leap ahead as fast as information technology, the risks of proliferation, or miscalculation, would continue to require constant monitoring and attention. The foresight study also drew attention to the potential of the Internet and cyber technology as a strategically disruptive force. The Canadian study pointed out that international trade, the movement of populations and instability in some countries enabled the spread of WMD expertise, and that there was little prospect of a corresponding improvement in proliferation control mechanisms. Whilst acknowledging the terrorist CBRN threat, it placed particular emphasis on the threat potential emanating from the activities of certain States.

The international community's response to the evolving CBRNe threats emanating from non-State actors has included a wide range of legislative, regulatory, enforcement and administrative measures. A prime example is UN Security Council Resolution 1540 (2004). Adopted under Chapter VII of the Charter, it created legally binding responsibilities for all states to take measures to prohibit and prevent the proliferation of WMD capabilities to non-State actors, in a comprehensive fashion.

The practical measures that this binding resolution compels States to adopt include (UNSC 2004):

- Refraining from providing any form of support to non-State actors in their attempts to acquire, transfer and use nuclear, biological and chemical weapons and their means of delivery;
- Adopting and enforcing appropriate effective laws which prohibit such acts by any non-State actor, as well as attempts to engage in any of the foregoing activities, participate in them as an accomplice, assist or finance them;

[1] The CoE Initiative was initially launched under the forerunner instrument of the IcSP – the Instrument for Stability (IfS). For the upcoming financing period starting in 2020, a new external financing instrument is being developed that is expected to continue funding the initiative.

- Taking and enforcing effective measures to establish domestic controls to prevent the proliferation of such weapons and means of delivery, including by establishing appropriate controls of related materials (through accountancy measures, physical protection measures, and effective border controls and law enforcement);
- Establishment and enforcement of effective national export and trans-shipment controls including appropriate laws and regulations to control exports, transit, trans-shipment and re-export, and controls on providing funds and services related to such export and trans-shipment.

This and subsequent Security Council resolutions encouraged States to review their situation with regard to measures to prevent the proliferation of WMD capabilities to and by non-State actors, to report to the Security Council about the measures they have taken, to adopt action plans to enhance their measures to prevent WMD proliferation by and to non-State actors, and they encouraged technical assistance to States that so required, including by other countries and by relevant International and Regional Organisations.

In parallel, efforts are being made to further strengthen the treaty system that has been constructed during and after the Cold War to reduce and where possible eliminate WMD threats. Treaties such as the Biological and Toxin Weapons Convention (BWC) of 1975 and the Chemical Weapons Convention (CWC) of 1997, although designed to address State-level WMD threats, contain provisions that compel their parties to enact and enforce domestic controls to prohibit and prevent the proliferation and use of such weapons and materials by individuals and legal persons subject to their jurisdiction or control. In addition, the international treaty system was further developed in such areas as countering terrorism, enhancing nuclear safety and security, strengthening transportation safety and security, ensuring biosafety and biosecurity, and in other relevant areas.

Natural events such as the 2014-2015 Ebola outbreak in West Africa, and major accidents caused by environmental disasters such as the 2011 meltdown of the Fukushima Daiichi nuclear power plant caused by the Tōhoku earthquake and tsunami, were further reminders of the destructive power inherent in CBRNe materials. But it is also recognised that many CBRNe materials are utilised in society as part of normal life, and that, consequently, there will always remain certain risks emanating from accidents or malevolent acts associated with their manufacturing, distribution and use for industrial, agricultural, medical and other peaceful purposes.

Efforts are being made to strengthen capacities to manage these risks, across the entire risk range (the whole of the CBRNe spectrum, as well as with regard to natural, accidental and hostile releases). Measures to this end involve strengthening civil defence organisations, the public health sector, first responder organisations, safety and security organisations within industry and the transportation sector, the development and adoption of safety and security measures and protocols by academic, research and other scientific communities and organisations, and many more.

It has been argued that arms control law provides a common legal framework for CBRN security (Myer and Herbach 2018). CBRN security, it is argued, was primarily about preventing non-state actors from obtaining, developing, using, and illicitly trafficking weapons of mass destruction (WMD) or related materials and technologies that may be used for hostile purposes. Although different from other instruments of arms control law that deal with controlling weapons and military capabilities of states, CBRN security arrangements deal fundamentally with the control of arms and related technology.

This may indeed be so at the level of international law dealing with CBRNe weapons and materials. Yet, there are no comprehensive legally binding international mechanisms or multilateral arrangements that deal with enabling technologies associated with CBRNe threats. The UN Convention on Certain Conventional Weapons (CCW) is currently used at the level of a UN Group of Governmental Experts (GoG) to discuss how to regulate lethal autonomous weapons systems (LAWS) – an issue that is taken up in several chapters of this book. A similar UN GoG might be launched to deal with the whole CBRNe threats spectrum and related enabling technologies for their employment.

At the practical level as well as with regard to policies and governance structures, however, finding a common approach to effectively mitigating CBRNe risks has remained a challenge. In practice, legislative, preventive, deterrence and response/recovery measures have often remained fragmented. For example, whilst States took steps under Security Council Resolution 1540 (2004), they also needed to respond to initiatives that took a more narrow, sectoral approach. International organisations (IAEA, OPCW, UNODA, UNODC, international organisations for the different transport modes, WHO, OIE, UNOCHA and many others) defined their respective contributions to the fight against terrorism within the context of their respective mandates and capabilities.

All these approaches are complementary, and coordination mechanisms such the UN's Counter-Terrorism Implementation Task Force (CTITF), as well as several UN Security Council Committees, were set up. Additional legal instruments were created, for example in the fields of counterterrorism and nuclear safety and security, and existing instruments were adapted to better address the evolving CBRN threats. Even legal instruments that on the surface might appear to have little to do with non-State actor CBRNe threats have been used to address certain aspects of the CBRN risk spectrum; an example is the International Health Regulation (2005) which requires States to develop core capabilities in their Public Health sectors – in the explicit understanding that these may be called upon to respond to health risks irrespective of whether they resulted from natural causes (disease outbreaks), accidents, or malicious acts.

The multitude of actors, the wide range of relevant technologies and their diversity in terms of maturity and risk potential, the fast and broad diffusion of some new technologies across the globe, and the widespread use of CBRN materials and technologies in society – all these factors complicate the development and application of a holistic and comprehensive CBRNe risk mitigation strategy. With regard to protection and response to incidents, such a "fuzzy" and ever-changing environment calls increasingly for a generic, resilience-based approach rather than the use of traditional (i.e., probabilistic) risk assessment and management tools. With regard to prevention and deterrence, at the same time, traditional arms control approaches are becoming out-dated and less likely to succeed. Nor are they likely to keep pace with the rapid advances in science and technology, the changes in industrial manufacturing, or the increasingly unpredictable global security environment. Some would argue that this signals the end of arms control as we have known it from previous decades.

The traditional arms control paradigm of the post Cold War period needs to be refocused in the twenty-first century since the so-called "strategic stability" can be affected by the advent of cyber and autonomous technologies, including artificial intelligence, and the close entanglement of nuclear and non-nuclear systems. An underpinning suggestion emerging throughout this book is the need to enhance the dialogue among different specialists in the CBRNe realm to keep a handle on the intangible dimensions of proliferation. Indeed, education, the development of professional ethics, and the development and application of tailored codes of conduct in the CBRNe field could enhance scientific responsibility globally. An appropriate framework for doing so is the Global Partnership Against the Spread of Weapons and Materials of Mass Destruction.

At the same time, in certain CBRN domains such as the life sciences, governments are no longer the primary producers and users of science. It has been noted that the primary producers and users of such technologies are becoming essential partners in preventing the misuse of technology for malevolent purposes (McLeish and Trapp 2011). For example, whilst traditional biosecurity policies have been government-instigated and top-down, the evolving world of science and technology calls for a multi-stakeholder governance approach. Furthermore, the increasingly global diffusion of certain enabling technologies (the Internet, cloud manufacturing, manufacturing at or close to the end user, point of care diagnostics and the like) is an indication that in some respect, we are already living in a "post-proliferation" world. It has been observed that "[in] such a world, traditional models of proliferation control are certain to fail, and the traditional top-down government approaches no longer seem appropriate. From a broader regulatory perspective, the role of governments is changing. The state alone is no longer able to control the way that life sciences discoveries are used. The circumstances beg instead for a governance system that brings together all stakeholders – science, industry, government, and the public – and broadens as well as deepens the basis for compliance with the safe and responsible conduct and utilization of science, thus supporting the norm against biological weapons" (Ibid p. 540).

As a consequence, future policies and mechanisms in the CBRNe domain are likely to be less reliant on top-down approaches such as global arms control treaties, and instead they will need to place more emphasis on arrangements between governments and other stakeholders, including measure taken by the developers and users of technologies such as voluntary compliance and control measures adopted by industry and traders, and the applications of soft tools such as codes of conduct and ethics. This signals a shift from attempts to control and prevent the spread of sensitive technologies and materials to certain actors, to a risk management approach that accepts that whilst prevention, prohibitions and deterrence will continue to play a role, they will not be able to prevent proliferation. What is needed instead is the development of a culture of non-proliferation and a stronger societal resistance to such threats. Chesney and Citron have argued, in the context of information warfare, that "democracies will have to accept an uncomfortable truth: in order to survive the threat of deepfakes, they are going to have to learn how to live with lies" (Chesney and Citron 2018). The same reasoning may apply to the risks posed by CBRNe materials and technologies. Society will have to accept that CBRNe risks cannot be eliminated altogether, and nor can the proliferation of CBRNe materials and technologies be completely prevented. Instead, society will have to learn how to live with these risks and how best to mitigate against them and develop resilience (Martellini et al. 2017).

The multitude of actors involved in such broader risk management strategies will also call for an approach that relies on effectively connecting hitherto unconnected networks of actors, to share information about the different mandates and capabilities, and to create platforms for coordination and collaborations. Such a "patchwork approach" of expanding and at the same time connecting different regimes and initiatives to create a broader framework to deal with the emerging CBRNe threat spectrum is gradually evolving. But the degree of fragmentation remains significant. Attempts to coordinate and synchronise the activities of the different actors have had only limited success. There have been attempts to develop common, more integrated platforms for practical measures in CBRN risk mitigation. For example, the EU's CBRN Centres of Excellence Initiative (a flagship programme under the EU's Instrument contributing to Security and Peace that provides technical support to non-EU partner countries in the field of CBRN risk mitigation) has taken a comprehensive risk mitigation approach, covering natural as well as man-made risks, accidental as well as hostile scenarios, risks associated with State as well as non-State actors, and risks from the entire spectrum of CBRN materials. More importantly perhaps, it builds on the context and the needs identified by the partner countries themselves, attempting to align technical assistance to these needs and conditions.

The CoE Initiative also has articulated the ambition to offer coordination and collaboration platforms that could be used by other actors, sponsors and benefactors alike. This has indeed been done in a few cases, but they remain few and far between. Despite some success, the initiative continues to struggle with effectively interfacing with other relevant initiatives and programmes that aim at strengthening CBRNe risk mitigation capacities.

This book is an attempt to help developing such a broader conceptual and practical framework towards a more holistic and comprehensive CBRNe risk mitigation strategy. It is structured into two Parts: Part 1 looks at key science and technology dimension of the changing CBRNe risk landscape, and considers the challenges that these developments create for governance approaches of emerging technologies and research, using the developments in the life sciences as a key example. Part 2 looks at some approaches to mitigating the evolving risks.

In Part 1, the book first develops and analyses key examples for advances in science and technology that have the potential of creating new CBRNe proliferation risks; Part 2 looks at technologies that have the potential to help mitigating these evolving as well as existing threats; and thereafter it discusses a number of practical measures that can be developed further to strengthen resilience and response to CBRNe threats.

With regard to advances that have the potential to change the level and nature of CBRNe threats, the book combines overviews of advances in science and technology with regard to the relevant agent groups and their manufacturing (toxic chemicals, explosives, biological agents, radiological/nuclear materials) with a survey of specific technologies that may change some of the fundamentals underpinning the CBRNe threat spectrum: new and increasingly widely distributed means of delivery (such as hypersonic missiles, drones and other UAVs, and autonomous weapons systems), vulnerabilities emanating from the possibility of cyber attacks on facilities and systems associated with CBRNe materials and technologies, and the potential for the misuse of artificial intelligence to develop novel types and means of warfare involving CBRNe materials and weapons. This section of the book also addresses evolving concepts of hybrid warfare that combine propaganda warfare with the threat or actual use of CBRNe weapons.

In the second Part of the book, Chaps. 11, 12, 13, 14, and 15 address technologies and scientific advances, as well as broader strategies and policy options, that are expected to help strengthening resilience against CBRNe threats. This includes trends in the detection of agent releases as well as medical countermeasures (using the medical response to nerve agent poisoning as a pertinent example). A field that has received increasing attention in recent years is forensics with regard to CBRN incidents. This is partly as the result of CBRNe threats associated with non-State actors; it also reflects the experiences of recent investigations of uses of chemical weapons in Syria, Malaysia and the UK. These investigations have underlined the importance of robust forensic and analytical methodologies as well as databases to be able to attribute responsibility to such incidents, based on scientific evidence. Finally, this Part offers practical guidance on training and education in the CBRNe risk mitigation field, as well as considers methodologies of assessing the effectiveness of such measures.

Whilst individual chapters draw their own conclusions in the context of their particular thematic scope, there was no attempt to put forward overall conclusions. However, as already indicated above, certain common themes emerged from several of the chapters, and we felt it might be useful to summarize them here to give the reader an overall perspective of what the current trends are that shape the CBRNe risk landscape today, and what concepts and strategies might be the most appropriate to manage these risks.

CBRNe risk mitigation, by its very nature, is multidisciplinary, science and technology based, and highly context/scenario dependent. At the same time, there are technical dimensions that are specific to the types of agent concerned, and there are also overarching and generic issues – these call for a holistic strategic approach.

What has become apparent is that within this complicated risk landscape, information management, processing and analysis are becoming more and more important. New tools are drastically enhancing our ability to process vast amounts of data and to analyse complex data sets in ways that the human brain itself is not able to master (including by means of deep machine learning / artificial intelligence).

This is about to change drastically the way in which CBRNe risks are evolving. At the same time, the growing global diffusion of technology and new manufacturing concepts (industry 4.0) are spreading capabilities globally, bringing manufacturing closer to the end user. All this is creating huge new opportunities to identify and manage CBRNe risks, but at the same time it also creates new vulnerabilities and complexities.

To us, it indicates the need for a shift in emphasis: from prohibitions, prevention and protection to strengthening resilience in society. To be clear, prevention and protection remain important and must not be neglected. In fact, strengthening resilience will overlap with preventive strategies (for example in the case of hardening critical infrastructure against the effects of CBRNe threats). But resilience will place a stronger emphasis on "softer", longer-term, strategies such as

- More strongly and explicitly fostering a culture of ethical behaviour and responsible foresight as part of the self-image and professional conduct of scientists, engineers and other professionals who deal with CBRNe materials and technologies
- Embedding these concepts and principles in educational systems – not as an add-on or a form of "securization" but as an integral part of the way in which science, technology and ethics are being taught and understood
- Promoting voluntary, self-regulatory compliance assurance systems in research and development institutions, industry and trade (in a manner perhaps comparable to safety as well as quality assurance systems widely employed today – voluntary but certifiable and seen as a bonus)
- Strengthening the role of international humanitarian law in international relations (thereby modulating the concepts of arms control and disarmament which traditionally function within narrowly defined sets of definitions and prescriptions rather than on a more holistic "do no harm" basis).

At the practical level, this will require outreach by governments and international organisations working in the field, engagement with a wide range of stakeholders, and the creation of suitable platforms – local, regional and global – that allow multiple actors to work together on analysing problems and devising and delivering solutions to CBRNe risks. This must be a bottom-up approach (aimed at strengthening indigenous capabilities that can take account of local conditions and needs), but it will also need an overarching strategic orientation and shared objectives of the different stakeholders.

Martellini and co-authors (Martellini et al. 2017) have argued that CBRN security is a sort of a new organizing principle of the international multilateral relations dealing with international security. Such an approach, even if theoretically sound, is difficult to operationalize and to harmonize with traditional arms-control arrangements.

A deep analysis of the new, cutting-edge technologies, as well as the re-emergence of well-established technologies applied today using new materials or being employed within a different industrial processes environment, including the management and utilization of mega-data, shows that despite these radical changes in the science, technology and industry environments, the fundamental architecture and the legal framework of WMD arms control and the related arms control treaties are still valid.

What is necessary is to "update" their definitions, technical concepts and implementation mechanisms to manage compliance, including national mechanisms to implement treaty requirements and verification processes used to provide confidence in compliance.

To give a few examples: in the BWC, the definition of a biological weapon should be understood to encompass "artificial DNA" (synthetic DNA using base pairs not expressed in nature) – consistent with but expanding on the common understandings adopted by BWC States Parties; in the CWC, the industry verification system in the domain of non-scheduled chemicals needs to be adapted to take account of the impact of chem-bio convergence (such as the changing nature of industrial manufacturing – bioprocesses, AM, cloud manufacturing, etc.); in the NPT the concept of minimal weapon-usable material ("significant quantity") should take into account that the sub-critical nuclear tests allow the five NW-States to design new thermonuclear weapons using sub-critical amounts of nuclear material, and that even the concept of ballistic missile as a vehicle of NWs delivery is too restrictive – indeed, the XXI Century is fast becoming the age of hypersonic gliders with nuclear payloads which would end any kind of nuclear strategic stability and deterrence as we know it from the Cold War. Furthermore, the brain-machine interface and AI augmented reality will create new ethical problems in International Humanitarian Law.

Furthermore, traditional approaches to the WMD arms-control verification processes need to be rendered sufficiently flexible and adaptive to account for the new disruptive technologies and materials that are going to shape the WMD landscape of the XXI Century. That adaptation, however, must not weaken the WMD arms-control norms and legal constraints. To say it in other words: the architecture and normative basis of the WMD arms control regimes should be perpetual whilst their definitions, implementation processes and review mechanisms (e.g., the WMD Treaty Conferences of States parties) should be revised whenever necessary, synchronised with the advances in science and technology as well as industrial application over time.

Of course, the diplomatic process itself would be enhanced by more actively engaging CBRNe scientists and engineers – as proactive partners of the diplomats that are in charge of the multilateral diplomatic gatherings that attempt to manage the new Prometheus S&T challenges. How to achieve concretely this task is not clear today, but some global reflection is needed if one wants to avoid the risk of a full collapse of XXI Century arms-control under the new S&T developments of our epoch.

References

Canada. 2016. Canadian Security Intelligence Service: 2018 Security Outlook – Potential Risks and Threats – a foresight project, Occasional Papers (June 2016).

Chesney, Robert and Danielle Citron. 2018. *Deepfakes and the New Disinformation War*, Foreign Affairs (11 December 2018). https://www.foreignaffairs.com/articles/world/2018-12-11/deepfakes-and-new-disinformation-war accessed on 12 January 2019.

Clapper James R. 2016. Director of National Intelligence: Worldwide Threat Assessment of the US Intelligence Community, US Senate Armed Services Committee, Statement for the Record on 9 February 2016. https://www.dni.gov/files/documents/SASC_Unclassified_2016_ATA_SFR_FINAL.pdf accessed 9 January 2019.

European Commission. 2017. Communication from the Commission to the European Parliament, the Council, the European Economic and Social Committee and the Committee of the Regions: Action Plan to enhance preparedness against chemical, biological, radiological and nuclear security risks, COM(2017) 619 final (18 October 2017).

GAO. 2011. U.S. Government Accounting Office (GAO), Report to the Committee on Homeland Security and Government Affairs, US Senate: *National Preparedness – DHS and HHS can further strengthen coordination for Chemical, Biological, Radiological, and Nuclear Risk Assessment*, GAO-11-696 (2011).

———. 2012. U.S. Government Accounting Office (GAO), Report to the Committee on Homeland Security and Government Affairs, US Senate: *Chemical, Biological, Radiological, and Nuclear Risk Assessments – DHS should establish more specific guidance for their use*, GAO-12-272 (2012).

Herzog, C. 2019. *Preparedness for low-probability incidents with high consequence potential – the example of bioterrorism, strategies and problems*, presentation at the annual meeting of the Working Party on CBW Non-proliferation, Berlin 2019, private communication.

Martellini, Maurizio, et al. 2017. *A reflection on the future of the CBRN security paradigm*. In *Cyber and chemical, biological, radiological, nuclear, explosives challenges – Threats and counter efforts*, ed. Maurizio Martellini and Andrea Malizia. Switzerland: Springer.

McLeish, Caitríona, and Ralf Trapp. 2011. *The life science revolution and the BWC*. The Nonproliferation Review 18 (3): 527–543.

Myer, Eric, and Jonathan Herbach. 2018. *Arms control law as a common legal framework for CBRN security*. In *Enhancing CNRNE safety and security: Proceedings of the SICC 2017 conference*, ed. Andrea Malizia and Maurizio D'Arienzo, 2018, 207–214. Switzerland: Springer Intl. Publ.

NRC. 2011. *Trends relevant to the biological weapons convention*. Washington D.C.: National Research Council of the National Academies (of the United States of America) in cooperation with the Chinese Academy of Sciences, IAP – the Global Network of Science Academies, the International Union of Biochemistry and Molecular Biology, and the International Union of Microbiological Societies, The National Academies Press.

OPCW. 2011. Note by the Director General: *Report of the Advisory Panel on Future Priorities of the Organisation for the Prohibition of Chemical Weapons*, Technical Secretariat document S/951/2011 (25 July 2011).

United Nations Security Council. 2004. Resolution S/Res/1540(2004), 28 April 2004.

United States. 2018. National strategy against weapons of mass sestruction terrorism, December 2018.

White, Scott. 2016. *CBRN Terrorism Risk Assessments – Methods and Applications*, Military Operations Research Society (MROS) 28 July 2016, available at https://www.pic.gov/sites/default/files/DHS%20Terrorism%20Risk%20Assessments%20July%202016_Part1.pdf. Accessed 13 February 2019.

Part I
The Changing CBRN Risk Landscape

The Twenty-first Century: The Epoch of Advanced Missile Systems and Growing Vulnerabilities

Matteo Frigoli

1 The New Strategic Layer

International security has been impaired by the fact that the underpinning material scenario to which it applies has undergone a technological evolutionary and revolutionary process which has boosted its complexity and the difficulty to manage it through exhaustive measures. The *complexity* element is linked to the continuous process of creation of new military capabilities which add new layers within the international security structure, requiring wholly fresh approaches for managing their destabilizing impacts.

Indeed, there has been a rise in the number of military technologies with strategic effects which are influencing nuclear postures[1]. These technologies do not necessarily belong to the nuclear realm, but, instead, are conventional capabilities able to exploit the vulnerabilities of nuclear deterrents and to disrupt the military systems on which the functioning of strategic arsenals is based. The evolving nature of threat to which key-defensive systems are exposed has brought a deep uncertainty on inter-state security relationships and, in particular, among the three great competitors, the U.S., China and Russia. Uncertainty has highlighted the need for more *flexibility* and *diversity* about the envisioned use of nuclear and advanced conventional weapons[2].

Relevant destabilizing elements of the current international security scenario are:

- New advanced strategic military architectures with intertwined conventional and nuclear elements vulnerable to asymmetric means of warfare (e.g. anti-satellite weapons, cyber-weapons).

[1] See, for example, the US Nuclear Posture Review (US DOD 2018); see also the Military Doctrine of the Russian Federation (Russian Federation 2015),

[2] *Ibid.*

M. Frigoli (✉)
Independent Researcher, Parma, Italy

M. Martellini, R. Trapp (eds.), *21st Century Prometheus*,
https://doi.org/10.1007/978-3-030-28285-1_2

- The development of new military technologies with strategic or long-term implications.
- The vulnerability of critical dual-use infrastructures on which the functioning of advanced military capabilities is based.

In the discussion about new advanced military architectures, the attention is focused on strategic stability relationships between the U.S., China and Russia. Strategic Stability relationships cannot be addressed exhaustively with a fixed and determined set of arms control measures. Arms control actions focused on reductions of strategic warheads, of delivery systems, fixed places of strategic forces deployment, encouraging survivability of strategic delivery platforms and strategic sensors would not be suitable for addressing the new concept of strategic stability open to include advanced conventional capabilities, i.e. cyber weapons, ASATs, advanced conventional missiles and the envisioned growing role of low-yield nuclear weapons. There is a set of military technologies which constitute a "*New Strategic Layer*" which needs to be managed.

This *New Strategic Layer* of military technologies is identified here as a fixed number of systems with strategic implications, but this concept is suitable for embracing new advancements in warfare technology outside the "nucleus" of issues which are managed with business-as-usual measures arms control instruments.

The most relevant destabilizing military capabilities taken into consideration in this paper are: boost-phase defence, hypersonic missiles, nuclear-powered missiles, ASAT weapons, development of new generation low-yield nuclear weapons, AI security implications.

The *New Strategic Layer* of military capabilities, it is to be understood both by looking at the single elements and at the interaction between the individual elements.

1.1 Boost-Phase Defence

Ballistic missile trajectories are typically divided into three phases. The boost-phase, the mid-course phase[3] and the terminal phase (Chen and Speyer 2008, p. 1). The Boost phase is the portion of flight immediately after launch when the booster accelerates to lift the munitions into the air which lasts from the missile's launch until the rocket booster shuts (*Ibid*). Boost phase is relatively short in duration. For ICBMs, which typically have two- or three-stage boosters, the boost phase lasts until the final stage burns out (IFPA 2009, p. 16). The second, or midcourse, phase lasts from the end of boost-phase until the warhead reenters the atmosphere (i.e. at more or less 100 km in altitude), during the midcourse phase the warhead has

[3] The Midcourse phase of an ICBM is the longest portion of flight lasting about 75% of all the flight. In this phase the munitions have separated from the booster and are flying un-powered. This phase offers the largest time window in which track and intercept the incoming warhead (Evers et al. 2015). See also Cepek (2005).

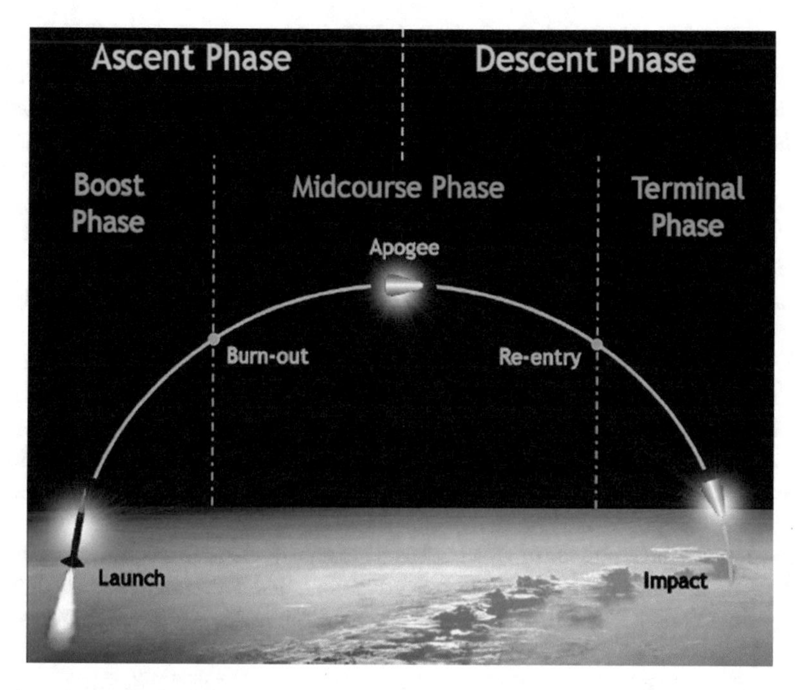

Fig. 1 Ballistic missile trajectory

separated from the booster and is flying un-powered (Regan and Anandakrishnan 1993, pp. 27–28)[4]. The *terminal phase* is the portion of flight when the warhead reenters the atmosphere and reaches its target, which lasts approximately 30 s (Chen and Speyer 2008, n. 5, p. 1) (Fig. 1).

While the objective of midcourse and terminal missile defenses is to intercept and destroy the warhead (i.e. a little target travelling at hypersonic speeds (Mallik 2004, p. 110)[5], boost-phase defense systems are aimed at destroying or damaging the warhead while its attached to a large rocket in the ascent phase. In the boost-phase, damaging the booster alone may prevent the warhead from achieving enough velocity to reach its target (Wilkening 2004, p. 1; Congressional Budget Office 2004, p. x.).

A reliable boost phase defense system has long been considered the "holy grail" of missile defense, as boosting missiles are much slower than missiles during the midcourse or terminal phase. For example, it is reported that an average liquid-fuel

[4] See Regan and Anandakrishnan (1993). The early portion of the midcourse phase, after the booster burns out but before the warhead is deployed, is sometimes called the ascent or early-ascent phase. See (Congressional Budget Office 2004).

[5] Noteworthy, once the warhead reenters from outer space, it would be greatly slowed down by the Earth's atmosphere. It is estimated that an average ICBM warhead would travel at more than 3 km/s (approximately Mach 8) at impact point (Wilkening 2004, p. 1; Congressional Budget Office 2004, p. x.).

ICBM after 30 s from launch has reached an altitude of about 3 km (a solid-fuel ICBM has, instead, a better acceleration ratio and can travel faster than liquid-fuel ICBM) (Congressional Budget Office 2004, p. 12). Moreover, during boost-phase, the hot exhaust gas put out by the booster rocket constitute a very visible target for infrared sensors and the rocket body itself is larger and more visible to radars than the much smaller warhead which will separate from the booster and travel unpowered when it reaches the midcourse phase. In addition, ICBM countermeasures such as decoys, which would be difficult to distinguish from the real warhead during the midcourse phase (Goodby and Postol 2018, p. 210), are complex to deploy during the boost-phase (Wilkening 2004, p. 45).

The challenges of boost-phase defense systems are linked to: the very short time available for the engagement and the access to the necessary deployment locations for interceptors which must be near to the ICBM launch point (whether the interceptors are based on kinetic or directed energy). The required velocity of a kinetic interceptor is dependent on these two variables (Congressional Budget Office 2004, p. 9; NRC 2012, p. 5; Wilkening 2004).

In concept, the architecture of a boost-phase missile defense system is composed of a set of sensors, infrared or/and radar systems, and a given kinetic or directed energy interceptor system which could be ground-based, ship-based, air-based or space-based (Congressional Budget Office 2004; NRC 2012). The system should be able to detect, track, fire and intercept the hostile ballistic missile before the end of the boost-phase, or, in the case of a kinetic interceptor, before the hostile ballistic missile reaches a speed higher than that of the interceptor (ibid.). In order to meet this timeline, the sensor system should be able to detect and track the target shortly after launch. It has been reported that ground-based radars should be positioned not farther than 550 km from the launch point, while airborne radar could allow for a boost-phase detection from as far as 800 km (ibid.).

Because of horizon limitations[6] ground-based radar systems could provide a full coverage against small countries, but an ICBM launched from the interior of a large country will not be visible to surface or airborne radars in time for a useful boost-phase detection (Congressional Budget Office 2004). Also, land-based or sea-based interceptor systems would be greatly constrained if the target is shot from deep inland of a large country, airborne interceptors could solve the problem where air supremacy has been achieved or where they can get close enough to the launch point of the hostile ballistic missile without being threatened by an effective air defense system.

Space-basing for boost-phase defense could overcome these disadvantages. Space sensors are not horizon-limited and can be oriented to guarantee a persistent coverage of a selected objective (NRC 2012, p. 6); space-based interceptors are not geographically limited. The disadvantage is that there would be the need of a very large constellation (hundreds or maybe thousands) of space interceptors in order to

[6] *OTH radars* are a type of radar systems which use the ionosphere to refract outgoing radar waves and return signals, enabling the system to detect and track targets that would otherwise be hidden by the curvature of the earth. See Air Combat Command (2012).

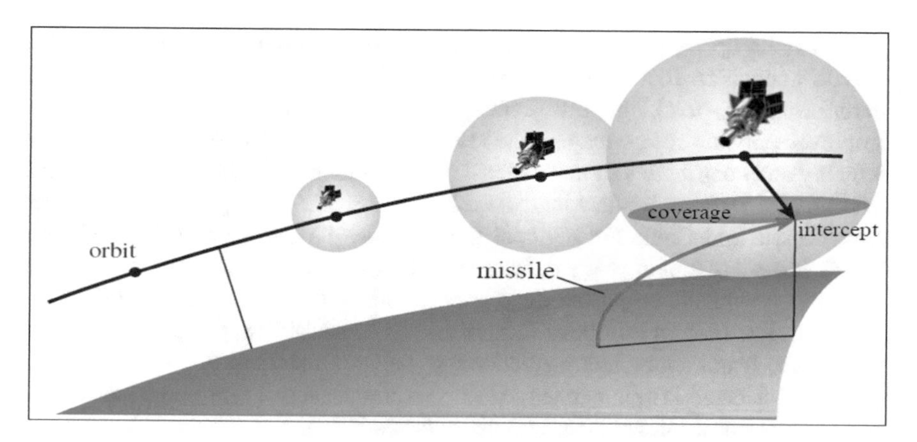

Fig. 2 Example of a space-based missile interception. (Source: Adapted from Barton et al. 2004)

have at least one interceptor always in range for a boost-phase engagement (Congressional Budget Office 2004, n. 10; NRC 2012) (Fig. 2).

At the time of this writing, no nation has a proven realistic capability to destroy a ballistic missile during the boost phase. The U.S. are currently developing the technology for boost-phase intercept of ballistic missiles, with the intent to counter threats from regional actors such as North Korea and (potentially) Iran (NRC 2012; Williams 2017). Indeed, the Pentagon has been required to study and formulate an initial plan to develop a boost phase missile defense capability according to the 2019 National Defense Authorization Act (US Congress 2018).

Either space-basing or ground-basing for boost-phase defense would entail that the interceptors could target missiles coming from other multiple countries (NRC 2012; Williams 2017). As regards the sensor architecture, a constellation of space sensors could be oriented to focus on specific latitude bands, they cannot be concentrated against individual countries (NRC 2012, p. 3). The U.S. have been working on boost-phase missile defense concepts to counter missiles fired from North Korea or Iran (Williams 2017, p. 2). It has been reported that a space-based system with orbits capable of covering North Korea could cover about 75% of the world's countries and about 90% of those that might now be considered potential threats of the U.S. (Congressional Budget Office 2004, p. xvi). It is noteworthy that even if the U.S. boost-phase defense would be committed to target Russia or China, such defense is not practical, given the size, sophistication, and capabilities of Russian and Chinese forces to respond to U.S. defense efforts, including by increasing the size of the attack to the point at which defenses are overwhelmed by numbers (NRC 2012); indeed, Russia and China could field warheads and decoys "at a dramatically less cost than the United States can add missile defense interceptors" (Rose 2018).

Nonetheless, a perfect defense is not necessary to introduce important uncertainties into an enemy's offensive power by devaluing its offensive missiles strikes capabilities. In addition, it is well-known how Russia feels threatened by U.S. missile defense in general.

The build-up of the envisioned space sensor architecture alone could feed Russia's concerns. Indeed, sensors are required across the entire intercept cycle: early warning, tracking, fire control, discrimination, and kill assessment. Improvements in sensors may, at the margin, be one of the best ways to improve lethality and contribute to a more robust existing defense (Williams 2017, p. xxii). The Missile Defense Review indicate that the U.S. are looking to counter threats from Russia and China "in regional theaters such as Europe and Asia" (Sonne 2018). Instead of looking for a defense from Russia or China strategic arsenals, the U.S. may be looking to expand the role of missile defense to defend NATO forces in regions like Eastern Europe or Eastern Asia (Fedasiuk 2018, p. 7), adding a layer to counter, if not short, intermediate-range ballistic missiles. In fact, one of the options considered is to integrate the F-35s aircraft as both a sensor and interceptor platform for boost-phase defense (Defense News 2019).

Nevertheless, the politics governing theater missile defenses and strategic missile defense are bounded, setting down a system capable to defend against one kind of weapon necessarily affects the strategy underpinning the use of other weapons (Fedasiuk, pp. 7–10), showing how delicate could be the equilibrium that sustains thrust and stability between competitors.

2 Hypersonic Missiles

The United States, China and Russia are by far the nations with the most developed hypersonic technologies. They are engaged in what can be regarded as a hypersonic arms race (Gubrud 2015, p. 1; Nagappa 2015, p. 9), these nations have conducted several tests of hypersonic missiles in the last decade and are near to deploying a limited hypersonic capacity (Lele 2019, pp. 71–74). Hypersonic missiles are now well-tested and near to be deployed. Hypersonic missiles are likely to impact on the international security scenario in the short-term. It is reported that Russia will deploy a limited number of hypersonic missiles during 2019 while the U.S. could do the same in a few years; China is strongly committed to the development of hypersonic missiles having undergone at least 7 tests between 2014 and 2017.

The new missile technologies referred to as "hypersonic missiles" consist of two new systems: boost-glide vehicles (HGVs) and hypersonic cruise missiles (HCMs). Before analysing the specific capabilities of each system, it is useful to mention their common characteristics.

Hypersonic missiles could reach and maintain hypersonic speeds, i.e. speeds above Mach 5. Speed is not the only element that characterizes hypersonic missiles, indeed even the existing re-entry vehicles (RV) mounted atop of ballistic missiles can reach hypersonic speeds in the terminal phase of their flight[7].

[7] An average ICBM can accelerate the RV to orbital velocities (about 7 km/s or Mach 20). Nevertheless, once the RV reenters from outer space it would be greatly slowed down by the Earth's atmosphere. It is estimated that an average ICBM RV would travel at more than 3 km/s (approximately Mach 8) at impact point (Mallik 2004, p. 110).

The game-changing capabilities of hypersonic missiles derive from the missile's speed, manoeuvrability and unusual altitudes that make them complex to detect and difficult to intercept by the most advanced missile defence systems.

Hypersonic missiles follow a non-ballistic trajectory, flying between 30 and 100 km in altitude, thus they would operate at altitudes above those of conventional aircraft, but significantly below those of ballistic missiles. Their maneuverability allows them to change their impact point and the associated trajectory throughout their flight time (Rand 2017, p. 8; CNS 2015, p. 8) and they can achieve a high degree of targeting precision. By contrast, existing cruise missiles offer good maneuverability but relatively low speeds, and ballistic missiles offer hypersonic speed but little or no maneuverability.

The unusual altitudes and unpredictable flight path of hypersonic missiles have implications for the detection: hypersonic missiles will likely be invisible to existing missile early-warning radars for much of their trajectory. It is reported that hypersonic missiles could be detected during their boost-phase[8] by satellite early-warning systems (or by surface radars sufficiently close to the launch point), but after the boost-phase they may disappear from view (Acton 2013, p. 118). After the "unobservable" phase, they could be detected once again if they fly 400–600 km from ground-based early-warning radars but, even if detected, there will be a high degree of uncertainty about their target since in-flight updates can program the missile to attack a different target from the originally planned (Rand 2017; NRC 2007, p. 7; NRC 2008, p. 128). In short, hypersonic missiles will leave to the target a few minutes to elaborate a reaction before the impact. By contrast, it is possible to predict the impact point of any given ballistic missile, allowing an opponent to calculate the warning-time at his disposal.

These characteristics make hypersonic missiles a suitable system for surprise long-range strikes. Thanks to their capabilities hypersonic missiles will:

- Highly compress target's reaction time (and the time at disposal to decision-makers to elaborate and communicate a response).
- Hold extremely large areas and targets at risk (given that the actual targets of a hypersonic missile attack might not be apparent until the last minutes of flight).
- Potentially overcome the most advanced missile defense systems.

As will be better discussed, HCMs and HGVs may be used as either strategic or tactical systems, armed with conventional or nuclear warheads, introducing not only flexibility but ambiguity of intent.

[8] Boost phase is the portion of flight immediately after launch when the booster accelerates to lift the munitions into the air. It lasts roughly 3–5 min for a long-range missile and as little as 1–2 min for a short-range missile (Chen and Speyer 2008, p. 1).

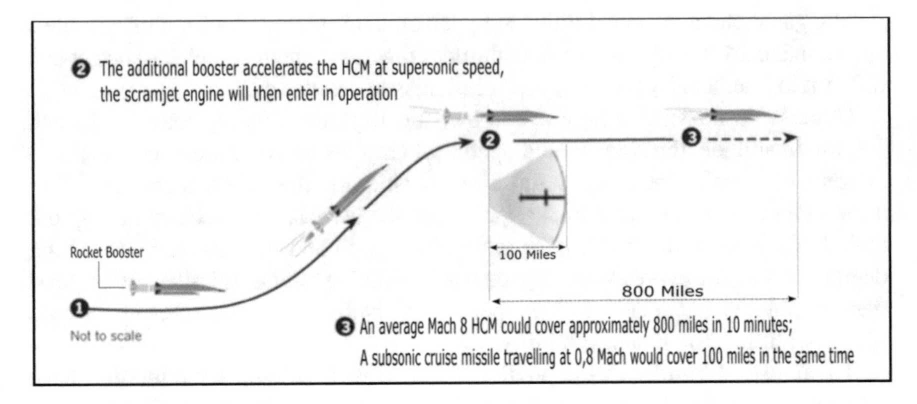

Fig. 3 Hypersonic Cruise Missile. (Source: Adapted from Boeing Graphics)

2.1 Hypersonic Cruise Missile

An HCM is a cruise missile capable of operating at hypersonic speeds, flying at 20–50 km in altitude (Rand 2017, pp. 8–10). In concept, these systems consist of two stages: the first-stage rocket booster and the second stage powered by a scramjet engine (i.e., supersonic combustion ramjet) which generates thrust from a supersonic airflow (Rand 2017, p. 12). The first-stage would accelerate the missile at the right supersonic speed needed for properly starting the second stage and begin to cruise at hypersonic speeds (Acton 2013, p. 68).

Existing subsonic and supersonic cruise missiles are difficult to defend against because they are hardly detectable and follow unpredictable trajectories. The additional speed provided by HCMs, relative to other cruise missiles, would pose a highly complex defensive challenge (Acton 2013, p. 73). They could be launched from land, air or sea and would provide a regional strike capability, being able to fly up to 1000 km in range (Fig. 3).

2.2 Hypersonic Boost-Glide Vehicles

An HGV is an unpowered vehicle capable of gliding on the upper atmosphere at hypersonic speeds for long-range distances. It is equipped with a small propulsion system (RCS thrusters) for orientation and directional control (Rand 2017, p. 9). The HGV is mounted atop of a large rocket, usually an existing type of ICBM, that will propel the HGV at hypersonic speeds (Wiener 2017, p. 142; Woolf 2019a, p. 17). The separation from the rocket will take place, depending on the target location, between 40 and 100 km above the earth's surface[9]. After the separation, the HGV

[9] In the HTV-2 DARPA experiment, the glider detached from the rocket after 270 s.

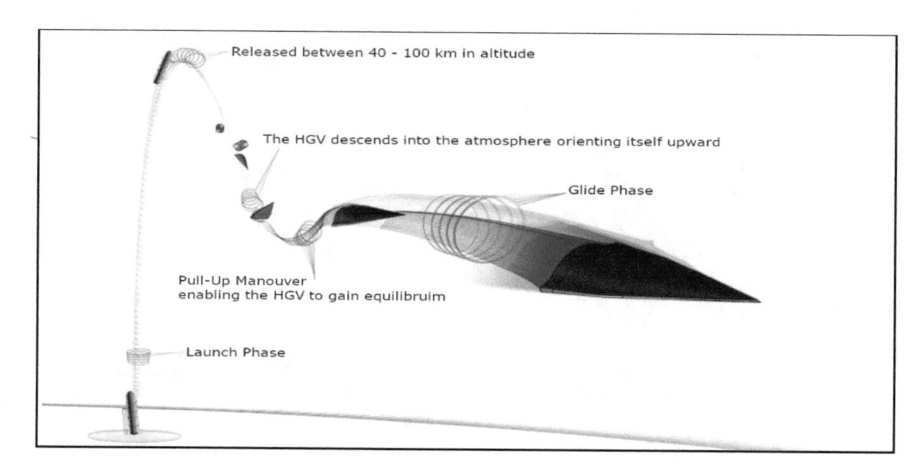

Fig. 4 Flight portions of a hypersonic glide vehicle. (Source: adapted from DARPA graphics)

descends into the atmosphere where a "pull-up" manoeuvre[10] is executed in order to enable the HGV to enter in the glide-phase of the flight. The HGV will then glide to its target along a relatively flat trajectory (Acton 2015, pp. 195–196). In contrast to existing ballistic missiles, HGVs would fly at much lower altitudes, they would follow a flat trajectory, and would be capable of maneuvering and changing the point of impact throughout their glide-flight, lasting for thousands of kilometers in range (Rand 2017, p. 3) (Fig. 4).

It is important to stress the difference between the HGV and the existing Maneuvering Re-entry Vehicle (MaRV) (Rand 2017, p. 9). It is true that HGV and MaRV are both able to glide, but there are decisive differences.

Indeed, MaRVs can offer a change of direction only in the terminal phase of their flight, and given that the majority of the movement trajectory is ballistic, they could be detected and tracked from thousands of kilometers of distance by early-warning radars. In addition, MaRVs follow a ballistic trajectory, they are vulnerable to mid-course ballistic missile defenses, while these last defenses would not be effective against HGVs (Rand 2017, p. 9). In other words, MaRVs have all the attributes and vulnerabilities of ballistic missiles, except for the opportunity to maneuver in the terminal-phase of the flight.

The figure below compares MaRV and HGV trajectories (Fig. 5).

[10] In the pull-up phase, the HGV will orient itself upwards using small thrusters and will enter in the gliding phase by using the lift force generated by its shape (Acton 2015, p. 194).

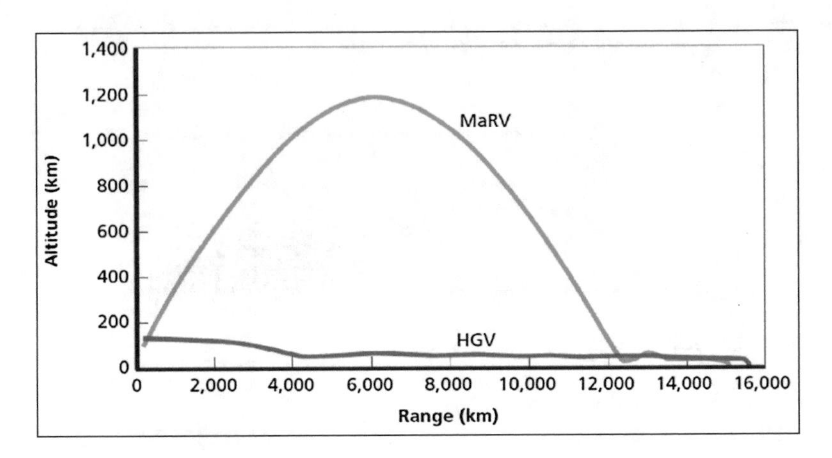

Fig. 5 MaRV versus HGV. (Source: RAND 2017)

2.3 Consequences for International Security and Strategic Stability Between U.S., China and Russia

The military characteristics of hypersonic missiles bring critical ambiguities likely to have a relevant impact on strategic stability relationships, especially among U.S., China and Russia which are near to deploy an operational hypersonic strike capability.

Warhead ambiguity: i.e. the impossibility of discerning whether a hypersonic missile is carrying a nuclear or a conventional warhead.

Target ambiguity: due to the susceptibility of HCMs and HGVs to multiple in-flight course corrections, a nuclear possessor-state could believe that its nuclear assets were at risk even though in fact conventional assets were the intended targets.

Destination ambiguity: an observing state could assume that HCMs and/or HGVs were targeting its territory even though the actual targets were located in the territory of a neighboring state.

These key-ambiguities could bring a scenario in which the risks of accidental war and the difficulties of escalation management will likely increase by quickly climbing the different steps of the “escalation ladder”. It is noteworthy to mention that the U.S. are developing hypersonic missiles only as a conventional weapon system, while, at the time of this writing, it is unclear if China and Russia are going to arm hypersonic missiles with nuclear warheads.

Indeed, hypersonic missiles may compress the warning-time that will follow the detection of a hypersonic missile fleet, the targets of a hypersonic strike will be unpredictable, and the nature of the threat posed by any given strike – conventional or nuclear – will be ambiguous. These factors will affect the strategic postures of states engaged in the hypersonic arms race and of other states which have reason to fear being in the “gunsight” of these weapons.

In fact, if nuclear-armed, hypersonic missiles could impair the nuclear equilibrium between competitors destabilizing one of the pillars of strategic stability: first-strike stability. Even if there would be a numerical equality in arsenals, the unique characteristics of hypersonic missiles will compound first-strike stability as both the possessor states would perceive each other as enhancing their first-strike capability (Acton 2013, p. 144).[11]

Indeed, the super-accuracy of these weapons could lead one state to believe that a surgical low-yield nuclear attack would be acceptable to an adversary. The surgical precision of these weapons could boost the confidence of a state leadership that strike the enemy's forces with precision and with a low-yield nuclear weapon could be a useful tool of "escalation control" (Leah 2017, p. 192). While in the event of a hypersonic low-yield nuclear strike, even with both states willing to accept a mutual exchange of low-yield nuclear weapons, the one who would then find himself at a disadvantage could prefer to escalate.

As has been said above, hypersonic missiles are being conceived as providing a conventional strike capability by the U.S.. Nonetheless, conventionally-armed hypersonic missiles represent as a dangerous issue for strategic stability and arms control as nuclear-armed hypersonic missiles. It is not by looking at 'first-strike stability' that it is possible to catch the core of the destabilizing effects of conventionally-armed hypersonic missiles as conventional hypersonic missiles are not suitable to conduct purely counter-nuclear missions. There is no assurance that conventionally-armed hypersonic missiles could effectively destroy small hardened targets such as silo-based nuclear missiles (Dvorking 2019; Acton 2013, pp. 84–87; Gormley 2015, p. 133). The destabilizing impact of conventionally-armed hypersonic missiles derives primarily from their employment as the leading edge of major combat operations, targeting early-warning radars, dual-use (nuclear/conventional) command and control nodes, air and missile defence assets (Acton 2018, pp. 61–62; NRC 2007, p. 2). Indeed, the so-called strategic conventional weapons are now part of the deterrence equation and are influencing nuclear postures. Both the Russian and the U.S. nuclear postures foresee the possible use of nuclear weapons in front of a purely conventional aggression to key-defense assets like command and control infrastructure and early-warning radars. China argues that a hypersonic conventional attack could put in a "disadvantaged, passive position" the Chinese nuclear counterstrike capability (Chinese Academy of Military Science 2013, pp. 170–171). Moreover, it is reported that the command and control infrastructure of nuclear weapon states are controlling both conventional and nuclear forces (a phenomenon named *entanglement*), even an accidental hypersonic strike on these infrastructures could degrade the communication systems required for operating nuclear weapons, triggering a potential nuclear response. This could bring a rise of the alert level of nuclear forces which will add further risks of accidental nuclear actions.

[11] See on this subject (Schelling 1970, p. 180).

3 Nuclear-Powered Engine for Cruise Missiles and Torpedoes: Burevestnik and Status 6

These two nuclear powered weapons system are being treated in the same paragraph as they have in common the fact that for both systems a compact nuclear reactor represents a pre-requisite; both are potentially conceived as weapons of pure "retribution" in the case of a nuclear attack, and for both systems there are doubts over their real deployment and capabilities.

The "Burevestnik" nuclear-powered cruise missile was unveiled during a speech by the Russian President Vladimir Putin before the Russian Federal Assembly in March 2018, he affirmed that "In late 2017, Russia successfully launched its latest nuclear-powered missile at the Central training ground" (GlobalSecurity.Org 2019a). At the time of this writing, the "Burevestnik", also known to U.S. observers as "Skyfall", has undergone at least 13 tests (Missile Defense Project 2019). The only two positive tests achieved a "moderate success" (Ibid.).

It is clear that the nuclear-powered cruise missile is a possible future weapon system, made feasible by the fact that Russia has created a new, compact, nuclear power unit that can be installed in a cruise missile (Cooper 2018). Russian analysts tend to agree that the nuclear unit acts as a heat source for what is essentially a ramjet engine (Hacker 1995. pp. 91–92).

This technology has a long history. In the early 1960s, the United States built a nuclear-powered missile known as "Project Pluto". The program begun in 1957 and terminated in 1964 (Hacker 1995, p. 90). The Department of Defense decided to rely on strategic bombers and ICBM as means for strategic delivery. In addition, there was no way to fly it without spreading dangerous levels of radiations. Analyst Stephen Schwartz noted that "the reactor was unshielded, emitting dangerous levels of gamma and neutron radiation. And as it flew, it would spew radioactive fission fragments in its exhaust, including over allies [states] en route to the U.S.S.R" (GlobalSecurity.Org 2019b). To be small enough to reasonably fit inside a missile, the nuclear ramjet which the United States developed for "Project Pluto" had no shielding to contain the dangerous radiation spreading. The exhaust plume also contained unspent fissile material that would have contaminated any area, enemy controlled or not.

In fact, it is reported that Norway's Radiation Protection Authority and the French *Institute for Radiological Protection and Nuclear Safety* detected small amounts of radioactive substances called ruthenium-106 tracing back to Russia in October 2017 (IRSN 2018, p. 10). Also, unusual amounts of Iodine-131 were detected in the same period (Nilsen 2018). The particles may have come from the test of the nuclear-powered missile (Gertz 2018). Further reports indicate that Rosatom (Russian State Nuclear Energy Corp.) was monitoring the site during at least one of the tests of the "Burevestnik-Skyfall", an indication that nuclear material was used in the test (Gertz 2018).

From a purely technical point of view, and despite these grave and critical issues, the "Burevestnik" nuclear-powered cruise missile would provide unique military capabilities. Indeed, it has been described by Russian officials as "A low-flying, barely noticeable cruise missile carrying a nuclear warhead with virtually unlimited range, an unpredictable flight path and the possibility of circumvention of interception lines is invulnerable to all existing and prospective systems of both missile defense and air defense". In fact, an unlimited-range cruise missile could fly intricate routes to exploit holes in enemy air defenses. Though, the exact missions of this system in the overall Russian deterrence strategy is not yet clear (Trenin 2019, p. 16). The "Burevestnik" or "Sky-fall" could be a bet with more to lose than to gain. Indeed, the spreading of dangerous radioactive material during the flight aside, some of the risks could regard: the consequences of the destruction (whether intentionally or accidentally) of these missiles' storage facilities during a combat operation; the implications of the interception of the missile while overflying a neutral state; the potential detection of the missile by neutral states and the consequent risk of accidental war as the same state could conclude to be the target (so called *target ambiguity*).

In addition, it is difficult to imagine what kind of purpose the nuclear-powered missile could have in war planning. What are the advantages of this missile over existing cruise and ballistic missiles in a real warfare scenario? It is true that with an unlimited thrust it could exploit holes in hostile missile defense systems and follow unpredictable long-routes where the hostile radar/sensor coverage is less focused. Will it be capable to destroy high value targets? Or there will be the need to build hundreds of these missiles to bypass point and area missile defense systems? Moreover, in case of the decision to carry out long-range nuclear strikes, strategic bombers, ICBMs, SLBMs will surely reach the targets faster than the Burevestnik[12], in these circumstances what could be the purpose of this missile? Another doubt could be the real "undetectable" capability, indeed if the missile will spew even a small amount of radioactive substances in the air, it could be easily detected by following its radioactive cloud (The Aviationist 2017). It seems more a weapon of "destabilization" rather than a system to increase the deterrence of the overall Russia nuclear arsenal.

The same concepts behind the allegedly developed Russian compact nuclear reactor used to propel the Burevestnik could work as a propulsion system for other relatively small Russian warhead delivery systems. Russia is developing an autonomous nuclear-powered torpedo designed to travel autonomously across thousands of miles to detonate a multi-megaton bomb near to a coastal objective, creating a massive radioactive tsunami (Sutyagin 2016, p. 1, GlobalSecurity.Org 2019c).[13] The torpedo, named Status-6 (also known as Kanyon), it supposed to be 1.6 m in diameter, and 24 m long (GlobalSecurity.Org 2019c). The existence of these weapon systems has been revealed for the first time by a video of Russia's state-owned "Channel One" shot at a meeting of Russian military officials held by President

[12] Note: Ballistic missiles could reach orbital velocities. See on this (Podvig 2006, pp. 83–85).

[13] See also Russia's Poseidon Thermonuclear Torpedo Is No Aircraft-Carrier Killer, available at https://nationalinterest.org/blog/buzz/russia%E2%80%99s-poseidon-thermonuclear-torpedo-no-aircraft-carrier-killer-40242

Vladimir Putin on November 9, 2015 (Sutyagin 2016, p. 1; Podvig 2015). The camera briefly focused on a slide showing diagrams of a nuclear-armed, nuclear-powered, torpedo-like device labeled as "Oceanic Multipurpose System Status-6". The leak has been labeled as a staged performance (Sutyagin 2016, p. 1).

The slide described the mission of the weapon as "Damaging the important components of the adversary's economy in a coastal area and inflicting unacceptable damage to a country's territory by creating areas of wide radioactive contamination that would be unsuitable for military, economic, or other activity for long periods of time" (Podvig 2015). The drone would reportedly have a maximum range of 5400 nautical miles (10,000 km) while traveling at a depth of 1000 m (GlobalSecurity.Org 2019c).

Nonetheless, there are doubts over its real capabilities (Cooper 2018). It is reported that even if the drone could reach the claimed top speed of 100 knots throughout a 10,000 km journey (the claimed range of the drone), the combat launch of the vessel would mean a 54-hour-long potential loss of effective control of a high-yield thermonuclear weapon by Russia's. Extremely low-frequency (ELF) communication, while possible in theory at a cruise speed much lower than 100 knots, does not provide an effective means of control of the submarine drone (Sutyagin 2016, p. 3). Despite this, the autonomous nuclear-drone has been mentioned in the U.S. Nuclear Posture Review, signaling that it is considered a real threat by the U.S..

It does not seem that these two destabilizing weapon systems are near to be deployed. Nonetheless, as has been said, these weapons could signal that Russia is determined to communicate that even after a crippling nuclear strike, greatly reducing Russia's counter-strike capability, Russia would still possess the option of inflicting an unacceptable level of nuclear annihilation (Cooper 2018).

4 Expanding "ASAT States' Club"

The outer space is an ideal environment for the acquisition of information of important strategic-military value, to the point that the development of space technology has been, since its origin, closely linked to its potential use for military purposes. As was correctly observed: "*Space has always been militarised. Military considerations were at the heart of the original efforts to enter space and have remained so to the present day*" (Ferreira-Snyman 2015, p. 495). The outer space has represented and represents what military commanders throughout the centuries have always sought: the "highest frontier", from which to acquire a complete strategic vision (Schmitt 2006, p. 94).

Military satellites play a fundamental role in telecommunications, reconnaissance, navigation, meteorology and early missile warning. They are an essential part of the command-and-control infrastructure, thus degrading spate assets will degrade the control of conventional (and nuclear) forces. Advanced armed forces, the one of the U.S. in particular, have deeply integrated space assets in theater battlefield operations. Space assets can guarantee instant communication between combatants engaged in an armed conflict, allow the observation of the movements of hostile troops, identify the objectives of possible attacks, warn of an impending missile attack against deployed troops. The U.S. military officials refer to space assets as a "force multiplier" (Aberle 2003, p. 41).

Operation Desert Storm, conducted in 1991 against Iraq by a coalition of states under the command of the United States, represents the cornerstone of the modern use of space assets in armed conflicts (Matthews 1996; Watts 2001; Chapman 2008, pp. 38–39), given the systematic integration of space technologies in all aspects of combat on the battlefield (Chapman 2008, p. 39).

Satellite systems have guaranteed, and continue to guarantee, military advantages that cannot be reached by any other technology, providing a decisive support for victory in armed conflicts (Rice 1996). As has been correctly said, in fact, "*Space today is, from a military perspective, fundamental to every single military operation that occurs on the planet today* (…). *Every operation, from humanitarian operations to major combat operations, is critically dependent on space capabilities* (…). *Every fighter aircraft, every bomber aircraft, every ship, every wheeled vehicle, every single soldier, every single airman, Marine and sailor is critically dependent on space to conduct their operations*" (Hyten 2017).

But while, on the one hand, space systems become more integrated with the mechanisms of warfare, on the other hand, the essential space support turned out to be also a manifest weakness which modern armed forces critically rely on. Militaries with developed space capabilities have grown a strong reliance toward space assets in association with the awareness that during a conflict, the adversaries will try to deny their linkage to space. Even non-state actors rely on space assets as a support for their military operation, for example, Palestinian al-Aqsa Martyrs' Brigades acknowledged using Google Earth for determination of rocket attack targets (Vittori 2011, p. 20).

In fact, satellite systems are considered soft targets: they are unshielded, they follow predictable orbits, they are still relatively few in number with the consequence that destroying or damaging even a handful of them could have a major impact, indeed, States and private corporations do not maintain standby fleets of spares to rapidly reconstitute a satellite architecture that was suddenly degraded by hostile action (Koplow 2009, p. 158; Klein 2006, p. 75).

NATO officials are aware of the heavy reliance of their armies on the support from satellites. NATO Parliamentary Assembly stated in 2017: "The twenty-first century will prove to be the race for space. Space-based systems are the key enablers of national and international infrastructures of today and tomorrow. The current speed of technological developments indicates the pace of diffusion of technology with some form of dependence on space-related hardware will only accelerate. Accordingly, outer space is becoming increasingly congested, contested, and competitive" (NATO Parliamentary Assembly 2017) (Fig. 6).

The major spacefaring nations have perceived the need to protect such a vulnerable and critical infrastructure; in fact, there is a renewed interest to the development of anti-satellite capabilities (ASAT) both by spacefaring countries and by

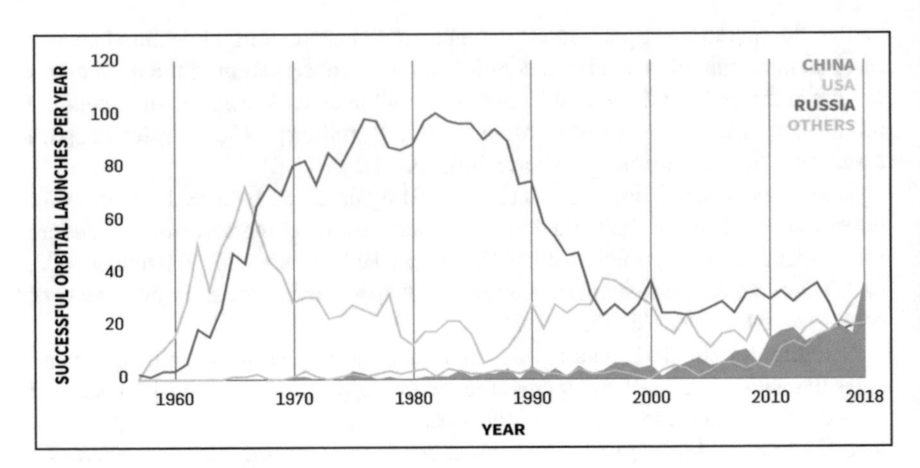

Fig. 6 Successful orbital launches per year. Note: In 2018, China conducted 38 orbital launches, surpassing the United States' 34 launches (CSIS 2019)

countries interested in signaling their capability to disrupt the use of space to potential competitors.

There are various kinds of ASAT systems[14], these weapons are usually divided in kinetic energy weapons and directed energy weapons. Kinetic energy ASAT weapons derive their destructive force from the speed of physical impact against the target, while directed energy weapons rely on direct transmission of energy to the target in order to temporarily or permanently damage it (Wolter 2005, p. 50; Zohuri 2016, pp. 80–81). These weapons could be fired from Earth to Space (via land, sea, air) or could be stationed in outer space (co-orbital ASATs). An example of kinetic energy weapons are standard missiles or killer satellites (i.e. a space vehicle which maneuver and explode near the target). Examples of directed energy ASAT weapons are high-energy laser weapons, radio-frequency weapons, subatomic-particle beam weapons, and weapons capable of generating an electromagnetic pulse (Koplow 2009, pp. 168–169; Kopp 1996).

To date, ASAT weapons are relatively easy to develop. Anti-ballistic missiles could be used as ASAT weapons without significant modifications (NRC 2012 p. 1; Grego 2012, p. 12). The proliferation of ABM missile with the ability to attack satellites represents a real threat to space activities (Baines 2003, p. 370). Furthermore, it has been reported that even less advanced but widespread missile systems, such as the Soviet-era SCUDs, could have the ability to attack satellites[15].

The "club" of states with proven ASAT capabilities is rapidly expanding, it seems that showing off ASAT capabilities has become a way to send a strong signal

[14] A system capable of disrupting, damaging, or destroying spacecraft in orbit from positions on land, sea, air, or space. (Chapman 2008, p. 370)+.

[15] The North Korean missile *Nodong* and the Pakistani missile *Ghauri*, both derived from SCUDs, could be used to attack satellites (Deblois et al. 2004, pp. 60–61; Hays et al. 2014, p. 73).

to opponents. Since intercepting a satellite is relatively easier than intercepting a missile (Johnson-Freese 2009, p. 39), the meaning of the signal is not related to showing to opponents a great military achievement, but it is linked to demonstrate one's own resolution and determination that in an eventual conflict enemy satellites will be targeted. The U.S., Russia and China have recently "showed" their anti-satellite capabilities. Currently, the main weapon systems to target and destroy satellites are direct-ascent conventional missiles. The U.S. used an Aegis SM-3 ballistic missile interceptor to destroy a satellite in 2008, proving that their missile interceptors of the Aegis Missile Defense system could be used as ASAT weapons. China utilized a modified version of its DF-21 ballistic missile and Russia has recently tested its new PL-19 Nudol ASAT missile (Grego 2012 p. 12).

India has also joined the "ASAT club" by testing on March 27th, 2019. The test involved an indigenously produced "Ballistic Missile Defense Interceptor" targeting a satellite in Low Earth Orbit (LEO), at a height of approximately 300 km (Shaikh 2019). The test could spark a dangerous local arms race with local powers interested in showing off the same capability. There are reports that Pakistan could be on the way to develop an Air-launched Satellite Launch Vehicle which could be used as a platform for an air-launched ASAT missile (Sace Threat Assessment 2019, p. 39).

It has been reported that a targeted satellite would only have an estimated 8–15 min warning time before impact (Global Counterspace Capabilities, p. 2–19) in the case of a direct-ascent missile, to counter such an attack during this timeframe, the ASAT launch must be detected, the targeted satellite identified, and the satellite maneuvered onto a new orbital path. However, fuel for maneuvering is finite and the maneuver itself could eliminate the satellite's future mobility.

ASAT weapons have always had a destabilizing impact on international security, indeed ASAT weapons increasingly render outer space a medium of warfare, pushing further states to develop ASAT technologies. At the time of this writing, Pakistan, Iran and North Korea could be the next players to enter the "ASAT club" (Ibid.).

Today ASAT weapons are more than destabilizing, the use of these assets could cripple the use of advanced military capabilities (Ibid.) and the intentional or accidental disruption of military satellites part of the command and control infrastructure could trigger a rapid nuclear escalation (AUSA 2017, NASIC 2019 p. 7). The U.S. have explicitly pointed out that damaging U.S. and allied early-warning infrastructure will cross a so-called nuclear red line (US DOD 2018). An imaginary warfare scenario could well explain the perils of damaging early-warning satellites. For example, during a low-intensity armed conflict between Russia and the U.S. in Eastern Europe, Russia's missile systems are largely intercepted by U.S. missile defenses crippling Russian land troops to rapidly advance, entering in a stagnating phase of the conflict. In order to degrade the rate of missile interception, Russian military officials decide to suddenly target U.S. ground-based and space-based early-warning systems. The U.S. capability to detect hostile missile launches would then be heavily degraded. The U.S. could think that the Russians are preparing an attack involving missile launched tactical nuclear weapons to destroy the U.S. theater command and control centers; temporarily deprived of effective early-warning capabilities and with command and control centers at risk, in the heat of the moment the U.S. could be tempted to escalate first.

As has been stated above, a large set of widespread missile systems could possess ASAT capabilities, their destabilizing effect are not isolated within the perspective of a war to be fought in outer space but, especially in a confrontation involving nuclear powers, ASAT missiles are an element that will make it difficult to manage a potential nuclear escalation.

5 Usable Nuclear Weapons: Low-Yield Nuclear Weapons

There has been a rapid evolution in nuclear weaponry since the 40's. An atomic device, the only sort available in 1945, relies upon fission – the release of immense quantities of energy from the splitting of the nucleus of atoms of a heavy element such as uranium or plutonium. In contrast, a nuclear ("thermonuclear" or "hydrogen") device operates via fusion – the energy is released from the forced joining of small elements such as atoms of hydrogen to create helium. The most sophisticated modern types of weapons utilize fission and fusion processes sequentially: an initial fission explosion (the so-called "primary"), which creates the heat and pressure conditions necessary to trigger a "secondary" fusion explosion, which accounts for most of the power of the bomb (Sublette 2019).

During the Cold War, both the U.S. and Soviet Union designed bombs with specialized effects to optimize their employment, bombs with reduced low-yield, for example, were conceived to be employed in a tactical warfare scenario.

The tactical use of nuclear weapons is defined as "the use of nuclear weapons by land, sea, or air forces against opposing forces, supporting installations or facilities, in support of operations that contribute to the accomplishment of a military mission of limited scope, or in support of the military commander's scheme of maneuver, usually limited to the area of military operations." (Woolf, Amy 2019b, p. 8).

In order to employ a nuclear weapon in support of tactical operations, the nuclear warhead should be characterised by small yields (Ibid.). After reductions in nuclear stockpiles and in the interest of maintaining tactical nuclear weapons during the 90s, the role of low-yield has been resurrected. Within this scenario, most attention is focused on the relations between Russia and the NATO as the most recent motives for developing low-yield nuclear weapons are driven by their weakened strategic (*in*)stability relationship. Indeed, in 1991 the U.S. and the U.S.S.R. agreed to dismantle their tactical nuclear weapons; while the U.S. dismantle process was quick, the Soviet and then Russian dismantling level has been uncertain (Adamsky 2013, pp. 7–8; Woolf, Amy 2019b, pp. 12–14). Moreover, the estimated locations of the Russian storage facilities of "non-strategic nuclear weapons" coincide with estimated theaters of their employment.

There at least three important elements which have led to the resurgence of the importance of low-yield nuclear weapon in military planning:

- The entanglement between conventional and nuclear deterrence in nuclear doctrines, which have led to foresee the use of nuclear weapons as a response to a conventional aggression.
- The contested "escalate to de-escalate" Russia's doctrine (there is a degree of uncertainty about this point in the arms control community), and the consequent need for the U.S. and NATO to rely on low-yield warheads for a credible deterrence.
- Improved accuracy of delivery systems which have boosted the "acceptability factor" of a low-yield nuclear strike.

Another argument which could sustain the development of low-yield nuclear weapons and a military doctrine governing their use could be that these devices could allow for nuclear escalation control (Krepon 2017). As will be discussed this argument is hardly sustainable.

Russia and the U.S. seems to now have fallen in to a vicious circle of nuclear confrontation which is bringing the calculations underpinning nuclear deterrence far from the goal of "avoiding war" and closer to "employable battlefield capabilities".

It is estimated that Russia possess between 1000 and 6000 non-strategic low-yield nuclear warheads (Adamsky 2013 p. 8; Kristensen and Korda 2019 p. 80; Woolf, Amy 2019b), and, according to reports, the Russian military continues to attribute importance to non-strategic nuclear weapons for use by naval, tactical air, and air– and missile-defense forces, as well as on short-range ballistic missiles (Kristensen and Korda 2019, p. 80); the rationale appears to be that non-strategic nuclear weapons are needed to offset the superior conventional forces of NATO and particularly of the United States (Ibid.). The U.S. Nuclear Posture Review states that "the United States will enhance the flexibility and range of its tailored deterrence options […] Expanding flexible US nuclear options now, to include low-yield options, is important for the preservation of credible deterrence against regional aggression."

Furthermore, the low transparency of nuclear doctrines contributes to add fuel to the "tactical nuclear confrontation". For example, some authors state that Russia's declaratory nuclear policy allows for a broader use of nuclear weapons, far from their official nuclear doctrine (Schneider 2019) (i.e. it allows for *preventive* nuclear strikes); other authors sustain that this is a wrong interpretation of Russia's nuclear intentions (Olike 2016).

Despite who is right and who is wrong on the debate about the Russian nuclear threshold, both the U.S. and Russia are developing low-yield nuclear weapons for tactical use (Kristensen and Korda 2019; Werner 2019a, b).

The doctrine behind low-yield nuclear weapons is related to render deterrence more credible. Despite this, low-yield nuclear weapons are something different to strategic nuclear weapons in terms of deterrence. A State which is disseminating low-yield tactical nuclear weapons in its armed forces it is not just deterring its opponents; that State is sending a strong signal of its readiness to employ such weapons as part of tactical operations in regional or local conflict (if not in limited military operations). Tactical nuclear weapons are intrinsically able to lower the bar for nuclear use as their role is to render nuclear weapons usable as part of a process to acquire victory in local military conflicts.

In the near future, military doctrines of employment of low-yield nuclear weapons will impact on deterrence theory, making it necessary to change the way of how deterrence is thought and managed. Today, even nuclear and non-nuclear capabilities have equal implications for deterrence and strategic stability relationships (Acton 2018). The same nature of deterrence will change if Russia and NATO will integrate low-yield tactical nuclear weapons as part of tactical warfare in order to prevail on hostile forces in limited military operations.

In this scenario, the "*contestable nature*" of conventional deterrence[16] will be accompanied by the potential use of low-yield nuclear weapons, making it "*uncontestable*". A clear division between nuclear and conventional deterrence will have relevance only in books, while there will not be a real distinction between conventional and nuclear deterrence in practical terms.

From a pure strategic view, it has no relevance which side will start first to foresee the use of low-yield nuclear weapons for defeating enemy forces in local warfare scenario, the other side(s) will be dragged in the same pattern.

Since low-yield nuclear weapons are not a tool of "escalation control" that will produce predictable and certain results[17], NATO should look at alternatives to ensure its deterrence even in front of hostile employment of low-yield nuclear weapons. In fact, the recognition that deterrence will benefit from the reliance on advanced non-nuclear military systems capable to cripple the enemy's key defence assets could be one of the bases to deter opponents keeping low-yield nuclear weapons outside tactical warfare scenario.

6 A Brief Look at the Military Applications of AI

AI gets viewed as 'simulation of human intelligence processes by machines, especially computer systems. These processes include learning (the acquisition of information and rules for using the information), reasoning (using the rules to reach approximate or definite conclusions) and self-correction. Particular applications of AI include expert systems, speech/voice recognition and machine vision' (Search Enterprise 2018). AI refers to any computer system that uses a logical process to learn and improve by factoring the surrounding environment and prior mistakes (Lele 2019, p. 140.).

Research on AI is happening for many years, both government and private industry are making heavy investments in this area. It is reported that almost 300%

[16] On the contestable nature of conventional deterrence see (Wirtz 2018).

[17] Indeed, if both opponents are willing to accept a mutual exchange of low-yield nuclear strikes, the one who would then find himself at a disadvantage could prefer to escalate.

increase in investment in AI took place during 2017 compared with 2016 (McCormick 2016). AI does not have a significance as a "stand-alone" military technology, its significance becomes more prominent when AI is applied as an element of a more complex weapon system (FAS 2018a, b p. 2).

AI systems are known to improve machine behaviour, and any military system which is based on electronic sensors or computer systems would achieve an enhanced effectiveness through AI. One of the relevant integration of AI is represented by its integration with so-called command, control, communications, computers, intelligence, surveillance, and reconnaissance (C4ISR) architecture.

AI is contributing significantly to the fields of ground, aerial, space and underwater ISR capabilities (Lele 2019). Noticeably, unmanned aerial vehicles (UAVs) are found using AI-based techniques for the purposes of intelligence, surveillance and reconnaissance.

The combination of artificial intelligence, persistent surveillance by unmanned aerial vehicles, airborne early warning systems, underwater drone sensors and constellations of imaging satellites continuously generate vast volumes of data and imagery, round the clock. Using such data, AI-based systems develop approaches to enhance active and passive sensors useful for identifying hidden threats in the battlefield with precision and accuracy.

AI development set the conditions for fully autonomous weapon systems (Fig. 7).

Fully autonomous military system has been increasingly integrating with existing weapon systems. If in conventional thinking fully autonomous weapons are yet to come, isolated cases of fully autonomous weapons go back decades (Fig. 8).

In concept, AI can be used to control systems that use a sensor to trigger an automatic military action. Fire-and-forget, once activated, select and engage targets on their own without any human intervention.

Autonomous weapons could be designed as offensive or defensive weapons.

Advanced anti-missile and anti air-craft system, like THAAD and S-400, are fully autonomous. As regards the THAAD systems, for example, an infrared satellite detects an incoming missile's heat signature, sends all collected real-time tracking data to the ground-based system through a communications satellite.

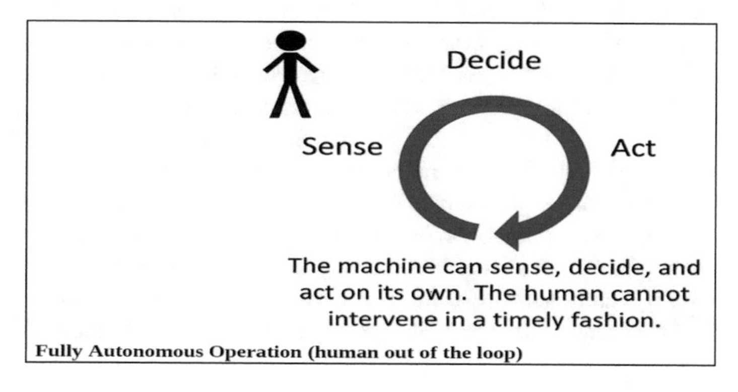

Fig. 7 Example of autonomous decision-making without human intervention

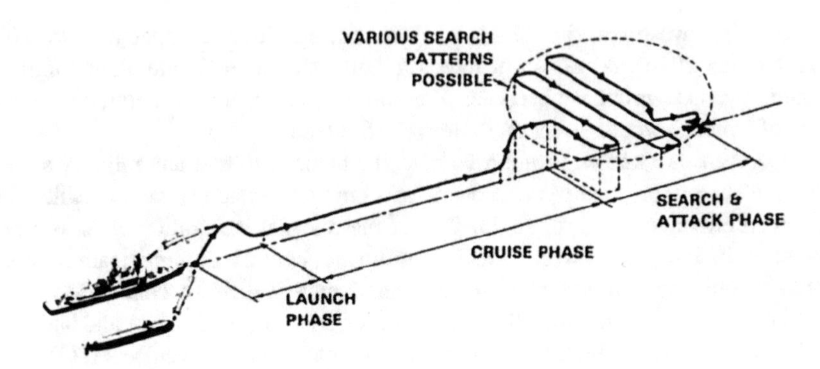

Fig. 8 Tomahawk Anti-Ship Missile, developed by the U.S. during the 80's

When the threat is confirmed, the weapon system turns into active mode. Subsequently, the long-range radar detects and tracks the missile for some time to further improve the accuracy. Among the group of batteries available to address the threat, the most effective interceptor battery is engaged and carries out the interception.

Near to be deployed offensive autonomous weapon system are increasingly making their appearance.

For example, the U.S. are developing armed drone swarms, unmanned flying units which fly in formation to achieve a given task. The Perdix project consists of autonomous drones operating in a formation represented by cooperative swarms of 20 or more flying units.

An already deployed advanced autonomous missile system is the Brimstone missile, developed for the Royal Air Force, an air-launched, fire-and-forget missile designed to destroy ground vehicles or small boats. The Brimstone could follow a target selected by a human operator with a laser but can also work in a "fire and forget" mode where the missile's software seeks a predesignated target type within an established kill box (FAS 2018b, pp. 144–15). Brimstone has reportedly been used against Islamic State targets in Syria (Ibid.). The Brimstone missile is similar to the U.S. Long Range Anti-Ship Missile (LRASM). An LRASM can find its own target autonomously by using its active radar homing to locate ships in an area, then using passive measures once on terminal approach (Gresham 2013; Pace 2016, p. 247). In the near future, armored robotic land-units will assist human operators on the battle field. Uran-9 and Vikhr are Russian semi-autonomous tanks which could serve as a platform for a full autonomous future system[18].

Hunting mobile targets in communications-denied areas represents one the most relevant military utility of autonomous weapon systems. In an ideal world, robotic fully autonomous weapons would attack only if a human operator authorized the engagement. If a communication link is not available, fully autonomous weapons will be used for search, select and engage targets on their own. In these situations,

[18] For a detailed list of Autonomous Weapon System see Scharre (2018).

there would be no ability to recall or abort the weapon if it failed into attacking the wrong target. There could even be the possibility to leave small "dormant" autonomous weapons where the enemy forces are expected to pass[19].

The ongoing evolution of AI and the capability to manage enormous amount of data constitute a potential enabler for autonomous weapon systems. From autonomous drones and sensors for persistent and continuous coverage and analysis of data enhancing command and control capabilities, to robotic weapons with a fully-autonomous decision making process, this arena is highly problematic in terms of ethics and military doctrine. It seems that the perils linked to robot military systems are being perceived by the same militaries, indeed, as was stated "The main rationale for building fully autonomous weapons seems to be the assumption that others might do so" (Scharre 2018, p. 317).

7 Conclusions

The conjunction of the technological process is accompanied by a low level of trust between the leading nations. This problem has affected the process with which States perceive other states' intentions in the international society, bringing a growing level of misperception and mistrust. Cooperation between U.S., China and Russia is mostly needed as they are the drivers for both new weaponry and the reasoning behind the need of new weaponry.

Without effective forms of discourse, the great powers are likely to develop a form of strategic stability relationship based on signalling their military readiness to meet mutual military threats. The modernization of strategic missile systems is increasingly accompanied by the development of advanced forms of military technologies which have the purpose to exploit the vulnerabilities of the other competitors. If this process will likely to continue, the "stability" between U.S., China and Russia and their allies will be unbalanced by a vicious circle of exploitation of mutual vulnerabilities. Russia's development of nuclear powered missiles and torpedoes as weapons of pure retribution in the case of a successful hostile first-strike highlight a dangerous malfunction of its strategic stability relationship with the U.S.. Does still effectively exist a will to pursue strategic stability? It seems that it has been substituted by the search for strategic superiority. Arms control measures must take a broader look to these issues as the nuclear and conventional realm are now, and will be growingly, intertwined. The broader scenario for an effective "stabilization" process should not be based on limited solutions; the recognition that national security will benefit from mutual restraint about pursuing new destabilizing warfare technologies could be one of the bases on which to start a process of strategic "re-stabilization".

[19]The Russian Uran-9 autonomous tank is expected to wait hidden for enemy forces to be in the nearing before engaging. See Scharre (2018).

References

Aberle, A. 2003. *Strategic force multiplier: The importance of space in the Army's future*. Army Space Journal 4. 41–42.

Acton, James. 2013. *Silver bullet? Asking the right questions about conventional prompt global strike*. Washington: Carnegie Endowment for international Peace.

———. 2015. Hypersonic boost-glide weapons. *Science and Global Security* 23 (3). 191–219.

———. 2018. Escalation through entanglement. *International Security* 43 (1). 56–99.

Adamsky, Dmitry (Dima). 2013. *Nuclear incoherence: deterrence theory and non-strategic nuclear weapons in Russia*. The Journal of Strategic Studies.

Air Combat Command. 2012. *Over the horizon backscatter radar: East and west*. Available at https://www.acc.af.mil/About-Us/Fact-Sheets/Display/Article/199120/over-the-horizon-backscatter-radar-east-and-west/. Accessed 11 June 1029.

Association of the United States Army (AUSA). 2017. *Satellite and ground communication systems: Space and electronic warfare threats to the United States Army*. Available at https://www.ausa.org/publications/satellite-and-ground-communication-systems-space-and-electronic-warfare-threats-united.

Baines, P.J. 2003. *Prospects for "Non-Offensive" Defenses, in Space*. In *New challenges in missile proliferation, missile defense, and space security*, ed. J.C. MOLTZ. Monterey: Monterey Institute of International Studies.

Barton, David K., Roger Falcone, Frederick K. Daniel Kleppner, Lamb, et al. 2004. *Report of the American Physical Society Study Group on boost-phase intercept systems for national missile defense: Scientific and technical issues*, 60. American Physical Society.

Center for Strategic and International Studies (CSIS). 2019. *Space threat assessment 2019*.

Cepek, Robert J. 2005. *Ground-based midcourse defense: Continue testing, but operational fielding must take a backseat to theater missile defense and homeland security*, Joint Forces Staff College 2005, p. 5. See also 'Mid-Course Phase" at https://www.globalsecurity.org/space/systems/mid-course.htm.

Chapman, B. 2008. *Space warfare and defense. A historical encyclopedia and research guide*. Santa Barbara: Bert Chapman.

Chen, Robert, and Jason Speyer (2008): *Terminal and Boost Phase Intercept of Ballistic Missile Defense,* AIAA Guidance, Navigation and Control Conference and Exhibit (18–21 August 2008).

Chinese Academy of Military Science. 2013. *The science of military strategy*. Beijing: Military Science Press.

CNS. 2015. *Hyper-glide delivery systems and the implications for strategic stability and arms reductions*. Washington DC: James Martin Center for Nonproliferation Studies.

Congressional Budget Office. 2004. *Alternatives for boost-phase missile defense*. Washington D.C.: United States Congressional Budget Office. Available at https://apps.dtic.mil/dtic/tr/fulltext/u2/1001718.pdf.

Cooper, Julian. 2018. *Russia's invincible weapons: Today, tomorrow, sometime, never?* Available at http://www.ccw.ox.ac.uk/blog/2018/4/30/russias-invincible-weapons-today-tomorrow-sometime-never. Acessed 11 June 2019.

Deblois, B.M., R.L. Garwin, R.S. Kemp, and J.C. Marwell. 2004. Space weapons, crossing the U.S. rubicon. *International Security* 29 (2004).

Defense News. 2019. *Pentagon considers an ICBM-killing weapon for the F-35, but is it affordable?* Available at https://www.defensenews.com/air/2019/01/17/pentagon-considering-an-icbm-killing-weapon-for-the-f-35-but-can-it-afford-it/.

Dvorking, Vladimir. 2019. *Preserving strategic stability Amid U.S.-Russian confrontation*. Available at https://carnegie.ru/2019/02/08/preserving-strategic-stability-amid-u.s.-russian-confrontation-pub-78319.

Evers, Lanah, Ana Isabel Barros, and Herman Monsuur. 2015. *The cooperative ballistic missile defence game*. In *Decision and gametheory for security: Proceedings of the 4th international conference*, ed. Sajal K. Das, Cristina Nita-Rotaru, and Murat Kantarcioglu, 85–86. Fort Worth. November 11-12, 2013: GameSec 2013.

Fedasiuk, Ryan. 2018. U.S. missile defense and the theater-strategic nexus. *New Perspectives in Foreign Policy, Fall* 2018 (16).

Federation of American Scientists (FAS). 2018a. *Emerging disruptive technologies and their potential threat to strategic stability and national security*, FAS Special Report. Available at https://fas.org/wp-content/uploads/media/FAS-Emerging-Technologies-Report.pdf?utm_source=FAS+General&utm_campaign=9b44d2bccd-EMAIL_CAMPAIGN_2017_02_21_COPY_01&utm_medium=email&utm_term=0_56a7496199-9b44d2bccd-222162373.

———. 2018b. *U.S. Ground Forces Robotics and Autonomous Systems (RAS) and Artificial Intelligence (AI): Considerations for Congress*. Available at https://fas.org/sgp/crs/weapons/R45392.pdf.

Ferreira-Snyman, A. 2015. Selected legal challenges relating to the military use of outer space, with specific reference to article IV of the outer space treaty. *Potchefstroom Electronic L.J.* 18 (3).

Gertz, Bill. 2018. *Pentagon: Russia tested nuclear-powered cruise missile twice*. Available at https://freebeacon.com/national-security/pentagon-russia-tested-nuclear-powered-cruise-missile-twice/. Global Counterspace Capabilities: An Open Assessment.

GlobalSecurity.Org. 2019a. *KRND Burevestnik [Petrel] SSC-X-9 Skyfall*. Available at https://www.globalsecurity.org/wmd/world/russia/krnd.htm.

———. 2019b. *KRND Burevestnik – Technology*. Available at https://www.globalsecurity.org/wmd/world/russia/krnd-tech.htm.

———. 2019c. *Status-6/Kanyon – Ocean Multipurpose System*. Available at https://www.globalsecurity.org/wmd/world/russia/status-6.htm.

Goodby, James E., and Theodore A. Postol. 2018. A new boost-phase missile defense system—and its diplomatic uses in the North Korea dispute. *Bulletin of the Atomic Scientists* 74 (4).

Gormley, Dennis M. 2015. US advanced conventional systems and conventional prompt global strike ambitions. *The NonProliferation Review* 22 (n. 2).

Grego, Laura. 2012. *A history of anti-satellite programs*, Union of Concerned Scientists (2012). Available at https://www.ucsusa.org/sites/default/files/legacy/assets/ documents/nwgs/a-history-of-ASAT-programs_lo-res.pdf.

Gresham, John D. 2013. *LRASM: Long range maritime strike for air-sea battle*. Available at https://www.defensemedianetwork.com/stories/lrasm-long-range-maritime-strike-for-air-sea-battle/

Gubrud, Mark. 2015. Going too fast: Time to ban hypersonic missile tests? A US response. *Bulletin of the Atomic Scientists* 71 (5).

Hacker, Barton C. 1995. *Whoever heard of nuclear Ramjets? Project pluto 1957-1964. Journal of the International Committee for the History of Technology* 1.

Hays, P., et al. 2014. *Space security index* # 56.

Hyten, John E. 2017. *Statement by General John E. Hyten, Commander of the US Strategic Command,* Stanford University Center for Security and Cooperation on January 24, 2017. Available at http://www.stratcom.mil/Media/Speeches/Article/1063244/center-for-international-security-and-cooperation-cisac/.

Institute for Foreign Policy Analysis (IFPA). 2009. *Missile defense, the space relationship, & the twenty-first century*. Available at http://www.ifpa.org/pdf/IWG2009.pdf.

IRSN. 2018. *Report on the IRSN's investigations following the widespread detection of 106Ru in Europe early October 2017*. Available at https://www.irsn.fr/FR/Actualites_presse/Actualites/Documents/IRSN_Report-on-IRSN-investigations-of-Ru-106-in-Europe-in-october-2017.pdf. Accessed 11 June 2019.

Johnson-Freese, Joan. 2009. *Heavenly ambitions: America's quest to dominate space*. Philadelphia: University of Pennsylvania Press.

Klein, John L. 2006. *Space warfare strategy, principles and policy*. Routledge.

Koplow, D.A. 2009. *Death by moderation: The U.S. military's quest for useable weapons*. Cambridge: Routledge.

Kopp, C. 1996. *The E-bomb: A weapon of electrical mass destruction*. Proceedings of InfoWarCon V, Washington 1996. Available at: http://www.dtic.mil/docs/citations/ADA332511.

Krepon, Michael. 2017. *Low yield nuclear weapons (again)*. Available at https://www.armscontrolwonk.com/archive/1202952/low-yield-nuclear-weapons-again/.

Kristensen, Hans M., and Matt Korda. 2019. Russian nuclear forces. *Bulletin of the Atomic Scientists* 75 (2): 2019.
Leah, Christin M. 2017. *The consequences of american nuclear disarmament strategy and nuclear weapons*. New York: Palgrave Millan.
Lele, Ajey. 2019. *Disruptive technologies for the militaries and security*. Singapore: Springer.
Mallik, Amitav. 2004. *Technology and security in the 21st century: A demand-side perspective*. Oxford: Oxford University Press.
Matthews, E. D. 1996. *U.S. space systems: A critical strength and vulnerability* (Newport).
McCormick, J. 2016. *Predictions 2017: Artificial intelligence will drive the insights revolution*. Available at https://go.forrester.com/wp-content/uploads/Forrester_Predictions_2017_-Artificial_Intelligence_Will_Drive_The_Insights_Revolution.pdf.
Missile Defense Project. 2019. *Russia tests nuclear-powered Burevestnik cruise missile*, Center for Strategic and International Studies, published February 6, 2019, last modified February 6, 2019. Available at https://missilethreat.csis.org/russia-tests-nuclear-powered-burevestnik-cruise-missile/.
Nagappa, Rajaram. 2015. Going too fast: Time to ban hypersonic missile tests? An Indian response. *Bulletin of the Atomic Scientists* 71 (5).
National Air and Space Intelligence Center (NASIC). 2019. *Competing in Space*. Available at https://media.defense.gov/2019/Jan/16/2002080386/-1/-1/1/190115-F-NV711-0002.PDF.
National Research Council. 2007. *Conventional Prompt Global Strike Capability*, Letter Report
———. 2008. *U.S. Conventional Prompt Global Strike: Issues for 2008 and Beyond*, Committee on Conventional Prompt Global Strike Capability.
———. 2012. *Making sense of ballistic missile defense: An assessment of concepts and systems for U.S. boost-phase missile defense in comparison to other alternatives* Committee on an assessment of concepts and systems for U.S. Vol. 2012. Boost-Phase Missile Defense in Comparison to Other Alternatives Washington, DC.
NATO Parliamentary Assembly. 2017. *The space domain and allied defense*, Defense and Security Committee, Draft Report – 068 DSCFC 17 E, 20 March 2017.
Nilsen, Thomas. 2018. *Is Putin's new nuclear systems source of mysterious radioactivity in the air?* Available at https://thebarentsobserver.com/en/ecology/2018/03/putins-new-nuclear-systems-source-mysterious-radioactivity-air.
Olike, Olga. 2016. *Russia's Nuclear Doctrine What We Know, What We Don't, and What That Means*, CSIS Reports, 2016. Available at https://csis-prod.s3.amazonaws.com/s3fs-public/publication/160504_Oliker_RussiasNuclearDoctrine_Web.pdf.
Pace, Steve. 2016. *The projects of skunk works 75 years of lockheed martin's advanced development program*. Beverly: Voyageur Press.
Podvig, Pavel. 2006. Reducing the risk of an accidental launch. *Science and Global Security* 14.
——— (2015): *Is Russia working on a massive dirty bomb?*, Russian strategic nuclear forces blog, 11 November 2015. Available at http://russianforces.org/blog/2015/11/is_russia_working_on_a_massive.shtml.
RAND. 2017. *Hypersonic missile nonproliferation: Hindering the spread of a new class of weapons*. Santa Monica: RAND Corporation.
Regan, Frank J., and Satya M. Anandakrishnan. 1993. Dynamics of atmospheric re-entry. *American Institute of Aeronautics and Astronautics* 1993.
Rice, Donald B. 1996. Global reach – Global power. In Peter L. Hays, Brenda J. Vallance, Alan R. Van Tassel (eds.) *American Defense Policy 7th edition*.
Rose, Frank A. 2018. *Will the upcoming missile defense review maintain the current course or plot a new direction?* Available at https://www.brookings.edu/blog/order-from-chaos/2018/06/11/will-the-upcoming-missile-defense-review-maintain-the-cur-rent-course-or-plot-a-new-direction/.
Russian Federation. 2015. The Military Doctrine of the Russian Federation. Available in English at https://rusemb.org.uk/press/2029. Accessed 11 June2019.
Scharre, Paul. 2018. *Army of None: Autonomous weapons and the future of war*. New York: W.W. Norton & Company.

Schelling, Thomas C. 1970. *Surprise Attack and Disarmament*. In *Theories of peace and security: A reader in contemporary strategic thought*, ed. John Garnett, 1970. New York: Macmillan St Martin's Press.

Schmitt, M.N. 2006. *International law and military operations in space*. In *Max planck yearbook of united nations law*, ed. A. von Bogdandy and R. Wolfrum, vol. 10. Leiden: Brill.

Schneider, Mark B. 2019. Russian nuclear "de-escalation" of future war. *Comparative Strategy* 37 (5).

Search Enterprise AI. 2018. *Definition AI (artificial intelligence)*. Available at http://searchcio.techtarget.com/definition/AI.

Shaikh, S. 2019. India conducts successful ASAT test, CSIS Missile Defefnse project. Available at https://missilethreat.csis.org/india-conducts-successful-asat-test/.

Sonne, Paul. 2018. *Pentagon looks to adjust missile defense policy to include threats from Russia, China* The Washington Post, March 02, 2018, https://www.washingtonpost.com/world/national-security/pentagon-looks-to-adjust-missile-defense-policy-to-include-threats-from-russia-china/2018/03/01/2358ae22-1be5-11e8-8a2c-1a6665f59e95_story.html?noredirect=on&utm_term=.45df834972e7.

Sublette, Carey. 2019. *Nuclear weapons frequently asked questions*. Available at http://nuclearweaponarchive.org.

Sutyagin, Igor. 2016. *Russia's underwater "doomsday drone": Science fiction, but real danger*, Bulletin of the Atomic Scientist.

The Aviationist. 2017. *U.S. Air Force deploys WC-135 nuclear sniffer aircraft to UK as spike of radioactive Iodine levels is detected in Europe*. Available at https://theaviationist.com/2017/02/19/u-s-air-force-deploys-wc-135-nuclear-sniffer-aircraft-to-uk-after-spike-of-radioactive-iodine-levels-detected-in-europe/.

Trenin, Dmitri. 2019. Russian views of US nuclear modernization. *Bulletin of the Atomic Scientist, 2019* 75 (1): 14–18.

United States Defence Department (US DOD). 2018. *The Nuclear Posture Review 2018*. Available at https://media.defense.gov/2018/Feb/02/2001872886/-1/-1/1/2018-NUCLEAR-POSTURE-REVIEW-FINAL-REPORT.PDF. Accessed 11 June 2019.

US Congress. 2018. *John S. McCain national defense authorization act for fiscal year 2019*. Available at https://www.congress.gov/bill/115th-congress/house-bill/5515/text#toc-H3B5BF0F78BDE4E55AD2A165E15881353.

Vittori, K. 2011. *Terrorist financing and resourcing*. New York: Palgrave Macmillan.

Watts, B.D. 2001. *The military use of space: A diagnostic assessment*. Washington, D.C..

Werner, Ben. 2019a. *STRATCOM commander wants to put low yield nuclear missiles on U.S. submarines*. Available at https://news.usni.org/2019/03/28/42238.

———. 2019b. *Pentagon developing low-yield nuclear cruise missiles for submarines*. Available at https://news.usni.org/2019/04/03/42398

Wiener, Rachel. 2017. *The impact of hypersonic glide, boost-glide and air-breathing technologies on nuclear deterrence*. In *Project on nuclear issues*, ed. Mark Cancian. Lanham: CSIS.

Wilkening, Dean A. 2004. Airborne boost-phase ballistic missile defense. *Science and Global Security* 12 (1).

Williams, Thomas Karakoian. 2017. Missile defense 2020: Next steps for defending the homeland (CSIS, 2017).

Wirtz, James J. 2018. How does nuclear deterrence differ from conventional deterrence? *Strategic Studies Quarterly* 12 (4).

Wolter, D. 2005. *Common security in outer space and international law*. Geneva: Detlev Wolter.

Woolf, Amy. 2019a. *Conventional prompt global strike (PGS) and long-range ballistic missiles*, Congressional Research Service Report R41464 (Washington D.C.).

———. 2019b. *Nonstrategic nuclear weapons*, Congressional Reports Service Report RL32572 (Washington D.C.).

Zohuri, B. 2016. *Directed energy weapons: Physics of high energy lasers (HEL)*. Albuquerque.

The Dark Side of Nuclear Energy: Risks of Proliferation from Domestic Fuel Cycle Technologies

Sharon Squassoni

1 Introduction

The peaceful and military uses of nuclear technology have coexisted uneasily for more than 70 years. The myth that civilian nuclear energy doesn't contribute to proliferation of nuclear weapons is exposed every few decades (e.g., India and Pakistan in the 1970s, Iraq in the 1990s, Iran in the 2000s) but countries cannot find the political will to completely cut off access to the most sensitive fuel cycle technologies for themselves or others. The most sensitive elements of the fuel cycle are uranium enrichment, which is necessary to manufacture the low-enriched uranium fuel that is used in 85% of commercial power reactors worldwide, and spent-fuel reprocessing, which is unnecessary for generating nuclear electricity, but is pursued by a handful of states seeking to ultimately close their fuel cycles by recycling spent nuclear fuel.

Both of these technologies also make fissile material for nuclear weapons and, in fact, were developed for bombs before their peaceful uses were explored. But they have gone largely unrestricted by treaties and agreements in order to promote the peaceful uses of nuclear energy.[1] Early assumptions about the difficulty of producing fissile material have given way to concern about the broad diffusion of technology and the inadequacy of controls. The example of Iran, and the additional

[1] The Nuclear Nonproliferation Treaty contains no prohibition on acquiring the means of producing fissile material for weapons, but rather the manufacture, acquisition or transfer of nuclear weapons, their control, or seeking or receiving any assistance in the manufacturing nuclear weapons. Negotiators realized the line was being drawn fairly high up in a long chain of decisions. The solution in the NPT was to impose an obligation, through comprehensive safeguards agreements, to ensure that all nuclear material in a state was used for peaceful purposes only.

S. Squassoni (✉)
George Washington University, Washington, DC, USA
e-mail: ssquassoni@email.gwu.edu

M. Martellini, R. Trapp (eds.), *21st Century Prometheus*,
https://doi.org/10.1007/978-3-030-28285-1_3

restrictions imposed upon its fuel cycle capabilities under the Joint Comprehensive Plan of Action (JCPOA) in 2015, illustrates the weaknesses of the current system of control. Even with additional monitoring and access to enrichment equipment and facilities and restrictions on uranium enrichment levels, it was thought necessary to include limits on stockpiles of materials (even at the low-enriched uranium level; high-enriched uranium was banned) to avoid a situation where Iran could quickly assemble enough material for several bombs. And yet, critics of the Iran deal have suggested that a state with questionable intentions ought not to have any fuel cycle technology at all.

This chapter discusses the risks of fuel cycle technology for the proliferation of nuclear weapons. It addresses how potential nuclear energy trends, including reactor development, might increase proliferation. It also explores how policies have adapted to those risks, suggesting that a more systematic approach to nonproliferation that specifically addresses fuel cycle risks should be developed, in contrast to responses tailored for specific, regional proliferation challenges.

2 The Dual Use Nature of Nuclear Energy

Nuclear technologies have abundant applications in civilian society, including industrial, medical and agricultural, and many of these pose no significant safety or security concerns apart from the appropriate handling of radioactive material. Research and power reactors and their supporting infrastructure, however, will always pose nuclear weapons proliferation risks.[2] As a U.S. Department of Energy official once remarked in candor, the only difference between civilian and military uses of nuclear energy was psychological (Gilinsky 2014). Whether it is the operation of the reactors themselves, the diversion of their fresh or spent fuel, or the operation of fuel cycle facilities like uranium enrichment and spent fuel reprocessing, there is no escaping the potential for misuse. Of course, most countries have no intentions to develop nuclear weapons, but the few that have considered that path and the even fewer that have succeeded, have either used civilian assets directly or used civilian programs or trade to mask their clandestine operations.[3] Nuclear weapon states have also blurred the lines between military and civilian facilities by using civilian facilities to support their weapons programs (Squassoni 2018a).

[2] In countries that already have nuclear weapons, one can argue that the risk of misuse is low because governments have little incentive to use civilian assets for military purposes, but all nuclear weapon states have historically ignored such boundaries. Even in such states, the risk of unauthorized diversion is not zero.

[3] The exception is South Africa, which conducted its nuclear weapons program largely disconnected from the outside world. South Africa is an exception in one other respect: it is the only country to have developed nuclear weapons and then dismantled them and joined the Nuclear Nonproliferation Treaty as a non-nuclear weapon state.

Unfortunately, there is no truly proliferation-resistant nuclear fuel cycle, despite decades of international studies, assessments, commissions and initiatives (National Research Council 2013). Although the Generation IV International Forum, begun in 2000 by the U.S. Department of Energy, incorporated proliferation resistance into its objectives, the fast reactors under development rely on spent-fuel reprocessing and in some cases, pyroprocessing.[4] Members of the forum chose to pursue four fast reactor technologies (gas-cooled, lead-cooled, sodium-cooled, and the molten salt reactor) as well as a supercritical water-cooled reactor and a very-high-temperature reactor. Two U.S. initiatives – the Global Nuclear Energy Partnership (GNEP) and the International Framework for Nuclear Energy Cooperation (IFNEC) – began with the objective of promoting more proliferation-resistant nuclear technologies, but they met with resistance from international partners.[5] According to the Department of Energy, a basic goal of GNEP was to "make it impossible to divert nuclear materials or modify systems without immediate detection." The idea was to improve International Atomic Energy Agency (IAEA) safeguards and recycling options that did not separate out pure plutonium, but GNEP also sought to "develop a fuel services program that would provide nuclear fuel and recycling services to nations in return for their commitment to refrain from developing enrichment and recycling technologies" (U.S. Department of Energy 2006). This last element never materialized. The follow-on program, IFNEC, made no pretense of engaging countries to refrain from developing their own enrichment and reprocessing, instead declaring in its mission statement that partner countries would not "give up any rights." So far, countries have been unwilling to compromise on nuclear research, plans, or future options to achieve greater security.

3 Risks from the Nuclear Fuel Cycle Technologies

Uranium enrichment and spent fuel reprocessing are the two fuel cycle activities that generate the highest proliferation concern because the material they are capable of producing can fall into the category of unirradiated, direct-use material by the IAEA (defined in the IAEA glossary as "nuclear material that can be used for the manufacture of nuclear explosives components without transmutation or further enrichment, such as plutonium containing less than 80% plutonium-238, highly

[4]Pyroprocessing, or electrometallurgical processing, separates plutonium from uranium, but not fission products. The resultant material can be used to produce fuel for fast reactors. As such, pyroprocessing is popularly considered by some to be more proliferation-resistant than the PUREX acqueous method of separating spent fuel. Official U.S. policy, at present, treats pyroprocessing as a form of reprocessing. This debate has been at the heart of proliferation concerns regarding South Korea's desired plans to pursue pyroprocessing as a way to condition its spent light-water reactor fuel to provide fuel for its future sodium fast reactors.

[5]The Bush administration prioritized the domestic portion of the GNEP program, which sought to revitalize reprocessing in the United States. This was abandoned by the Obama administration in 2009.

enriched uranium (HEU) and uranium-233"). Such material can be weapons-ready within a matter of days. Most uranium enrichment facilities today enrich at or slightly below 6% low-enriched uranium – about the level needed for commercial nuclear power plant fuel. Only Russia is producing HEU at the Electrochemical Plant in Zelenogorsk (Glaser and Podvig 2017).

Proliferation risks include diversion of declared material, undeclared production at a declared facility, and parallel, undeclared facilities. In the case of undeclared production at a declared facility, the operator could introduce undeclared inputs (e.g., operating an additional, undeclared shift at the plant) or falsify records to suggest lower production levels.

Downstream facilities that use HEU or mixed oxide fuel (a mixture of plutonium and uranium in oxide form) such as fuel fabrication plants also pose diversion risks. Power reactors and research reactors pose two kinds of risks: diversion of fresh fuel and clandestine production of plutonium (or U-233) in their cores. Such undeclared production could provide feed for undeclared reprocessing plants.

Reactors are typically treated as less sensitive in terms of proliferation risk than uranium enrichment and spent fuel reprocessing. However, the plutonium that is created when natural uranium captures a neutron in a reactor (or in an accelerator) is considered direct-use by the IAEA (defined in the IAEA glossary as "nuclear material that can be used for the manufacture of nuclear explosives components without transmutation or further enrichment, such as plutonium containing less than 80% plutonium-238, highly enriched uranium (HEU) and uranium-233."). The only difference between irradiated direct-use material and unirradiated direct-use material is that it would take several weeks instead of days to convert the material (i.e., separate it from fission products) into a weapon-ready state.[6]

Thus, the United States and Russia have understood that the HEU fuel and research reactors they exported from the mid-1950s pose a risk, both for the potential diversion of fresh HEU fuel, and for the clandestine production of plutonium. As early as 1980, the International Nuclear Fuel Cycle Evaluation study concluded that enrichment of reactor fuel should be limited to no more than 20% U-235. The tradeoff is clear: at higher enrichment levels, the fresh fuel could pose a diversion risk; at low levels, the fuel could produce more plutonium in the reactor. In its 2016 report on *Reducing the Use of Highly Enriched Uranium in Civilian Research Reactors*, the National Academy of Sciences noted "For example, a 40-MW natural-uranium-fueled reactor makes about 8 kg of plutonium per year, enough for at least one nuclear weapon, while its (fresh) uranium fuel is of very little concern. At the other extreme, a similar 40-MW reactor fueled with W-HEU makes almost no plutonium (less than 100 g per year), but requires about 30 kg of fresh HEU fuel per year" (National Academies of Sciences, Engineering, and Medicine 2016).

[6]This is also true for other fissile materials like U-233, americium and neptunium. The IAEA has officially designated U-233, which is produced when Thorium-232 captures a neutron, as a special fissionable material but has not officially designated americium and neptunium-237 as special fissionable material.

The solution has been, since 1978, to convert the fuel of civilian research reactors to low-enriched fuel. As of 2017, 97 research reactors had been converted to use low-enriched uranium (LEU) fuel precisely for this reason (IPFM 2018, IAEA 2018). As of 2016, there were at least 74 civilian research reactors and close to 40 military and civilian propulsion reactors using HEU fuel (National Academies of Sciences, Engineering, and Medicine 2016).

Commercial nuclear power reactors, on the other hand, are widely dismissed as contributing to proliferation risk. One reason for this is the dominance (more than 85% of total reactors) of light-water reactors (LWRs) in the commercial nuclear power industry. LWRs use LEU fuel and light water as a moderator. Typically, the need to produce electricity consistently and cost-effectively means high operating capacities, which also means higher burn-up of fuel. With long periods of irradiation, more Pu-240 is created in the reactor.[7] This so-called "reactor-grade" plutonium is thought to be less desirable for weapons because the spontaneous fissioning of the Pu-240 can lead to pre-initiation and lower-than-anticipated yields in nuclear explosives. However, U.S. weapons designers achieved a 20-kiloton yield with a device using reactor-grade plutonium during a test in 1962, a fact revealed by the Department of Energy in 1977 (Jones 2018; Jones 2019; U.S. DoE 1977). Another consideration for advanced nuclear states is the possibility of enriching plutonium from reactor-grade to weapons-grade, as the U.S. Department of Energy intended to do in a proposed Special Isotope Production plant in the early 1980s (Squassoni 2018b).

The other 15% of commercially operating power reactors in the world are heavy water and graphite reactors. These designs were deployed early on in weapons programs and then commercialized by a handful of countries (Canada, UK and India). They are generally considered to be less proliferation-resistant than light water reactors because their natural uranium fuel produces more plutonium that is compositionally more attractive for weapons than its light-water cousins, for the reasons described earlier. There are about 31 CANDU reactors (between 500 and 600 MWe in size) operating now in Canada, Argentina, China, South Korea, India, Pakistan and Romania. A drawback to CANDUs is the amount of spent fuel they produce, which is significantly more than LWRs. India, which has developed its own heavy-water-moderated pressurized reactors (PHWRs), may seek to export its designs in the future. For some countries, such reactors might be attractive because they are smaller (300 MWe in contrast to 1000–1500 MWe reactors).

Figure 1 below depicts typical steps in a nuclear fuel cycle. For heavy water fuel cycles, the step of enrichment is not necessary.

Until now, only a few countries have operated both uranium enrichment and spent fuel reprocessing. These are the nuclear weapon states – the United States, United Kingdom, France, China and Russia – and a few others like Japan and Germany.[8] A handful of other states outside of the Nuclear Nonproliferation

[7] The US Department of Energy differentiates between weapons-grade (<7% Pu-240); fuel-grade (7-19% Pu-240) and reactor-grade (>19% Pu-240).

[8] Belgium briefly ran the Eurochemic reprocessing plant for 12 other European states from 1966 to 1974.

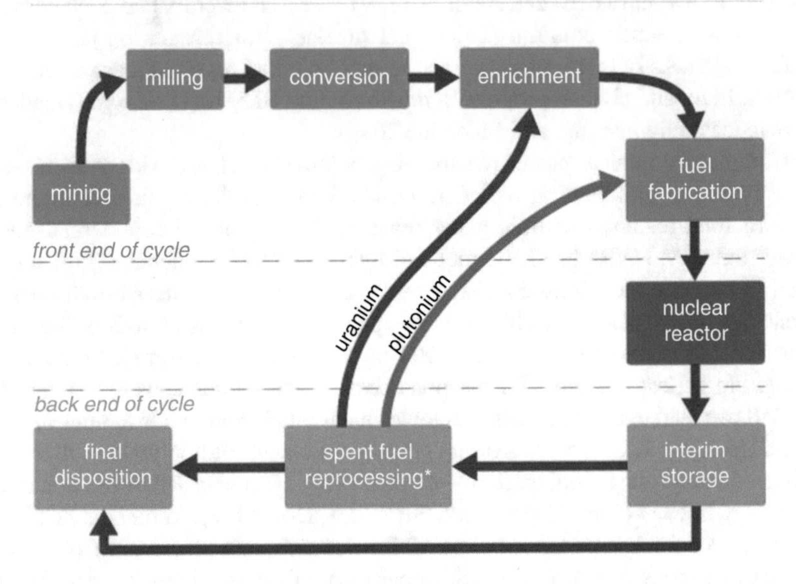

Fig. 1 The nuclear fuel cycle. * Spent fuel reprocessing is omitted from the cycle in most countries, including the United States. (Source: U.S. Department of Energy, Energy Information Administration)

Treaty (NPT) – India, Pakistan, Israel and North Korea – have used enrichment and reprocessing for military purposes primarily (India enriches and reprocesses also for commercial purposes). Most countries have not developed their own fuel cycle capabilities because they are too costly, but the NPT contains no prohibitions against doing so.

Proliferation risks associated with the nuclear fuel cycle have changed dramatically over time, both in terms of the technologies that are perceived as posing significant risks and the countries (or non-state actors) that may pursue fuel cycle capabilities. When the NPT was negotiated in the Eighteen Nation Disarmament Conference, experts and diplomats assumed that the tremendous costs, energy requirements and physical footprint of uranium enrichment plants (based on gaseous diffusion technology) would make clandestine enrichment very difficult, if not impossible. Today, gaseous diffusion technology has been phased out in favor of gas centrifuge technology and it is clear from the recent examples of Iran and North Korea that clandestine gas centrifuge operations are difficult to detect. Material advances – particularly carbon fiber – have greatly increased the efficiency of centrifuges. Whereas early centrifuges had low separative work capacity, current generation URENCO machines are now more than 20 times more efficient. As the individual separative work capacity per machine increases, the required number of machines per enrichment cascade declines, improving the chances that clandestine cascades will not be detected.

Many enrichment techniques have been explored since the early weapons programs, but few have been commercialized. Technical developments that decrease cost, enhance effectiveness, and decrease detectability could increase proliferation risk. Thus, critics contend that laser enrichment technology, if it can be commercialized, could increase proliferation risk because it will be more efficient and cost less than gas centrifuge. Moreover, like gas centrifuge facilities, laser enrichment facilities do not require a large footprint or significant electricity, nor do they provide other signatures that would facilitate remote detection. Although more than twenty countries have researched laser enrichment, none has yet commercialized the technique, despite some significant efforts like Global Laser Enrichment's attempt to commercialize SILEX technology acquired from Australia.[9]

For reprocessing, technologies and techniques have not changed dramatically. To date, around 400 000 metric tons of heavy metal have been discharged from nuclear power plants and about 25% of discharged fuel is being reprocessed. The IAEA expects that percentage will rise to 30% by 2020 (IAEA 2018). PUREX is still the industry standard, although France has promoted COEX as the next-generation reprocessing technology. Developed jointly by AREVA and the French Atomic Energy Agency (CEA), COEX would keep uranium and plutonium together in one stream, thus avoiding separation of pure plutonium. Many other variations have been developed and in some cases, a few have been developed specifically to handle Generation IV fast reactor fuel (Gray et al. 2015).

Particle accelerators are also capable of producing U-235, U-233 and Pu-239 (Scott Kemp 2005; Nusbaum 2013) but these have been used primarily for research and development rather than commercial purposes. Increased interest in partitioning and conditioning of waste from nuclear power reactors may increase the proliferation potential from accelerators. In addition, partitioning and conditioning activities could produce other fissile materials present in spent nuclear fuel that are useful for weapons – americium and neptunium. Since 1999, the IAEA has been monitoring the production and export of americium and neptunium in about 33 states and noted in its Safeguards Implementation Report for 2017 that the quantities have been small enough not to pose a significant proliferation risk (IAEA 2017).

4 Impact of Other Technical Developments

In addition to proliferation risks emanating from fuel cycle facilities, some technical developments may affect security risks to and from fuel cycle facilities. Computing technologies and software used for information warfare, system infiltration (hacking), and development of artificial intelligence can defend against or create incidents that expose people and the environment to radiation. With regard to the

[9] GE-Hitachi decided to back out of the GLE project in 2016 and sought to sell its entire share to SILEX. Cameco and SILEX are now poised to jointly own (49/51 split) GLE, with the purchase price only being paid to GLE after the project surpasses its first $50M in revenues.

vulnerability of internal control systems at such facilities, conventional wisdom holds that sensitive nuclear facilities are kept isolated from the internet and thus have an "air gap" that protects them. The infamous example of the Stuxnet virus, which infected Iranian centrifuges in 2010 using social engineering (preying on unsuspecting employees to introduce the virus on a flashdrive), demonstrated potential vulnerabilities even where air gaps exist, however. Operators may now recognize that cybersecurity could require a separate regulatory framework (PIR 2016; Van Dine et al. 2016). The risks produced by the information and computing revolution will likely continue to evolve in an unpredictable fashion. At the far end of the spectrum, the continuing evolution of artificial intelligence may create new potential risks. For example, AI could enable sabotage risks that are difficult to protect against, like swarms of armed drones (Kallenborn and Bleek 2018).

5 Supply and Demand for Nuclear Energy

Increased civilian nuclear commerce can facilitate proliferation at the margins. In the 1970s, many countries turned to nuclear energy as an alternative to dependence on oil and at the same time, several countries also sought to acquire nuclear fuel cycle capabilities (e.g., Brazil, Pakistan, South Korea). A truly determined proliferator may choose a completely clandestine route, quite separate from any civilian applications, but proliferation risks can grow in the context of robust nuclear trade, particularly when civilian nuclear programs can be used to justify acquiring fuel cycle capabilities.

Today, the desire to reduce greenhouse gas emissions has led to increased support for nuclear energy. At one point, more than 80 countries were interested in acquiring a commercial nuclear energy capability for the first time. Prior to the 2011 Fukushima Daiichi nuclear accident, the nuclear industry itself proclaimed a "nuclear renaissance." After the accident, the predictions have become a little more sober, but two trends are clear. The first is that the concentration of nuclear energy is likely to shift from advanced economies in Europe and North America to less advanced economies in Asia, the Middle East, Africa and Southeast Asia. The second is that the next wave of nuclear power plants will have to differ considerably from the ones that dominate nuclear energy capacity today if they are to be sustainable in the long term. The desired improvements include cost, safety and waste management.

Thirty-one countries (plus Taiwan) – mostly those with developed economies – host the 413 commercial nuclear power plants now in operation (Schneider and Froggatt 2018).[10] The reactors generate about 363 GWe of electricity, or about 11% of the world's total electricity generation. Five countries – the United States, France, China, Russia, and South Korea – generate over 70% of all nuclear electricity.

[10] Note that this number does not include scores of commercial power reactors in Japan that are not yet back in operation.

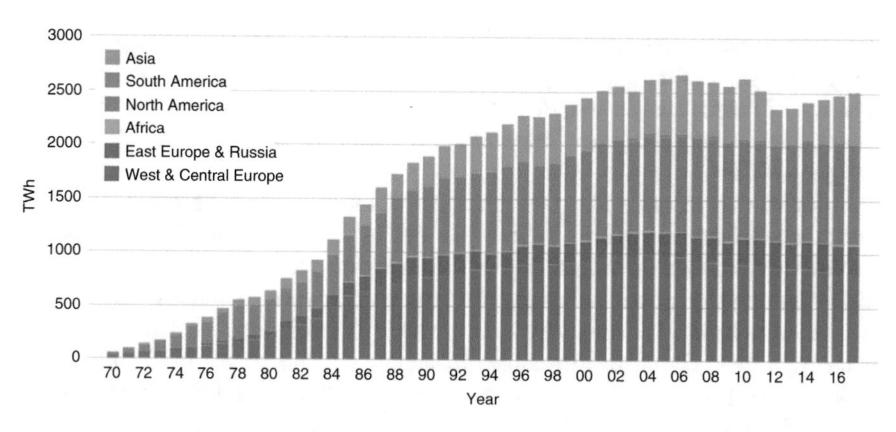

Fig. 2 Nuclear electricity generation by region. (Source: IAEA PRIS)

(Previously, Japan and Germany were included among the big nuclear states, but this is no longer the case). The United States operates 98 power reactors, generating about a third of the world's nuclear-sourced electricity in 2016; countries in OECD (Europe) generate another 30%, and the rest of the capacity has been dominated by Japan, Russia, and South Korea (see Fig. 2).

China and India are, by far, the most active in nuclear power plant construction. Of the 59 reactors under construction, 43 are in China, India and Russia. China has 26 reactors under construction at present. A handful of other countries in Asia with aspirations to build nuclear power plants, such as Vietnam, Indonesia, and Malaysia, will likely move slowly if at all. Thailand, Philippines and Singapore appear to have postponed nuclear power plans for the indefinite future.

In the Middle East, Iran has been operating the Bushehr reactor since 2009, but it is difficult to predict the future of commercial energy in that country as long as there is great uncertainty about how the Iran nuclear deal (known by its acronym JCPOA), will proceed. The United Arab Emirates recently completed construction on the first of four nuclear power reactors; the Barakah plant, built by a South Korean consortium, is expected to connect to the grid in 2020. Jordan, Morocco, Saudi Arabia, Oman, Qatar, and Bahrain in the Middle East have all expressed interest in nuclear power, but apart from Saudi Arabia, none of these is likely to build more than one or two reactors. It is striking that Saudi Arabia's decision to pursue nuclear power came after Fukushima. Saudi Arabia is pursuing a variety of vendors and technologies, including small modular reactors.

A few realities may impinge upon the plans of developing countries. The first is that many countries may not have the electricity grid infrastructure to support the size of nuclear power plants that are most commonly marketed (1000–1500 MWe range). Although small modular reactors have become a buzzword in the nuclear industry, there are few licensed and ready to be built. The second is the intensive process required to put in place the necessary regulatory and legal framework to operate safe, secure nuclear power. A third is competition from renewable electricity generation sources like solar and wind. According to the IAEA, "A steady increase in hydropower and the rapid expansion of wind and solar power has cemented the position of renewables as an indispensable part of the global energy mix; by 2035, renewables account for almost one-third of total electricity output" (Schneider and Froggatt 2018).

Another development that may change the complexion of proliferation is the emergence of new suppliers. Before the end of the Cold War, Russia supplied nuclear reactors and materials to its East Bloc allies in a fairly consistent way (Duffy 1978) and the U.S. and its allies supplied the Western bloc of countries. In the last two decades, Russia has sought to capture a large share of global nuclear exports. Its strategy, which combines favorable financing, flexible terms like Build-Own-Operate, and promises of integrated fuel services, have appealed to nuclear newcomers. Rosatom reported in 2018 that it had 36 projects in 12 countries. Of these, it was actively constructing 7 units in China, Belarus, India and Bangladesh (Schepers 2019). In contrast, bankruptcies and restructurings among the major U.S., French and Japanese vendors have weakened their competitiveness. Strong government support for exports, which Russia, China and South Korea can muster, will likely continue to be a key element to success. According to the IAEA, China plans to export 30 reactors overseas by 2030 (IAEA 2018). Advocates for stronger U.S. government support for nuclear exports argue that a shift away from traditional suppliers will result in lower nonproliferation standards, but this remains to be seen.

6 Policy Responses

Efforts to control the fuel cycle are almost as old as the bomb itself. Beginning in late 1945, the United States, Canada and the UK proposed a commission under the United Nations to consider exchanges of scientific information about nuclear energy, controlling atomic energy to ensure its peaceful purposes, eliminating nuclear weapons, and establishing effective safeguards and inspections (Donovan 1977). The next year, the State Department's Acheson-Lilienthal report recommended international ownership and operation of all "dangerous" nuclear activities, including virtually the entire nuclear fuel cycle, beginning with uranium and thorium mines. The report stressed that dangerous nuclear activities could not be left in "national hands" and that national rivalries in nuclear energy that could be readily converted to military uses were risky. Three months later, the U.S. representative to the UN Atomic Energy Commission, Bernard Baruch proposed the creation of an International Atomic Development Authority (Rydell 2006). Such an authority would "be entrusted all phases of the development and use of atomic energy, starting with the raw material and including:

- Managerial control or ownership of all atomic-energy activities potentially dangerous to world security.
- Power to control, inspect, and license all other atomic activities.

Table 1 System wide responses to proliferation shocks

Year	Proliferation Shock/Prompt	Action	Wider adoption
1974	India's Peaceful Nuclear Explosion	Tighter export controls	Formation of Nuclear Suppliers Group (NSG)
		US 1978 Nuclear Nonproliferation Act	Full-scope safeguards requirement by NSG (1991)
1970s	Attempts by Pakistan, Brazil and South Korea to acquire enrichment, reprocessing	Bilateral diplomacy by US with West Germany, France	Incorporation of enrichment, reprocessing transfer restrictions into NSG guidelines
1991	Iraq's clandestine nuclear weapons program	Strengthened safeguards program	Additional Protocol INFCIRC/540
1992	North Korea's program	Agreed Framework 1994	Adoption of satellite imagery analysis by IAEA
2002	Iran's program		JCPOA 2015 Enhanced monitoring
2004	AQ Khan network	Bush proposal for enrichment, reprocessing moratorium	El Baradei recommended 7 steps

- The duty of fostering the beneficial uses of atomic energy.
- Research and development responsibilities of an affirmative character intended to put the Authority in the forefront of atomic knowledge and thus to enable it to comprehend, and therefore to detect, misuse of atomic energy. To be effective, the Authority must itself be the world's leader in the field of atomic knowledge and development and thus supplement its legal authority with the great power inherent in possession of leadership in knowledge."

Less than 10 years later, the International Atomic Energy Agency adopted some but not all of these roles. However, nuclear energy capabilities remained firmly in national hands.

Historically, efforts to halt countries' attempted nuclear proliferation have relied on diplomacy, threats, and luck. However, some system-wide improvements have been possible to implement after specific proliferation crises. These are listed in Table 1.

A short 4 years after entry into force of the NPT, India's nuclear explosion constituted a major shock to the regime. Many states were still outside of the treaty at the time, but the example of India, which had pledged peaceful uses for the Canadian-supplied CIRUS research reactor and the U.S. heavy water that moderated it, starkly demonstrated that ostensibly peaceful activities could produce bombs. Suppliers of nuclear material and equipment acted quickly to restrict the spread of sensitive nuclear fuel cycle technology.[11]

[11] Even though India was granted an exemption from NSG rules that require NPT membership for nuclear trade, transfers of enrichment and reprocessing technology are still off-limits because of tighter restrictions.

The discovery of Iraq's mini-Manhattan Project in 1991 also prompted system-wide changes. Despite comprehensive safeguards inspections and monitoring, the IAEA did not detect Iraq's clandestine activities, a severe blow for the Agency and for the exporters that unwittingly supplied Iraq's nuclear weapons program. NPT members developed the Model Additional Protocol (INFCIRC/540), and NSG members adopted a comprehensive safeguards requirement for exports, and further revised the NSG guidelines.

Although less of a shock to the system, the difficulties in verifying North Korea's initial declaration in the early 1990s led to a showdown in which satellite imagery played an important verification role. North Korea, like Iraq, was a proving ground for new verification techniques like environmental sampling, much of which was adopted under the Additional Protocol. North Korea's withdrawal from the NPT in 2003, arguably the biggest shock of all, had no systemic effect. Although states continue to discuss ways to avoid another NPT defection, there are no agreed solutions.

The discovery in 2004 that Pakistani scientist A.Q. Khan had been selling nuclear technology and equipment to Libya, Iran, and North Korea constituted another serious shock. For the first time, suppliers realized that technology advances in manufacturing were allowing many more countries to play a role in supplying nuclear technology including those with deplorable export controls. UNSC Resolution 1540 sought to close the gaps by requiring states to prioritize export controls and to criminalize sub-state activities to evade export controls.

In response to the Khan network, the IAEA's Director General El Baradei in 2005 proposed seven steps to enhance the nuclear nonproliferation regime. These included a five-year moratorium on construction of uranium enrichment and plutonium reprocessing facilities; conversion of nuclear reactors using highly enriched uranium to low-enriched uranium fuel; universalizing adherence to the Additional Protocol; revisiting U.N. Security Council actions in response to a state's withdrawal from the NPT; universal implementation of U.N. Security Council Resolution 1540; acceleration of Article VI actions by nuclear weapons states (toward nuclear disarmament); and resolution of regional security tensions that give rise to proliferation, including a Middle East nuclear-weapons-free zone (ElBaradei 2005). Of these proposals, only two have progressed – conversion of reactors to LEU fuel (NSS 2016) and progress in getting countries to sign on to the Additional Protocol.

7 The Special Case of Iran

It remains to be seen whether the shock of Iran's noncompliance with its IAEA safeguards agreement over clandestine uranium enrichment activities (as well as failure to report imports, facilities, material and some purported weaponization activities) will provoke significant improvements for the nonproliferation regime or just short-term ameliorative measures.

The JCPOA, signed in 2015, implicitly recognized that IAEA safeguards alone are insufficient to provide assurances of the peaceful nature of fissile material production facilities. When it became clear that Iran would not foreswear uranium enrichment (as it did for reprocessing), negotiators added stockpile restrictions, caps on enrichment levels, restrictions on research and development and procurement and a cap on the number and sophistication of centrifuges in operation. Continuous enrichment monitoring technology, a procurement monitoring and approval system and physical alteration of the Arak heavy water reactor to limit its proliferation potential were also key elements. The JCPOA added explicit procedures for challenge inspections and specific methods for compliance disputes.

Critics of the Iran deal signed in July 2015 contend the agreement is flawed because it did not irreversibly eliminate Iran's uranium enrichment capacity. Such critics, however, tend to argue the point from a regional perspective, claiming that Iran poses national security threats to the region and to the United States and Israel. What they should be arguing against, however, is the systemic weakness in the NPT, which does not restrict fuel cycle capabilities. Any non-nuclear weapon state – not just Iran – is able to acquire enrichment and reprocessing equipment and produce and stockpile weapons-grade fissile material.

However, the nonproliferation system implicitly assesses proliferation risk as a combination of capability and intent, favoring those countries whose intentions are trusted. This is a recipe for continued inequity within the nonproliferation regime and may ultimately backfire. For example, safeguarded enrichment facilities in Germany and Japan, two non-nuclear weapon state parties to the NPT are not viewed as proliferation risks. While some may criticize Japan's plan to open the Rokkasho reprocessing plant, given its current stockpile of more than 47 tons of separated plutonium (most of which is in the UK and France), few call into question Japan's nonproliferation credibility.

Although the agreement to limit Iran's nuclear program is historic in its restriction of fuel cycle activities and material stockpiling, it is unlikely to be a model going forward. The JCPOA specifically rejects a broader application of the measures contained in the agreement, most likely requested by Iran to underscore its perceived right to fuel cycle capabilities. Iran might also have wanted to avoid a situation in which restrictions under the JCPOA extended beyond the terms of the agreement because other members of the NPT chose to apply them. However, Iran might be willing to agree to the wider application of such restrictions if all member states – non-nuclear weapon states and nuclear weapon states – agreed to such restrictions in their civil nuclear fuel cycles. Total elimination of HEU in the civil sector, for example, might be one such measure.

8 Future Developments

The case for new commercial uranium enrichment, from an economic perspective, is gloomier than ever. Both uranium production and uranium enrichment capacity already exceed demand for the next five or so years and unless a good portion of Japan's power reactors come back into service, the glut of supply is likely to persist.

Ten years ago, South Korea tried to persuade the United States that it needed a uranium enrichment capability to support its reactor export business, but no country is likely to attempt to use economic arguments to justify new enrichment capacity in light of current market trends.

However, a gradual unraveling or complete collapse of the Iran deal could trigger a race in the Middle East to acquire nuclear assets. Saudi Arabian officials have spoken publicly in the last few years about their intention to match Iran's capabilities, but it is unclear whether this referred just to a fissile material production capability or nuclear weapons themselves. A complete collapse of the JCPOA could give Saudi Arabia a strong incentive to pursue parallel civilian and military capabilities. It could thus continue to benefit from civil nuclear cooperation with a variety of states while accruing specific military expertise, most likely from Pakistan.

On the plutonium side of the ledger, wider adoption of spent fuel conditioning techniques – whether reprocessing, partition and transmutation, pyroprocessing, etc. – seem more imminent than before. Arguments for reprocessing based on higher fuel utilization are again being made, and now they are being paired with the argument that such fuel should be recycled in fast reactors to reduce the volume of waste (by one reckoning to 20%) (Gray et al. 2015). The United States is planning to build a Versatile Test Reactor specifically to test fast reactor fuels, in the hope of facilitating fast reactor commercialization in the United States. Some of these next-generation reactors will likely incorporate reprocessing into their fuel cycles to enhance their cost competitiveness and fuel efficiency. But even if reprocessing is not explicitly planned, engineers may find that some fuels could require conditioning before final disposal (Krall and Macfarlane 2018).

A U.S. embrace of reprocessing (or spent fuel conditioning) could have ripple effects on its partners. Japan would feel more confident in opening up the Rokkasho reprocessing plant and South Korea would press the United States harder to allow it to pyroprocess its fast reactor spent fuel. In the past, the United States concluded that pyroprocessing presents the same proliferation risks as reprocessing. As Russia and China step into the lead of exporting civilian power reactors, their pro-reprocessing orientation may influence other states to regard reprocessing as necessary for the nuclear fuel cycle.

Overall, the role of advanced reactors in promoting fuel cycle capabilities has been underestimated, primarily because fast reactors have not heretofore been commercially competitive. However, China, France, Germany, South Korea, India, Japan, the US and Russia have constructed or are developing fast reactors. The United States and South Korea are the only two among those countries without a reprocessing capability.

The proliferation risks of these reactors range from their using HEU, plutonium or U-233 in their initial fuel loads, like the China Experimental Fast Reactor and Russia's BN-800, to their potential requirement for reprocessing for safety reasons. Even though proposed fuel cycles vary from once-through (no recycle) to non-traditional reprocessing (e.g., pyroprocessing), or traditional reprocessing, some technologies may require reprocessing before disposal because the fuel or coolants can be reactive, toxic or corrosive. For example, molten salt-based fuels, when exposed to water, form highly corrosive hydrofluoric acid. Therefore, reprocessing (or some form of "conditioning" the waste) will likely be required for safety reasons before disposal. The same is true for other fast reactors. Sodium coolants are reactive, while the uranium metal in sodium fast reactors poses criticality and pyrophoric risks.

One unintended outcome of wider implementation of partitioning and conditioning waste is the potential separation of americium and neptunium from spent fuel. As noted earlier, the IAEA has been monitoring activities in states that could potentially separate these materials from spent fuel since 1999 (Albright and Kramer 2005). New reprocessing techniques developed to separate these actinides for waste conditioning purposes could potentially trigger an adjustment within the IAEA safeguards system to institute material accounting and control for americium and neptunium.

9 The Prospects for Systemic Solutions

There are several explanations for why countries have failed to confront the challenge of nuclear fuel cycle risks directly. A stubborn optimism that the next generation of technology will be more proliferation-resistant infuses the scientific community despite lack of evidence to support this. Several countries with good nonproliferation intentions (like the US and South Korea) are pursuing fast reactor technology although it is clear that some of these reactors can be configured to breed plutonium and some technologies require their fuel to be reprocessed before storage. Of course, such countries do not see themselves as risks for proliferation, but other countries may view them as models of behavior. The fact that reprocessing technology has not spread much beyond countries that have nuclear weapons may be a source of optimism and supplier countries therefore may feel that they can control trade. Pressures for more lenient trade agreements, however, may arise from stiff competition for limited reactor sales. Conversely, the prospect of significant nuclear energy growth has also been used as a rationale for leaving recycling options open in the future. For its part, the International Atomic Energy Agency (IAEA) has, at times, championed multilateral approaches to the nuclear fuel cycle to partly resolve the risk of proliferation from acquisition of fuel cycle technologies, but the Agency is bound to follow the lead of its member states and its mandate to promote nuclear energy.

At the same time as the suppliers are loathe to limit their cooperation agreements and contracts, so too are the recipients. There is a traditional resistance to stronger nonproliferation measures among non-nuclear-weapon-state parties to the NPT, which tirelessly point out that their own nonproliferation commitments have not been balanced by nuclear disarmament progress under the treaty. Therefore, they refuse to compromise on their access to peaceful nuclear technology. Geopolitical factors also doubtless enter into this complicated equation. Debates about fuel cycle capabilities in North East Asia and in the Middle East cannot help but factor in North Korean and Iranian capabilities.

The bottom line is that the nonproliferation community is probably doomed to inventing specific responses to cases of proliferation as they arise, rather than fixing the system as a whole. If countries (let alone terrorists) are able to acquire fuel cycle technologies more freely and clandestine activity becomes more difficult to detect, the significance of these proliferation episodes would grow. What's more, if the world eventually moves towards having small numbers of nuclear weapons, the significance of such proliferation would increase exponentially.

10 Conclusions

The nonproliferation regime has managed since 1968 to limit the acquisition of nuclear weapons by additional states to just four – India, Pakistan, Israel, and North Korea. South Africa developed weapons and then dismantled them, and other states pursued elements of a nuclear weapons program but not to completion. A strong push to reduce nuclear weapons stockpiles, secure, reduce and eliminate nuclear material stockpiles, strengthen the mechanisms to detect illicit activities, and strictly regulate the further spread of sensitive technologies used in the civilian nuclear energy will all be vital improvements to an effective nuclear nonproliferation regime.

Cherished myths about the safety and security of peaceful nuclear energy are likely to persist into the future until the next big proliferation crisis. Saudi Arabia, if it acquires uranium enrichment capabilities, would quickly challenge the conventional wisdom that such proliferation risks can be controlled. Yet, truly systemic adjustments that would really reduce proliferation risk across the board are unlikely to win approval without major compromises by nuclear weapon states. At the moment, collaboration among nuclear weapon states to reduce, rather than raise, nuclear risks, is scarce.

Nuclear vendors will continue to downplay nuclear proliferation risks out of self-interest. Risks from advanced nuclear technology are not likely to appear before another 20 years, since fast reactors are unlikely to be commercialized until then and any efforts to open new reprocessing plants would require at least that amount of time from the point of decision to operation (in the US, this could take even longer). At that point, it will be apparent that nuclear energy cannot be deployed fast enough to make much of a contribution to mitigating climate change but countries might still desire to keep nuclear energy in their portfolios to enhance diversity.

Until nuclear weapons are eliminated and probably for a long time after that, humans will have to guard against the risks of nuclear weapons proliferation from peaceful nuclear energy. In a world without nuclear weapons, the risks from nuclear energy may finally be evaluated appropriately and countries may consider that the benefits of multilateral control of certain elements of the nuclear fuel cycle will outweigh the costs in sovereignty. It may be possible to move towards an outright ban on highly enriched uranium, as long as states can come up with a solution for naval fuel (Lobner et al. 2018). Or, countries may consider that it is time to restrict commercialization of some technologies, such as breeder reactors. If previous cases of noncompliance are instructive, attempts to plug gaps across the board rather than for one country are necessary. Only then can some of the inequities that have plagued the nonproliferation regime begin to fade.

References

Albright David and Kimberly Kramer. 2005. Neptunium 237 and Americium: World Inventories and Proliferation Concerns, available on-line. Revised 22 Aug 2005 http://www.isis-online.org/uploads/isis-reports/documents/np_237_and_americium.pdf.

Donovan, Robert J. 1977. *Conflict and crisis: The presidency of Harry S. Truman, 1945-1948*. New York: Norton & Co.

Duffy, Gloria. 1978. Soviet nuclear exports. In *International security*, vol. Vol. 3, No. 1 Summer. Cambridge, MA: MIT Press.

ElBaradei Mohammed. 2005. Seven steps to raise world security. *Financial Times*, February 2, 2005.

Gilinsky, Victor. 2014. Nuclear weapons, nuclear power – clarifying the links. In *Moving Beyond Pretense*, ed. Henry Sokolski. Carlisle: The Strategic Studies Institute Publications Office, United States Army War College, June 2014. Available at: http://www.npolicy.org/books/Moving_Beyond_Pretense/Ch5_Gilinsky.pdf.

Glaser Alexander and Pavel Podvig. 2017. Production of new highly enriched uranium in Russia for the FRM-II in Germany. IPFM Blog, November 8, 2017.

Gray L. W., K. S. Holliday, A. Murray, M. Thompson, D. T. Thorp, S. Yarbro, T. J. Venetz. 2015. Separation of plutonium from irradiated fuels and targets, Lawrence Livermore National Laboratory, LLNL-TR-677668.

International Atomic Energy Agency. 2017. Safeguards Implementation Report 2017, Vienna, Austria

———. 2018. Nuclear Technology Review 2018, GC/62/INF/2, Vienna, Austria, August 2018.

International Panel on Fissile Materials. 2018. IPFM Blog. All HEU Removed from Nigeria. 7 Dec 2018. Available at: http://fissilematerials.org/blog/2018/12/all_heu_removed_from_nige.html.

Jones Gregory S. (2018) *Reactor-Grade plutonium and nuclear weapons: Exploding the myths*, Nonproliferation Policy Education Center, Arlington, VA. 16 April 2018. Available at https://nebula.wsimg.com/3fd1e3cfbbf101d6c4f562e17bc8604c?AccessKeyId=40C80D0B51471CD86975&disposition=0&alloworigin=1.

Jones, Gregory S. 2019. Reactor-grade plutonium and nuclear weapons: Ending the debate. *The Nonproliferation Review* 26: 61–81.

Kallenborn, Zachary, and Phillip C. Bleek. 2018. Swarming destruction: Drone swarms and chemical, biological, radiological, and nuclear weapons. *The Nonproliferation Review* 25: 523–543.

Krall, Lindsay, and Allison Macfarlane. 2018. Burning waste or playing with fire? Waste management considerations for non-traditional reactors. *Bulletin of the Atomic Scientists* 74 (5): 326–334. https://doi.org/10.1080/00963402.2018.1507791.

Lobner Peter, George M. Moore, Laura Rockwood, Matias Spektor, Sharon Squassoni, Frank Von Hippel. *Reducing Risks from Naval Nuclear Fuel*, IISTP Occasional Papers, IISTP-WP-2018-10, Volume 1, October 2018.

National Academies of Sciences, Engineering, and Medicine. 2016. *Reducing the use of highly enriched uranium in civilian research reactors*. Washington, DC: The National Academies Press.

National Research Council. 2013. *Improving the assessment of the proliferation risk of nuclear fuel cycles*. Washington, D.C.: The National Academies Press.

NSS. 2016. Gift basket on minimizing and eliminating the use of highly enriched uranium in civilian applications. Available at: https://static1.squarespace.com/static/568be36505f8e2af8023adf7/t/56febac0b654f939134d97d1/1459534530157/HEU+Minimization+Gift+Basket+for+NSS+2016.pdf.

Nusbaum David. 2013. Smashing atoms for peace: using linear accelerators to produce medical isotopes without highly enriched uranium. Policy Brief, Managing the Atom Project, Belfer Center, October 2013.

PIR Center. 2016. Cybersecurity of civil nuclear facilities: Assessing the threat, mapping the path forward, a policy memo. Available at http://pircenter.org/media/content/files/13/14758399064.pdf.

Rydell Randy. 2006. LOOKING BACK: Going for Baruch: The nuclear plan that refused to go away. *Arms Control Today*, June 2006. Available at http://www.armscontrol.org/act/2006_06/LookingbackBaruch.

Schepers Nevine. 2019. Russia's nuclear energy exports: Status, prospects and implications. EU non-proliferation and disarmament consortium, non-proliferation and disarmament papers, No. 61, February 2019.

Schneider Mycle and Antony Froggatt. 2018. *World nuclear industry status report 2018*, published on-line. Available at: https://www.worldnuclearreport.org/IMG/pdf/20180902wnisr2018-hr.pdf.

Scott Kemp, R. 2005. Nuclear proliferation with particle accelerators. *Science and Global Security* 13: 183–207.

Squassoni Sharon. 2018a. Not separate and not equal: Co-mingling defense and civilian nuclear activities. Published on-line. Available at https://cpb-us-e1.wpmucdn.com/blogs.gwu.edu/dist/c/1963/files/2018/10/Not-Separate-and-Not-Equal-1ocx95x.pdf.

———. 2018b. Towards a fissile zero future: Final report. Published on-line. Available at https://cpb-us-e1.wpmucdn.com/blogs.gwu.edu/dist/c/1963/files/2019/05/Prof.-Squassoni_Towards-a-Fissile-Zero-Future_Report_2019.pdf.

U.S. Department of Energy. 1977. Additional information concerning underground nuclear weapon test of reactor-grade plutonium, Washington, D.C. Available at: https://www.osti.gov/opennet/forms?formurl=https://www.osti.gov/includes/opennet/document/press/pc29.html.

———. 2006. Factsheet on global nuclear energy partnership. Available on-line at: https://www.energy.gov/sites/prod/files/edg/media/GNEP/06-GA50035i_2-col.pdf.

Van Dine Alexandra, Michael Assante, Page Stoutland. 2016. Outpacing cyberthreats: Priorities for cybersecurity at nuclear facilities. Published on-line at: https://media.nti.org/documents/NTI_CyberThreats__FINAL.pdf.

Chemical and Biological Risks in the Twenty-first Century

Ralf Trapp

1 Introduction

Hazardous chemical and biological materials have been known to, and used by, humans throughout history. Disease has plagued civilisations through the centuries, and at times significantly decimated populations. Hunters have used poisons extracted from flora and fauna to kill animals. As our understanding of the natural world grew, scientists developed methods to manufacture materials not found in nature, such as explosives, synthetic drugs and poisons, vaccines, and genetically modified organisms. Some of these developments have found application as weapons of war or terror, but up to the end of the nineteenth century, with few exceptions, our ability to manufacture such materials in large industrial quantities remained limited.

The beginning of the twentieth century experienced a qualitative shift in the manufacturing and use of chemicals, with the emergence of a chemical industry that enabled the large-scale and economic manufacturing of synthetic chemicals. Chemistry became an important source of materials used in a wide range of industrial and agricultural applications, from food production to fuels for transportation, energy production, as materials used in construction and building work, as medicines for treating diseases, in the manufacturing of consumer products, and much more.

But industrial-scale manufacturing of chemicals also was a precondition for waging the types of wars that characterised the last century: military forces consumed huge amounts of fuels, lubricants, explosives and other chemicals; and the twentieth century experienced the massive-scale use of toxic chemicals as weapons of war, from the use of the poison gas during World War I to the use of Cyclon B in Nazi concentration camps in World War II, from the build-up of huge stockpiles of

R. Trapp (✉)
Independent Disarmament Consultant, Chessenaz, France

M. Martellini, R. Trapp (eds.), *21st Century Prometheus*,
https://doi.org/10.1007/978-3-030-28285-1_4

nerve and blister agents during the Cold War to the recent uses of chemical weapons in conflicts in the Middle East, in Syria in particular. The use of toxic chemicals was not limited to killing or injuring human beings: agent orange is an example for the large-scale use of plant poisons to defoliate large areas of forests and destroy crops, at the same time causing long-lasting effects in humans and the environment. Biological weapons were used on a large scale by Japanese forces in China during the Second World War.

But the past century was also characterised by efforts to mitigate and control these risks. Protections and countermeasures were developed, safety practices were introduced in industry, agriculture and society at large, and legal frameworks were created to mitigate chemical and biological risks. With regard to hostile uses of hazardous chemical or biological agents in armed conflict, conventions and other regulatory frameworks were adopted to control and where possible prohibit weapons that employed such agents. That included such instruments as the 1925 Geneva Protocol, the 1975 Biological and Toxin Weapons Convention, the 1997 Chemical Weapons Convention, the Anti-personnel Landmines Convention (Ottawa Convention) of 1999 and the Convention of Cluster Munitions (Oslo Convention) of 2010, amongst others.

At the beginning of the twenty-first century, a technological shift similar to the one that enabled the industrial-scale manufacturing of chemicals is expected in the biological sciences. This is often cast in terms of the convergence of sciences at the intersection of chemistry and biology, underpinned by advances in enabling technologies such as the Internet which is being used as a platform for information sharing, distributed manufacturing of goods and services, and collaboration; rapidly advancing information technology with vastly expanding data stores, processing speed and computational capacity; artificial intelligence with a shift from simple machine learning algorithms to deep learning, combined increasingly with the use of Big Data generated in life science research and elsewhere; advances in material sciences, materials engineering and nanotechnology; and new manufacturing technologies such as the use of micro-reactors/micro-processing equipment and additive manufacturing ("3D printing"). Some of these developments are still relatively young, but there are early indicators that they may profoundly change the way in which chemical and biological materials are developed, produced, marketed/traded and used. This will affect the economic relations between countries as well as the way in which chemical and biological risks will be perceived, and must be managed.

As these new scientific, technological and economic realities emerge, the policy and security environment of the world, too, is changing. When the twentieth century could be characterised by big-power rivalry leading into two World Wars and the Cold War, the first part of the twenty-first century seems to be characterised by greater multi-polarity and fragmentation, with a multitude of actors set into asymmetric conflict scenarios (not only governments and state-sponsored groups but also terrorist organisations, organised crime, and private enterprises). The same can be said about the actors involved in the mitigation of the risks emanating from chemical and biological materials and technologies. Whilst the role of States remains essential, other actors including industry, the science and technology communities, educators, and other non-state actors including civil society are increasingly called upon to shoulder responsibility in the development and implementation of strategies and practical measures to manage these risks.

This chapter will first look at how the advances in science and technology at the intersection of chemistry and biology are affecting the chemical and biological risk landscape. It will identify developments that have particular relevance with regard to our increasing understanding of biology and of how biological systems can be manipulated (for good or bad), and for the development of new materials and products with hitherto unknown properties and applications. Some of these advances are truly revolutionary (in a scientific sense), and some involve technologies that are expected to be disruptive (in a technological and economic sense).

The chapter will then look at how these advances may affect the potential for chemical and biological warfare. Which of these advances need to be studied and monitored closely to ensure that their application will not undermine existing regimes to control high-risk agents, equipment and technologies, or weaken established prohibitions of certain types of weapons.

But this emerging risk landscape is not only framed by new scientific advances and emerging technologies. Existing, sometimes fairly old, technologies, too, must not be ignored. This is particularly so in a security and political environment that is more divers and fluid than that of the past century, with a wider range of State, State-sponsored, and non-state actors involved. The chapter will recall some of the recent experiences in the hostile use of old CBW technology in armed conflict, and reflect on what that may mean for the future risk mitigation approach.

At the same time, the advances in science and technology will also have direct benefits, enable new risk mitigation strategies, and deliver new tools to detect, identify, characterise, protect against and treat the effects of chemical and biological agents. This chapter will reflect in particular on some of the advances that affect capabilities to investigate suspected uses of chemical and biological weapons, including with regard to establishing their origin and attributing responsibility for their uses.

2 Convergence at the Intersection of Chemistry, Biology and Technology

A number of studies and symposia have been devoted in recent years to review and assess the impact of scientific and technological advances at the intersection of chemistry and biology on chemical and biological security, and on the arms control system developed after the Second World War. These science and technology advances have often been conceptualised as "convergence" – a term described as "integrative and collaborative trends in the life sciences that bring together theoretical concepts, experimental techniques and knowledge of different science and engineering disciplines at the intersection of chemistry and biology. Such interdisciplinary approaches often revolutionise scientific discovery and open up new areas of application of science and technology in society. The benefits of convergence can be huge, but it can also create new risks to safety and security, including the existing arms control regimes" (Spiez Laboratory and ETH Zürich 2014, p. 11).

The OPCW's Scientific Advisory Board (SAB) set up a temporary working group to study the impact of convergence on the Chemical Weapons Convention, and published its report in June 2014. The group looked at convergence through the lens of Chemical Weapons Convention implementation, noting how the science and technology environment within which the treaty was functioning was changing under the influence of technical, commercial and economic factors. Biological and biologically medicated processes were becoming more prominent in industrial manufacturing of chemical products. At the same time, considerable progress was also being made in the chemical synthesis of complex biomolecules and biological systems, from DNA to peptides and other biomolecules to the reconstruction of viruses and the rational design of viral capsids. The SAB observed that "[new] production processes, combined with developments in drug discovery and delivery, could be exploited in the development of new toxic chemicals that could be used as weapons. Such developments would still be covered by what is known as the 'general purpose criterion', but this highlights the importance of monitoring developments in science and technology" (SAB 2014, p. 4)

The SAB was cautious, however, not to overstate the immediacy of the risks emanating from these advances, pointing out that "[although] there are concerns that biotechnology could be applied to the production of new toxic chemicals, bioregulators and toxins, the temporary working group (TWG) assessed that potential applications to scheduled chemicals are currently limited. Scaling up a new biological process will continue to take a considerable investment of capital, resources and time…" (SAB 2014, p. 3).

The SAB formulated a series of recommendations, which largely dealt with the need to continue monitoring particular trends in science and technology as well as to ensure more effective outreach to and communication with other communities – academia, industry, as well as the policy and security community dealing with the Biological Weapons Convention.

Subsequently, the SAB followed up with a number of more specialised studies, which resulted in:

- A report on emerging technologies that amongst others identified a set of technologies that may find applications in non-routine (contingency) and/or assistance and protection operations, in investigations, or that may enhance the laboratory capabilities of the OPCW and its Member States. It also identified the integration of informatics tools, mobile devices and remote sensing with an expanding range of capabilities as an area of interest to support the implementation of the Chemical Weapons Convention (SAB 2017a).
- A report on trends in chemical production, which confirmed the convergent nature of twenty-first century technology development well beyond chemistry and biology. As a consequence, the scientific review process must engage broad scientific communities and look for opportunity in technological change

to ensure that implementation of the Convention remained fit-for-purpose (SAB 2017b).
- The establishment of a SAB Temporary Working Group on investigative science and technology; the group met first in February 2018, when it reviewed experiences from past investigations and fact-finding missions, discussed the role of laboratories in investigations, and discussed crime scene management, data management, legal issues and operational considerations (SAB 2018). The work of this group continues, and a second report was issued in January 2019 (SAB 2019) which developed some of these themes further.

In a complementary process, Spiez Laboratory of Switzerland organised a series of so far three Convergence Workshops, held between 2014 and 2018. These workshops were designed as "horizon screening" exercises and brought together experts from research and development organisations, laboratories, industry, and the arms control and security communities. As observed in its latest report, the Spiez Convergence series set out to discuss how proofs of concept, technological breakthroughs, and scientific game changers might affect our understanding and perception of chemical and biological weapons (Spiez Laboratory and ETH Zürich 2018, p. 6).

These workshops provided a platform for conversations between divers scientific, industrial and policy communities about the possible impact that these advances may have on arms control in the chemical and biological fields and on the scientific and industrial environment within which the existing arms control treaties and mechanisms need to function. They confirmed the general trend in the life science and technology domain from descriptive to predictive approaches, a more refined understanding of the functioning of biological systems and subsystems, and a shift from analysing biological materials and structures to information, modelling and simulation approaches.

The Royal Society of Chemistry in London, too, published a study that amongst others reviewed the science convergence in chemistry and biology and its impact on chemical and biological weapons arms control (Crowley et al. 2018). This study was conducted in the run-up to the Fourth Chemical Weapons Convention Review Conference in 2018, but it addressed the much broader range of instruments and arrangements in the chemical and biological arms control field that have been set up to address and manage risks associated with these technologies.

The technology areas that have drawn particular attention in these studies mentioned above included:

- Genome editing (in particular CRISPR technology)
- Directed evolution
- Gene drives
- Synthetic biology
- Chemical and biology-based manufacturing technologies for chemicals (including highly potent chemical ingredients of pharmaceuticals) and complex biomolecules
- Automation of manufacturing processes and cloud manufacturing

- Additive manufacturing
- Nanomaterials for a range of applications
- Particle engineering including smart participles for targeted drug delivery
- Bioinformatics, omics, big data, and systems biology
- Artificial intelligence (deep machine learning).

Many of these technologies have the potential of revolutionising the life sciences, leading to new products and applications as well as resulting in different global dissemination patterns of science, technology and manufacturing capabilities. This, however, may not itself affect the functioning of arms control and non-proliferation systems. Some of these new technologies, however, may also be disruptive – they may alter the industrial landscape, leading to changes in markets and value networks. Such changes can directly affect how arms control and non-proliferation measures function.

It is worthwhile in this context to recall some the "take-home points" from the Third Spiez Convergence workshop in 2018:

- Since 2014 the focus of discussions has shifted from materials and equipment, to information, automation, and remote manufacturing.
- With manufacturing moving closer to the point of use, the role of the end-user is changing; access to data as well as intangibles transfers are becoming more relevant from a regulatory and control perspective.
- Novel CBW production facilities would have a smaller footprint and different technological features compared to what is known from past state programmes (Spiez Laboratory and ETH Zürich 2018, p. 39).

These trends pose the question, firstly, of how these new technologies would fit into a contemporary CBW programme should a State decide to acquire new types of CB weapons. Asked differently: if there were a State CBW programme today, is it conceivable that it would *not* exploit these new concepts, materials, equipment and technologies? But if it did make use of these new capabilities, what kind of CBW programme and threat would that be? Would current control system recognise it and be able to effectively curtail it? In short: is there a risk that significant "latent break-out capabilities" might emerge in the field of chemical and biological security, and would the barriers that have been erected during the last century (legal, political, ethical, technical and institutional) be strong enough to prevent such latent capabilities from turning into real-world threats?

3 Novel CB Agents

One might ask whether this is a valid question given the near-universal adherence to the Chemical Weapons Convention, as well as the Biological and Toxin Weapons Convention. But firstly, both treaties are not as yet fully universal, and among the States that have yet to join the regimes are some that have posed concerns about capabilities and intentions with regard to chemical and biological weapons in the past. Secondly, adherence to the Chemical Weapons Convention has not stopped the use of chemical weapons altogether, and concerns are being raised about smaller-scale uses of toxic chemicals for sabotage or assassination purposes. Such smaller-scale, non-battlefield uses of chemical weapons would require capabilities quite different from those of past military chemical weapons programmes. In past military programmes, factors such as low manufacturing costs, long-term agent stability in storage, ability to disseminate the agent over large areas on the battlefield, and the potential to cause mass casualties, amongst others, where important constraints that drove the selection and acquisition of agents. There seems to be a broad consensus today that this type of chemical (or biological) warfare may be largely a thing of the past.

As the recent chemical attacks in Malaysia and the United Kingdom have shown, however – attacks against individuals[1] that involved the use of the nerve agents VX and Novichoc in public spaces – some States may well have both the intent and the capability to wage new types of CB warfare, despite the global conventions which comprehensively ban such weapons.

In such different CBW scenarios, many of today's advances in the life sciences and related enabling technologies are perhaps more relevant than they would be in traditional chemical/biological warfare concepts as we know them from the two World Wars and the Cold War. A shift from large-scale weapons procurement for battlefield use, to small-scale CB agents used for targeted attacks on specific individuals may make types of agents and dissemination methods more attractive that would have been too expensive or otherwise not fit for purpose in a traditional military scenario. Examples could include chemical agents that have higher lethality than traditional CBW agents that would have made them problematic for large-scale manufacturing, storage and use; or that have lower lethality compared to traditional nerve or blister agents but cause more subtle physiological effects such as interferences with certain functions of the central nervous system; or types of lethal agents that are difficult to detect and treat.

The use of such types of chemical agents as weapons would, of course, constitute a violation of the Chemical Weapons Convention, which uses a General Purpose Criterion in its definition of chemical weapons.[2] But there remain concerns that States may decide to exploit (deliberately or inadvertently) the uncertainties inherent in the Convention's rule which allows the use of toxic chemicals for law enforcement including domestic riot control purposes, for developing, acquiring and using new types of toxic chemicals and related delivery systems for conducting law enforcement operations that at times may be difficult to distinguish from the use of toxic chemicals in warfare.

---

[1] For a discussion of the implications of the use of a Novichoc agent in the assassination attempt on the Skripals, see for example (Trapp 2018). For more information on the assassination of Kim Jong-Nam see, for example, (Nakagawa and Anthony 2018).

[2] For a more detailed discussion see (Krutzsch et al. 2014, pp. 76–85).

Related to these concerns are also developments in the field of targeted drug delivery – see for example (Crowley et al. 2018, pp. 201–204). Work is progressing on engineering nanoparticles, modifying viruses or developing drug delivery platforms modelled on certain toxins or proteins that have the ability to penetrate biological membranes – to be used as delivery systems that can carry biologically active molecules such as bioregulators or toxins across biological barriers (for example the blood brain barrier) to specific target sites (for example certain parts of the brain), thereby resulting in more subtle and specific physiological effects than traditional CW agents.

The speed with which some of these concepts are progressing from proof of concept in the laboratory to practical application can be illustrated by looking at DNA origami. DNA origami structures are two- or three-dimensional nano-scale objects that are self-assembling from single strand DNA and short "staple" DNA strands. The technique exploits the rigid complementarity rules of DNA base pairing; size, shape and other properties of the nanostructures are predetermined by the DNA sequence of the long single-DNA strand (typically a viral DNA) and the synthetic DNA staple strands. The staple strand sequence required to fold the long DNA strand into the desired shape is modelled by specialised software that uses the base pairing rules to match the required sequence of each staple strand to specific segments of the long DNA strand, thereby causing the long strand to fold as desired.

In the context of CBW arms control, DNA origami was characterised in 2016 as a technology in its early exploratory stages, and initial concerns related largely to the longer-term prospects of developing molecular machines at the nanoscale that could mimic physiological functions. However, after a mere 2 years, it had become clear that there was significant potential in applying this technology for the development of targeted drug delivery systems, and work was already under way to improve the economy of the process by drastically reducing the cost of manufacturing the DNA staple strands at industrial scale (Spiez Laboratory and ETH Zürich 2018). In parallel, animal testing of DNA origami delivery methods of cancer medicines was under way, and concepts of using DNA origami for the development of molecular biosensors were being discussed. There remained a number of problems to be resolved, including low DNA origami stability *in vivo* and high manufacturing costs. But similar to the development of other engineered nanoparticles, fast progress should be expected given the strong demands for better drug delivery systems in chemotherapy, cancer treatment, and the treatment of other diseases that would benefit from the ability to deliver complex biomolecules such as proteins to specific targets in the body.

It was the potential for the emergence of new agent concepts and targeted delivery methods, amongst others, that prompted discussions about so-called "incapacitating chemical agents" (today usually described as CNS-acting chemicals disseminated as aerosol). Many countries have made formal statements to the OPCW about their policies with regard to not using toxic chemicals that act on the CNS and are delivered as aerosols, for law enforcement purposes. In 2017, 39 States Parties of the CWC coming from all 5 regional groups of the OPCW joined a paper, which concluded that "CNS-acting chemicals cannot be dispersed by aerosol in a completely safe manner in law enforcement settings. This also raises concerns that CNS-acting chemicals could be used as chemical weapons" (OPCW 2017, p. 2). However, it remains unclear whether all States share this view. Should some States decide to acquire such agents and delivery system for what they may perceive or portray as legitimate law enforcement purposes, we would face a situation where the very purpose of the Chemical Weapons Convention – not only to prohibit the use of chemical weapons in armed conflict but also eliminate the capability to do so – would be put into question.

4 Preventing the (Re)-acquisition of CB Weapons

The trends in science and technology summarised above also pose the question of whether the national and international implementation systems of the two Conventions, as well as the export control measures adopted by many countries to prevent the proliferation of chemical and biological warfare capabilities to State as well as non-State actors, can remain effective as barriers against the (re)acquisition of novel CB warfare capabilities.

The shift from shipping materials and equipment to transferring and sharing information, as well as the trends in chemicals and biologicals manufacturing to use cloud services to bring manufacturing closer to the end-user, will have profound implications for the ability of traditional export control systems, as well as arms control treaties' requirements for transfer controls and prohibitions. These measures have been adopted to prevent the proliferation of proliferation-critical materials, equipment and technologies. Export controls have, of course, included the licensing and control of transfers of "intangibles" in the past, as well as catch-all clauses to catch transfers of items not included in control lists. But in practical terms, enforcement and prevention with regard to such transfers turn out to be complicated.

Also, the symbiosis of new manufacturing technologies with advanced information processing and artificial intelligence – such as additive manufacturing or the automated synthesis of complex molecules with biological activity using cloud-based manufacturing services and design based on artificial intelligence – is predicted to create a new industrial landscape in which traditional non-proliferation measures, as well as the national implementation measures States apply under the Chemical Weapons Convention and the Biological Weapons Convention, may no longer match the real world. It remains to be seen which particular direction these advances will take and at what scale they will manifest themselves in industrial manufacturing. But as these new technologies move from the lab bench into the real world of industrial manufacturing, a longer-term and more fundamental rethinking of proliferation controls may become necessary.

Some technological developments, however, may call for faster adaptation of existing control regimes. An example in case is additive manufacturing (AM).

The Australia Group control lists[3] include, amongst others, high-performance production equipment that is corrosion resistant (this type of equipment is needed to manufacture certain CW agents), or that can withstand the high temperatures and pressures used in steam sterilisation (needed in the manufacturing of biological agents). Early assessments of the proliferation potential of AM raised significant concerns about the possibility that the relative ease with which 3D-printers became available might undermine the export controls of such sensitive production equipment (Spiez Laboratory and ETH Zürich 2014, pp. 32–33). Subsequently, some of the physical limits of certain AM processes became more apparent, for example with regard to powder bed AM using electron or laser beam technology. This is a continuous welding process, which unavoidably involves the creation of welding defects, leading to material fatigue. Such defects are difficult to detect and cannot be predicted, making the technology suitable for fast prototyping but not for industrial-scale manufacturing of critical high-performance components of manufacturing equipment for CBW agents (Spiez Laboratory and ETH Zürich 2016, p. 7).

As the technology matures, however, some of these limitations are being overcome, and alternatives are being explored. Progress is being made with regard to minimizing defects, ensuring better reproducibility, and improving post-processing steps including surface finishing. This will increase the stability of printed pieces and improve their corrosion resistance – 3D-printing of equipment parts made of high-performance metal alloys or ceramic materials is now state of the art (it does require technical know-how, however, as well as a thorough process understanding). In addition, new printing materials are becoming available, including polyfluorinated polymers. This enables the 3D printing of corrosion-resistant equipment parts that may be suitable for key process steps in the manufacturing of nerve agents (Spiez Laboratory and ETH Zürich 2018, p. 30).

Nevertheless, considerable professional know-how and technical competence remain significant hurdles, and at this stage, only industrial AM systems can supply high-quality parts to a standard that can compete with incumbent industrial processes. It remains unlikely that 3D printers that are capable of manufacturing highly corrosion resistant equipment (or components thereof) would be available to individuals or consumers soon (Ibid.).

There is also room for self-regulation by the industry. As of today, high performance AM systems tend to be located in service centres that manufacture parts to customer specification. This configuration is similar to producer-customer relationships in other industries that have posed proliferation concerns in the past (such as the custom manufacturing of specialty chemicals, or DNA synthesis to customer specification). Outreach to these industries has helped create awareness of the proliferation risks that may be associated with these activities, and the industry responded with the adoption of voluntary measures including customers and order screening to detect suspect order. Similar measures may also be adopted by the AM

[3] The current control lists of the Australia Group can be found at https://australiagroup.net/en/controllists.html.

industry. But this will require outreach by regulatory authorities to raise awareness in the industry for the proliferation risks inherent in their activities, and for the need to adopt voluntary control measures.

At the same time, the shift in manufacturing from shipping materials and equipment to transferring information can also provide new opportunities for oversight, compliance monitoring, and verification. The data that are collected and processed for process and quality control, if made available and authenticated, could equally be used to verify (or demonstrate *post ante*) regulatory compliance with regard to security-driven limitations and criteria. To do so, however, would require a regulatory framework that would allow the use of these data for compliance verification purposes, and the cooperation of the industry in creating such a framework in ways that would be acceptable as well as effective.

5 Old Versus New Technologies

The above discussion leads to the assumption that a State programme in the field of chemical and biological weapons today – should there be such a programme – would likely make use of the new materials, equipment and processes developed in and around the life sciences. Most of the developments described above, however, do not as yet alter dramatically the risks emanating from non-State actors.

At least for the time being, non-state actors' attempts to acquire CBW capabilities are likely to remain opportunistic and constrained, amongst others, by access to requisite knowledge and technical skills, relevant materials and equipment, and time and money to sustain extensive development and testing programmes and the establishment of manufacturing capabilities for both agents and effective dissemination systems.

The uses of chemical weapons in Syria and Iraq, on the other hand, have demonstrated the utility of improvised chemical weapons that exploit easily available materials such as chlorine – a toxic industrial chemical widely used for legitimate purposes – and equipment such as compressed gas cylinders adapted to air or rocket delivery. The Islamic State mastered the technical challenges of weaponising and using mustard agent on the battlefield. Even the Syrian government used improvised chemical weapons such as chlorine barrel bombs to great effect, after it had accepted to join the Chemical Weapons Convention in 2013 and pledged to dismantle its military CW programme. It also continued to use Sarin on a number of occasions. For a detailed description of the uses of chemical weapons by Syria and ISIS, see for example (Bunker 2019).

The use of chlorine in the Syrian conflict is testimony to the fact that "improvised" does not mean "ineffective" when it comes to adapting easily available materials and equipment for use as chemical weapons. A recent study of the uses of chemical weapons by Syrian government forces concluded that "the Syrian regime's persistent and widespread use of chemical weapons is best understood as part of its overall war strategy of collective punishment of populations in opposition-held areas.

Chemical weapons are an integral component of its arsenal of indiscriminate violence, alongside sieges and high-explosive weapons such as 'barrel bombs'" (Schneider and Lütkefend 2019).

It is important to appreciate that the impact of chemical and biological weapons use in such scenarios does not necessarily measure in battlefield casualty numbers. In the Syrian context, it appears that the use of improvised as well as military chemical weapons did not aim at inflicting maximum casualties on military targets, but to target civilians living in opposition-held areas, in an attempt to force them to assert pressure on the military opposition groups and thereby to break their will to resist. In more general terms, the psychological impact of the use of chemical and biological weapons must not be underestimated. Even "ineffective" improvised chemical or biological devices can have a devastating impact on civilian populations, crippling morale and their will to resist.

It should also be recalled that for non-State actors, explosive devices remain the chemical of choice given the relative ease with which the necessary materials for bomb making can be acquired and given the comparative simplicity of manufacture.

6 New Technologies for Risk Mitigation

Advances in science and technology do not, however, only increase the risk potential in the CBRN domain. Chemical warfare and chemical defence are examples for how science and technology have been exploited for offensive as well as defensive purposes, creating a competition between offensive and defensive applications of science and technology where the development of new and more effective weapons only ever produced a temporary advantage until effective means of protection against them had been found.

Medical countermeasures as well as physical protections against toxic and infectious agents to prevent them from entering the body are obvious examples (Crowley et al. 2018, pp. 204–6 and 211). Whilst advances in drug discovery and delivery may lead to new types of CBW agents, the very same methods can also be exploited to develop new prophylactics and antidotes as well as other medical countermeasures to neutralise the effects of toxic chemicals or infectious biological agents. Nanotechnology, for example, cannot only be utilized to develop more targeted drug delivery systems, but also to develop materials with properties that can be brought to use in air filtration systems, for decontamination purposes, as auto-decontaminating or agent-repelling coatings for clothing or sensitive equipment parts, or in detection systems.

The OPCW SAB has identified a number of technologies that have utility for personnel operating in dangerous environments, in the context of contingency operations such as when providing assistance in case of CW uses, or during investigations of alleged CW incidents and other non-routine operations. The technologies reviewed by the SAB in this regard included (SAB 2017a):

- The use of aerial platforms including UAVs for reconnaissance, scene documentation, sample planning and agent detection
- Space-based imaging and sensor technologies that combine high spectral, spatial and temporal resolution
- Hyperspectral imaging and analysis as well as optical sensors for detecting biophysical and biochemical changes in plants, as indicators of chemical changes in the environment
- Data fusion of satellite observation data and data from dispersion models and ground data to monitor ongoing events and compute 3D concentration fields of chemical emissions (this has been done, for example, to compute SO_2 dispersion during a fire at the sulphur production site Al-Mishraq, which had been started on 20 October 2016 by Daesh (Islamic State) during the battle of Mosul)
- Novel sensors based on targeted catalytic degradation of organophosphates, for example by using graphene or carbon nanotubes functionalised with metal nanoparticles
- eNose sensor technology combined with wireless communications systems for air monitoring of facilities/areas, or for use in clinical applications such as diagnosis of certain cancers
- Mobile, wearable and point-of-care sensing and analysis technologies, such as paper-based electronics, smartphones interfaced with sensor technology, electrochemical devices, sweat sensors, micro-needles, sensors for measuring certain biomarkers or enzyme activities (e.g., wearable devices to monitor cholinesterase activity), personal dosimeters, or colorimetric toxic gas sensors for external exposure monitoring
- Wearable digital health technology to monitor body functions (heart rate, skin temperature, blood oxygen levels, physiological activity), which could be combined with Big Data technology and artificial intelligence for health diagnosis and exposure monitoring[4]
- Unmanned airborne platforms for use of sensors as well as analytical instruments such as mass spectrometers for *in situ* analysis in harsh, inaccessible environments, or for the detection, identification and monitoring of toxic chemicals for area and perimeter measurement of toxic agents
- Marine unmanned systems for sample collection and processing, using tools and techniques derived from biomedical diagnostics and research such as low-density DNA probe and protein assays, 2-chanel real-time PCR, or sample collection and preservation for off-site analysis
- Modular robotic platforms fitted with CBRN countermeasures including forensic tools
- Modelling tools for use in response to or investigation of chemical incidents (computer aided engineering tools, dispersion models).

[4] In a recent pilot competition for innovation teams that were developing CBRN risk mitigation products and solutions with market potential, in the framework of the EU's CBRN Centres of Excellence Initiative, such a wearable device for real-time ECG monitoring and the monitoring of other parameters, combined with AI-supported prediction of cardiac incidents caused by, for example, extreme stress, was selected as the winning competition entry.

Although the SAB reviewed the utility of these techniques under a specific perspective – their possible use in OPCW contingency operations such as assistance delivery after the use of toxic chemicals, or investigations of alleged uses of chemical weapons – they are equally relevant for enhancing preparedness and response systems dealing with a broad range of chemical and biological incidents (civil defence organisations, incident response structures in industry and transport organisations, preparedness structures in academic and research institutions, and others).

Not all these technologies will be sufficiently rugged to function as required under field conditions. As the SAB emphasised, their suitability for field use requires evaluation of fieldable capabilities that meet operational requirements and fit within mission specific modalities (SAB 2017a). Practical issues that may arise in the field may include power sources availability, availability of consumables, limitations in terms of time and space (access), transportation requirements related to weight or transportation safety restrictions, limitations with regard to the ability to transfer data from the field, restrictions emanating from the need to preserve chain of custody, requirements related to data authentication and protection, data management and storage limitations, and more.

But the combination of different emerging platforms and technologies clearly will increase key capabilities that underpin protection, detection/identification, response and recovery capabilities to manage chemical and biological incident. They are expected to offer new and more effective solutions in such areas as incident response planning and management, protecting and monitoring of the health status of responders operating in highly hazardous areas, allowing access to otherwise dangerous or difficult to reach areas, the use of stand-off detection systems and other (wearable, personalised) devices that provide early warning of exposure to threats, faster data collection, and additional confidence in regard to data collection in the field.

In addition to developing better countermeasures against chemical and biological threats, science and technology advances can also be helpful to enhance verification capabilities of international arms control treaties and mechanisms, and of control and enforcement measures implemented by national agencies.

In the biological field, there is today no international enforcement agency comparable to the OPCW, and no international verification system for routine or special inspections. There is, however, a mechanism under the UN Secretary-General for the investigation of allegations of the use of chemical, biological and toxin weapons that was set up at the end of the 1980s – the UN Secretary General's Mechanism or UNSGM. It was procedurally updated in 2007 and last used to investigate allegations of the use of chemical weapons in Syria in 2013. This UNSGM gives the UN Secretary-General authority to conduct investigations when a UN Member State brings a report to him/her indicating that a CBT weapon may have been used. But in terms of operational capacity and available resources, the mechanism largely depends on what UN Member States are able and willing to contribute to it. Its operational basis includes a roster of qualified experts for the conduct of field investigations, expert consultants nominated by Member States to the UNSG to provide expert advice and guidance for the conduct of such investigations, and a roster of laboratories designated to the UNSGM.

Concerted efforts have been made in recent years to strengthen the operational capacity of the mechanism. This includes the operational training of experts, the development of operating procedures for the conduct of investigations to confirm the deliberate release of chemical or biological agents, and first steps towards the setting-up of an international network of UNSGM designated laboratories, in particular in technical areas that are not covered by the existing OPCW network of Designated Laboratories. That emerging new network covers the off-site analysis of biological agents as well as toxins (the OPCW has recently begun moving into the field of toxin analysis and is a partner of this process of strengthening the laboratory capacity of the UNSGM).

The current status of setting up this laboratory network under the UNSGM was summarised in the report of the 4th workshop on UNSGM Designated Laboratories organised by Spiez Laboratory in September 2018 (Spiez Laboratory 2018). Spiez Laboratory has providing a platform for conducting regular workshops to share ideas and information among participating experts and laboratories. As part of this overall process, Germany is implementing a project called “RefBio 2017–2020” that has set out to better understand the roles and capabilities expected of UNSGM designated laboratories. The project provides a framework for self-evaluation of participating laboratories, and improvement through interlaboratory quality assurance exercises and workshops. The project will provide exercises for viruses, bacteria and toxins, with levels of difficulty increasing over time. The initial exercises have already shown that investigative analyses will require in-depth sample analysis and characterisation that goes well beyond everyday clinical diagnostics, if the objective is to discriminate between natural disease outbreaks and man-made incidents. Validated analytical procedures, access to reference standards and data, and careful documentation are important for data analysis and interpretation in such investigations.

Denmark and Sweden have started a project of confidence building exercises aimed at detection and characterisation of a biological agent and its associated genetic markers using genomic analysis. These are bioinformatics (dry-lab) tests that so far have focused on the identification of target species and of genomic modifications of virulence factors, using genome analysis.

The OPCW, too, has conducted three exercises in toxin analysis (as of January 2019). It is continuing to expand the capabilities of its network of Designated Laboratories beyond the current scope of environmental as well as biomedical sample analysis, to also include toxins. Initial confidence building exercises have focused on Ricin and Saxitoxin.

This is part of a broader effort by the OPCW to enhance its verification systems and tools – in part to take account of changes in the chemical industry and in chemical trade that the verification system needs to take into account, in part to make better use of new opportunities offered by advances in science and technology. In 2013, the Director-General convened a SAB temporary working group on verification that, amongst others, looked at new and emerging technologies that might add value to the existing verification tools of the OPCW. Many of the recommendations developed by the SAB address the verification system in general, including such issues as how to verify the completeness of information submissions to the OPCW, how to ensure a more holistic and analytical approach in CWC verification and make use of a broader range of complementary data available (including open source data), and how to improve information management for verification purposes. But the group also reviewed new technologies, and highlighted the desirability to pay attention to such developments as remote/automated monitoring equipment, satellite imagery and the use of new types of publicly available information (e.g., geospatial data). It also noted the developments of 3D modelling and virtual reality applications that may be useful tools to enhance inspector training. More generally speaking, the group emphasised that these emerging technologies generate increasing magnitudes of different types of data that need to be integrated (visual, temporal, analytical, and geospatial data from diverse sources). There is thus a need for specialised equipment, such as integrated information management solutions, and trained staff to produce meaningful data summaries (SAB 2015, pp. 28–9).

The SAB verification report also drew attention to the need to fill capability gaps of the OPCW in the area of investigations, including with regard to toxin analysis, and more generally speaking in chemical forensics. This has become particularly relevant after the OPCW adopted, in June 2018, a decision on how to address the threat from chemical weapons use. This decision was prompted by recent uses of chemical weapons in Syria, Iraq, the United Kingdom and Malaysia, and whilst it specifically authorised arrangements to identify perpetrators of chemical weapons uses in Syria, it also affirmed that the perpetrators, organisers, sponsors or otherwise involved parties of *any* CW use should be identified, and it recognised the value of the Technical Secretariat conducting an independent investigation of an alleged use of chemical weapons with a view to facilitating universal attribution of all chemical weapons attacks (OPCW 2018). In response to this decision, the OPCW is setting up a dedicated office to investigate alleged CW uses with the objective of collecting data that would allow attributing responsibility for these acts, or support thereto (the Investigation and Identification Team or IIT).

This process is supported by a new working group of the SAB that specifically deals with investigative science and technology. At the time of writing, the group had conducted two meetings (SAB 2018, 2019). It had started looking specifically at chemical attribution methodologies such as impurity profiling and stable isotope ratio determination to link chemical signatures of agents or degradation products to their possible origins/sources. The group also discussed how quickly chemical evidence may degrade and what practices may be needed to maintain sample integrity. Other issues that were reviewed included digital evidence authentication, remote sampling using ground based and aerial platforms including the work of the European Defence Agency Improvised Explosive Device Detection Programme, the use of satellite imagery (including the services available from UNOSAT), the experience available from forensic attribution analysis of artwork, the analysis of characteristic biomarkers or reaction products for certain chemicals with natural background (such as chlorine), and methods from waste water epidemiology to identify biomarkers of human activity.

In short, chemical forensics for chemical weapons investigation purposes is making some progress, but there remain many gaps with regard to methodologies, validated and generally accepted procedures and standards, as well as robust and comprehensive reference data. A separate chapter of this book discusses the situation in this field in more detail.

7 Conclusions

When chemical weapons were first used in industrial quantities in warfare, some proponents praised them as "a higher form of killing", thereby pointing to the scientific and technological advancement that their mass production, long-term storage and effective dissemination on the battlefield required. The history of chemical and biological warfare during the twentieth century knows of many cases where it has been the technologically advanced State that chose these types of weapons, often to compensate for lack of manpower or to deny the enemy access to strategically or tactically important areas. Chemical as well as biological warfare require a degree of sophistication – from production to effective use – that only countries with a developed base in science, technology and industry could master. Also, the protection against chemical or biological weapons – which, if effective, is a strong deterrent against their use – involves a range of technologies (rapid detection, unambiguous identification, physical means of protection against exposure by breathing or physical contact, means of effective and fast decontamination, medical countermeasures from prophylaxis to the management of mass casualties) that not all countries are able to afford.

This dependence on a strong science and technology base has made chemical and biological weapons strong candidates for types of weapons that advanced countries may consider, whilst putting underdeveloped countries at a high risk and a clear disadvantage.

As the progress in the life sciences and in related technologies is rapidly advancing, this logic may still hold in State-to-State conflict scenarios, at least to a degree. However, the international regimes against these types of weapons, in particular with reference to the Biological Weapons Convention of 1975 and the Chemical Weapons Convention of 1997, have created barriers which at least in words, essentially no State is challenging today. Despite the numerous uses of chemical weapons in Syria, not a few of which have been attributed to the Syrian government, no State has come out supporting or justifying the use of chemical weapons in this (or any other recent) armed conflict.

The pressure that the regimes that prohibit chemical and biological warfare face today, in a State context, emanate less from outright challenges to the prohibitions themselves, but more from the use of such weapons in ways and with tactics that make attribution of responsibility difficult.

At the same time, these regimes do have certain grey areas that States can exploit. One such area is the defence against CB weapons – a safeguard against regime failure (treaty violations) as well as lack of universality (absence of certain presumed possessor States from the treaties) – that had to be included in the Conventions to ensure the continued security of its parties. But up to a point, protective research and development overlaps with what an offensive programme would also look at. Deciding from where on a programme has an offensive character rather than a purely defensive orientation is a contextual assessment that can be difficult and at times controversial – it is certainly open to interpretation and challenges.

Another grey area relates to the right of States to use chemical (as well as biological) agents for legitimate purposes, in particular with regard to using them in law enforcement including riot control. Certain physiologically active chemicals (in the past often labelled as "incapacitating chemicals", today more often referred to as CNS-acting chemicals) have drawn attention as States might be tempted to include such agents into their law enforcement tool kit, under the notion of advanced riot control agents. Whilst this may appear to some to be consistent with the letter of the law, it has the potential to undermine the objectives of the disarmament regime and thus to violate the spirit of the Conventions – note that the CWC and the BWC are not merely agreements to prohibit the use of CB weapons in armed conflict (which the Geneva Protocol already does) but also to eliminate the physical means to wage CB warfare.

A third area has to do with what might be called "latent standby capabilities" in the chemical and biological industries of States parties to the treaties. Whilst the verification system of the CWC in the chemical industry has been set up to keep such legitimate capabilities in check, its adaptation to new scientific and technological developments has been slow. Certain types of facilities that contain technological and operational features similar to Schedule 1 facilities have already emerged – they may be liable to receive on-site inspections, but the data declared on them would make it difficult for the OPCW Technical Secretariat to actually identify them from the declared data sets so as to align inspections in industry with the risks these facilities pose to the objet and purpose of the Convention.

All these risk domains are of course under the purview of the States Parties of the two Conventions, and their policy-making organs (the Executive Council and the Conference of the States Parties of the CWC, and the Review Conference of the BWC, respectively). What will be important for the dependability and stability of the regimes in the long run is the collective political will of the States Parties to manage these risks, uphold the prohibitions and respond to situations when the norms are put at risk.

With regard to non-State actors, the advances in science and technology at the intersection of chemistry and biology may lead into a different kind of shift in the risk landscape. With knowledge and tools becoming more widely distributed, and more accessible including to non-specialists, and despite the continuing barrier function of tacit knowledge, the risk potential that determined non-State actors (terrorist groups with sustained funding; criminal organisations; insiders with access to sensitive knowledge, materials and equipment) could pose is increasing. Given enough time, this is likely to make scenarios of asymmetric warfare involving chemical or biological agents both more attractive for determined actors with sustained funding, and easier to instigate.

At the same time, the experience in Syria and Iraq has shown that non-State actors are quite capable of using old chemical (less so biological) technology in conflict scenarios, to considerable effect.

Traditional control measures such as export controls or international arms control regimes are an increasingly ineffective response to such threats. As has been argued before, what is needed is a combined and sustained effort to strengthen and implement international norms and controls, enact and enforce national laws, encourage compliance measures taken voluntarily by industry, traders and science and technology organisations, increase educational efforts in schools and institutions of higher education, and to engage civil society. Revill and colleagues argue that new thinking is required to boost confidence in compliance (Revill et al. 2019). This would entail, amongst others, stocktaking of what actually works in the current system, and to share lessons-learned across the entire spectrum of regimes and institutions. They also call for a broad stakeholder involvement (including engaging a wider range of countries, taking account of their particular interests and capacities), and a horizon-screening with regard to capabilities that could augment the WMD-related treaties in the future.

Such a more comprehensive approach is aimed at increasing resistance and resilience in society rather than attempting to eliminate the security risks emanating from chemical and biological agents, equipment and technologies altogether – something that will remain out of reach given the nature of the threat and of the diffusion of the sciences into, and use by, society.

References

Bunker, Robert L. 2019. *Contemporary chemical weapons use in Syria and Iraq by the Assad regime and the Islamic State*. Carlisle: US Army War College, Strategic Studies Institute.

Crowley, Michael, Malcolm Dando, and Lijun Shang, eds. 2018. *Preventing chemical weapons: Arms control and disarmament as the sciences converge*. Cambridge: The Royal Society of Chemistry.

Krutzsch, Walter, Eric Myjer, and Ralf Trapp. 2014. *The chemical weapons convention – A commentary*. Oxford: Oxford University Press.

Nakagawa, Tomomasa, and Anthony T. Yu. 2018. *Murders with VX: Aum Shinrikyo in Japan and the assassination of Kim Jung-Nam in Malaysia.* Letters to the Editor, Forensic Toxicology (Springer Nature). https://link.springer.com/article/10.1007/s11419-018-0426-9. Accessed 6 June 2019.

OPCW. 2017. Joint paper – Aerosolisation of central nervous system – Acting chemicals for law enforcement purposes, OPCW document C-22/Nat.5.

———. 2018. Decision – Addressing the threat from chemical weapons use, document C-SS-4/Dec.3.

Revill, James, John Borrie and Augusta Cohen. 2019. Global WMD risks are rising. It's time to do something about it. *The Diplomat* (14 June 2919).

Schneider, Tobias and Theresa Lütkefend. 2019. *Nowhere to hide – The logic of chemical weapons use in Syria.* Global Public Policy Institute study (February 2019). Available at https://www.gppi.net/2019/02/17/the-logic-of-chemical-weapons-use-in-syria. Last accessed 7 Mar 2019.

Scientific Advisory Board SAB. 2014. *Convergence of chemistry and biology*. Report of the Scientific Advisory Board's Temporary Working Group, OPCW document SAB/REP/1/14.

———. 2017a. Report of the Scientific Advisory Board's Workshop on emerging technologies, OPCW document SAB-26/WP.1.
———. 2017b. Report of the Scientific Advisory Board's Workshop on trends in chemical production, OPCW document SAB-26/WP.2.
———. 2018. Summary of the first meeting of the Scientific Advisory Board's Temporary Working Group on investigative science and technology, OPCW document SAB-27/WP.1.
———. 2019. Summary of the second meeting of the Scientific Advisory Board's Temporary Working Group on investigative science and technology, OPCW document SAB-28/WP.2.
Spiez Laboratory. 2018. UNSGM Laboratories Workshop Report, Spiez 9–11 September 2018.
Spiez Laboratory and ETH Zürich. 2014. Spiez Convergence – Report on the first workshop 6–9 October 2014, Spiez.
———. 2016. Spiez Convergence – Report on the second workshop 5–8 September 2016, Spiez.
———. 2018. Spiez Convergence – Report on the third workshop 11–14 September 2018, Spiez.
Trapp, Ralf. 2018. *Novičok, die Skripal Affäre und das Chemiewaffen-übereinkommen* (Novichoks, the Skripal Affair, and the Chemical Weapons Convention), SIRIUS 2018; 2(3), pp. 219–238. https://www.degruyter.com/downloadpdf/j/sirius.2018.2.issue-3/sirius-2018-3002/sirius-2018-3002.pdf. Accessed 6 June 2019.

Neuroscience-Based Weapons

Tatyana Novossiolova and Malcolm Dando

1 Introduction

In 2009, the US National Research Council (NRC) published a report titled *Opportunities in Neuroscience for Future Army Applications*. The report noted that:

> Emerging neuroscience opportunities have great potential to improve soldier performance and enable the development of technologies to increase the effectiveness of soldiers on the battlefield. Advances in research and investments by the broader science and medical community promise new insights for future military applications. These include traditional areas of interest to the Army, such as learning, decision making, and performance under stress, as well as new areas, such as cognitive fitness, brain–computer interfaces, and biological markers of neural states. [...] Opportunities exist for the Army to benefit from research in neuroscience by applying and leveraging the results of work by others (including academic research and R&D by other federal agencies and the commercial sector) or by making selective investments in Army-specific problems and applications (NRC 2009).

The report covered the potential implications of neuroscience research for several domains related to the army activity, such as training and learning; optimising decision-making; sustaining soldier performance; and improving cognitive and behavioural performance. The report further assesses high-risk, high-payoff neuroscience technology opportunities in terms of their potential importance to the army, the likelihood of their development by others, and the time frame for initial operational capability, as well as trends in neuroscience research that are likely to yield future opportunities for the Army and should therefore be monitored by a suitable

T. Novossiolova (✉)
Research Fellow, Law Program, Center for the Study of Democracy, Sofia, Bulgaria
e-mail: tatyana.novossiolova@csd.bg

M. Dando
Leverhulme Emeritus Fellow, Division of Peace Studies and International Development, University of Bradford, Bradford, UK

M. Martellini, R. Trapp (eds.), *21st Century Prometheus*,
https://doi.org/10.1007/978-3-030-28285-1_5

and continuing mechanism (NRC 2009). Among the overarching recommendations of the report was that:

> The Army should establish a group consisting of recognized leaders in neuroscience research in both the academic and private sectors to track progress in non-military neuroscience R&D that could be relevant to Army applications. To ensure that the monitoring group remains sensitive to and abreast of Army needs, the membership should also include Army civilians and soldiers whose backgrounds and interests would suit them for meaningful participation in the group's deliberations (NRC 2009).

In 2012, the Royal Society in the UK issued a report entitled *Neuroscience, Conflict and Security* which considered some of the potential military and law enforcement applications arising from key advances in neuroscience (Royal Society 2012). The report was published within the framework of the Brain Waves project which explored the potential and the limitations of neuroscience insights for policy-making, as well as the benefits and the risks posed by applications of neuroscience and neurotechnologies (Royal Society 2019). The report highlighted a number of research areas with possible military implications broadly divided into two categories: performance enhancement and performance degradation. Among the reviewed areas in which advances in neuroscience might confer performance advantages in military context were neuroimaging techniques; brain stimulation techniques; neuropharmacological agents; neural interface systems; and neural enhancement technologies. In the context of degradation, the report focused on the potential weaponisation of advances in neuroscience such as developments in anaesthetics and neuropharmacological drug research, including targeted drug delivery. It also referred to the development of incapacitating chemical agents and the effects of directed energy weapons (Royal Society 2012).

The report further contained a list of recommendations directed at different stakeholders. With regard to the scientific community, the report recommended that:

> There needs to be fresh effort by the appropriate professional bodies to inculcate the awareness of the dual-use challenge (i.e., knowledge and technologies used for beneficial purposes can also be misused for harmful purposes) amongst neuroscientists at an early stage of their training.

With regard to the UK Government, the report recommended that:

> The UK government […] should improve links with industry and academia to scope for significant future trends and threats posed by the applications of neuroscience.

With regard to the international community, the report recommended that:

> The implementing bodies of the Biological Weapons Convention (BWC) and CWC should improve coordination to address convergent trends in science and technology with respect to incapacitating chemical agents.

And that:

> Neuroscience should be considered a focal topic in the science and technology review process of the BWC because of the risks of misuse for hostile purposes in the form of incapacitating weapons.

Both the NRC and the Royal Society reports have recognised that advances in neuroscience can have legitimate military application. At the same time, the risk that cutting-edge neuroscience developments can undermine the established international legal rules by facilitating the development of novel biological and/or chemical weapons targeted at the nervous system is real and requires attention (Dando 2003; Tucker 2008). The purpose of this chapter is to examine the efforts to strengthen the international biological and chemical prohibition regime. In particular, the chapter seeks to offer fresh insights into the role that the scientific community can play in fostering a culture of responsible science that is keenly responsive to the multifaceted dual-use conundrums that challenge the existing international arrangements for preventing the development, stockpiling, acquisition, retention, and use of chemical and biological weapons. The chapter begins by an overview of key trends in neuroscience research that raise dual-use concerns. It then reviews the key international regulatory mechanisms in the area of chemical and biological disarmament and examines the ongoing efforts to reach out and engage the neuroscience community with issues of responsible science and dual use. The chapter concludes by looking into the role of the neuroscience community in strengthening the biological and chemical prohibition regime.

One area that remains beyond the scope of the present chapter is the development of artificial intelligence (AI). Advances in neuroscience contribute to an enhanced understanding of human psychology, moods, and behaviour. At the same time, computing science and digital technologies heighten the potential risk of the misuse of such insights for malicious purposes including deceit, manipulation, surveillance, and eventually control on an unprecedented scale (Brundage et al. 2018). Whilst the following sections do not address this conundrum in detail, the key message that this chapter aims to convey about the value of a responsible science culture in ensuring the legitimate use of science and technology is of relevance to the rapidly growing field of artificial intelligence and machine learning.

2 Neuroscience Advances of Concern

The industrial application of the growing capabilities of civil chemists in the second half of the nineteenth Century eventually led to the largescale use of chemical weapons during the First World War. Some of the chemical agents used, for example sulphur mustard (Wright et al. 2009), can have an effect on the nervous system, but not by a well-defined mechanism. Our understanding of the nervous system lagged behind our understanding of chemistry and it was only towards the turn of the nineteenth Century that the studies of Cajal demonstrated that the nervous system was made up of separate cells – neurons – rather than a reticulum, and only in 1897 that Sherrington suggested that the junction between such cells formed a synapse (Squire et al. 2013). It was not until the 1920s that Otto Loewi demonstrated that information transfer between neurons and target cells was by means of chemical transmitters (Meyer and Quenzer 2013). Subsequently, it has been shown that chemical transmission is the predominant means of information transmission between cells in the nervous system. The first chemical transmitter to be identified was acetylcholine and since then, an ever-growing list of other such chemicals has been discovered.

Yet by the 1930s the nerve agents that acted at specific sites within the nervous transmission system were discovered. In 1936, Gerhard Schrader was developing new pesticides. He sprayed insects with "a solution of one of his new chemicals that he diluted 200,000-fold, which killed them." He also was contaminated by the agent and developed "pinpoint-constricted pupils, sensitivity to light, shortness of breath, and giddiness." The agent was later called tabun (GA) the first of the nerve agents (Henderson et al. 2016).

The difference between these new nerve agents and the chemical weapons agents of the First World War was, of course, in the specificity of their action on the operations of the central and peripheral nervous system. All of the organophosphorus nerve agents disrupt the function of the major acetylcholine neurotransmitter system. In this cholinergic system the acetylcholine neurotransmitter is released from the presynaptic neuron terminal and attaches to receptors on the postsynaptic neuron or muscle to exert its effect. The acetylcholine is then broken down rapidly by an acetylcholinesterase enzyme to bring its effects to a close. The nerve agents act by inhibition of this enzyme and thus the synapse is flooded with acetylcholine with the well-known potentially lethal consequences (Sidell 1997). Similarly, botulinum toxin which was weaponised in the last century (Test and Evaluation Command 1982) operates on the pre-synaptic neuron to prevent the release of neurotransmitter into the synaptic cleft (Middlebrook and Franz 1997). It is this specificity of action that makes such agents so dangerous at low concentrations (Franz 1997). Botulinum toxin having a LD_{50}(ug/kg) in mice of 0.001 and VX nerve gas of 15.0. Even Staphylococcus Enterotoxin B, also weaponised in the last century (Pinchuk et al. 2010), but acting indirectly via the immune system (Kaempfer et al. 2002), has an LD_{50}(ug/kg) of 27 when used in an aerosol against Rhesus monkey. Moreover, we are gaining more and more understanding of the modes of operation of the viruses that cause encephalitis, a number of which were also weaponised during the last century (Bale 2015).

Fortunately, the agreement of the 1925 Geneva Protocol, The Biological and Toxin Weapons Convention and the Chemical Weapons Convention during the last century provides a firm international framework for the prohibition of such weapons (Crowley et al. 2018a). The question now is how to ensure that such weapons do not re-emerge during the present period of instability in the international security system and very rapid scientific and technological change in relevant areas (Crowley et al. 2018b). The regulation of the use of incapacitating agents for law enforcement purposes is a case in point. Whilst Article II.9(d) of the Chemical Weapons Convention recognises law enforcement, including domestic riot control, as a potentially acceptable purpose for the use of certain toxic chemicals (provided that the types and quantity used are consistent with this purpose), the range of potentially permissible chemicals has not been established (Shang et al. 2018). The process of turning a new candidate agent into a legitimate incapacitant is very similar to that of developing a viable chemical weapon. For example, it requires the identification of formulations that are stable in storage, effective in dissemination, and economically feasible in manufacturing, as well as the development of suitable dissemination systems/equipment, doctrines for use, and protective measures for ensuring the safety of one's own forces. Undertaking these types of developments under the banner of a legitimate activity amounts to legitimising a break-out capability.

The issue is further complicated by the fact that "contemporary definitions of incapacitating chemical agents have tended to emphasise the rapid onset of action and short duration of effects" (Royal Society 2012). Such agents are suitable for large scale military operations, but it is worth noting that this choice of effect was also forced on weaponeers by their lack of understanding of the central nervous system. Their original intentions during the early Cold War period was to seek means to cause a much wider range of 'selective malfunctions of the human machine', and it may not be too farfetched to imagine that the present rate of advances in our understanding of civil neuroscience, given the huge investments being made in State-level brain projects, could now present an increasing spectrum of such wider-ranging dual-use opportunities (Dando 2018; Nixdorff et al. 2018). What we are likely to see is an ever more detailed mechanistic understanding of the circuits of the central nervous system that can be put to good uses and unfortunately also malign purposes.

The dangers have not escaped the notice of the States Parties to the Chemical Weapons Convention in recent years. A total of 43 States Parties presented a joint paper on *Aerosolisation of Central Nervous System- Acting Chemicals for Law Enforcement Purposes* to the 4th Review Conference in November 2018. The paper stated that:

> Toxic (and potentially lethal) chemicals that target the central nervous system (CNS) so-called 'incapacitating chemical agents or ICAs', and their potential use in certain law enforcement scenarios, have been discussed in numerous forums. We believe these chemicals pose a serious challenge to the Convention (Albania et al. 2018).

The paper went on to note that "CNS-acting chemicals are not riot control agents (RCAs)" highlighting that "individuals exposed to aerosolised CNS-acting chemicals face inherent safety risks that include potential long-term health effects". The paper thus concluded that "CNS-acting chemicals cannot be dispersed by aerosol in a completely safe manner in law enforcement settings" which in turn "raises concerns that CNS-acting chemicals could be used as chemical weapons."

The problem, as some commentators have observed, is that as States become more concerned about the possible development of such agents and conduct research into how to deal with their use, others may well see potential offensive possibilities and react in turn. Thus, an erosion of the prohibition of chemical and biological weapons could come about by intent or mischance in what has been called a "degradation market" as new means of incapacitation and counters to novel agents are sought (Dando 2015).

3 The International Biological and Chemical Prohibition Regime

Table 1 gives an overview of the principal international regulations concerning the prohibition of chemical and biological weapons.

Within the framework of the Biological and Toxin Weapons Convention (BTWC), review of scientific and technological developments that are relevant to the

Table 1 International regulations prohibiting CB Weapons

Regulation	Year of adoption	Entry into force	Membership (as of December 2019)	Scope
Geneva Protocol	1925	1928	142 States Parties	Prohibition of the use of asphyxiating, poisonous, or other gases, and of bacteriological methods of warfare.
Biological and Toxin Weapons Convention	1972	1975	183 States Parties 4 Signatory States	Article I Each State Party to this Convention undertakes never in any circumstances to develop, produce, stockpile or otherwise acquire or retain: 1. microbial or other biological agents, or toxins whatever their origin or method of production, of types and in quantities that have no justification for prophylactic, protective or other peaceful purposes; 2. weapons, equipment or means of delivery designed to use such agents or toxins for hostile purposes or in armed conflict.
Chemical Weapons Convention	1993	1997	193 States Parties 1 Signatory State	Article I 1. Each State Party to this Convention undertakes never under any circumstances: To develop, produce, otherwise acquire, stockpile or retain chemical weapons, or transfer, directly or indirectly, chemical weapons to anyone; To use chemical weapons; To engage in any military preparations to use chemical weapons; To assist, encourage or induce, in any way, anyone to engage in any activity prohibited to a State Party under this Convention. 5. Each State Party undertakes not to use riot control agents as a method of warfare. Article VI 1. Each State Party has the right, subject to the provisions of this Convention, to develop, produce, otherwise acquire, retain, transfer and use toxic chemicals and their precursors for purposes not prohibited under this Convention.
United Nations Security Council Resolution 1540	2004	2004	UN Member States	Acting under Chapter VII of the Charter of the United Nations, 1. Decides that all States shall refrain from providing any form of support to non-State actors that attempt to develop, acquire, manufacture, possess, transport, transfer or use nuclear, chemical or biological weapons and their means of delivery; 2. Decides also that all States, in accordance with their national procedures, shall adopt and enforce appropriate effective laws which prohibit any non-State actor to manufacture, acquire, possess, develop, transport, transfer or use nuclear, chemical or biological weapons and their means of delivery, in particular for terrorist purposes, as well as attempts to engage in any of the foregoing [illegible]

Convention is carried out at an annual Meeting of Experts (MX). Throughout the Intersessional Process 2018-2020, the MX on Review of developments in the field of science and technology related to the Convention will consider the following topics:

(a) "Review of science and technology developments relevant to the Convention, including for the enhanced implementation of all articles of the Convention as well as the identification of potential benefits and risks of new science and technology developments relevant to the Convention, with a particular attention to positive implications;
(b) Biological risk assessment and management;
(c) Development of a voluntary model code of conduct for biological scientists and all relevant personnel, and biosecurity education, by drawing on the work already done on this issue in the context of the Convention, adaptable to national requirements;
(d) Any other science and technology developments of relevance to the Convention and also to the activities of relevant multilateral organizations such as the WHO, OIE, FAO, IPPC and OPCW" (MSP 2017).

Over the recent years, States Parties to the BTWC have given attention to the need for engaging the life science community with issues of relevance to the Convention. In particular, States Parties have noted the value of national implementation measures to:

(a) "implement voluntary management standards on biosafety and biosecurity;
(b) encourage the consideration of development of appropriate arrangements to promote awareness among relevant professionals in the private and public sectors and throughout relevant scientific and administrative activities;
(c) promote amongst those working in the biological sciences awareness of the obligations of States Parties under the Convention, as well as relevant national legislation and guidelines;
(d) promote the development of training and education programmes for those granted access to biological agents and toxins relevant to the Convention and for those with the knowledge or capacity to modify such agents and toxins;
(e) encourage the promotion of a culture of responsibility amongst relevant national professionals and the voluntary development, adoption and promulgation of codes of conduct" (Eight Review Conference 2017).

These provisions are pertinent to all those engaged in the life sciences regardless of whether they work in government institutions, industry, or academia.

In 2017, Ukraine, Japan and the United Kingdom tabled a Working Paper titled "Awareness-Raising, Education, and Outreach: Recent Developments" to the BTWC Meeting of the States Parties. The paper noted that:

> One key factor that will help facilitate the responsible conduct of biological science is ensuring that biosecurity education and awareness is a key component of scientific training early on in the career progression of life scientists and those in other relevant disciplines. It is, however, not just the responsibility of scientists to ensure that life science research is

> not directed to misuse: in particular, policy makers, other government officials and industry all have a role to play (Ukraine et al. 2017).

Another relevant initiative within the framework of the BTWC is the joint effort of China and Pakistan to develop a model code of conduct for biological scientists (China and Pakistan 2018).

The Organisation for the Prohibition of Chemical Weapons (OPCW) oversees the implementation of the Chemical Weapons Convention (CWC) at an international level. The OPCW Technical Secretariat assists the Conference of the States Parties and the Executive Council in performing their tasks and carries out the CWC verification measures (OPCW 2019a). The Scientific Advisory Board (SAB) is an OPCW subsidiary body that is tasked with the provision of specialised advice in science and technology (OPCW 2019b). The role of the SAB is to assess developments in scientific and technological fields that are relevant to the Convention. In its report that was prepared for the Fourth Review Conference of the CWC in 2018, the SAB analysed different trends of science and technology relevant to the Convention and gave recommendations on addressing issues that may impact the implementation of the Convention (SAB 2018). With regard to the need for closer cooperation with the Biological and Toxin Weapons Convention, it was recommended that:

> 17. The SAB and the Secretariat should continue to work across areas of overlap between the Chemical Weapons Convention and the Biological Weapons Convention and promote joint discussions amongst international experts in these areas.

When considering incapacitating or central nervous system (CNS)-acting chemicals, the SAB reached the following conclusions:

> 97. As previously noted, [riot-control agents] RCAs are expected to produce an immediate disabling effect on personnel. They act primarily on the sensory nerves of the peripheral nervous system in the eyes, nose, respiratory tract, and skin, having limited or no effect on the central nervous system (CNS). However, other types of disabling chemicals, historically referred to as incapacitating chemical agents (ICAs) differ from RCAs as they act primarily on the CNS. ICAs also differ from RCAs in that their effects are not usually confined to sensory irritation of a temporary nature. These compounds can induce incapacitation including cognitive impairment, loss of motor function and ultimately unconsciousness.

And that:

> 101. The SAB is of the view that technical discussions on CNS-acting chemicals remain exhausted; this issue is important to the Convention and is now in the policy domain. The SAB sees no value in revisiting this topic as the scientific facts remain unchanged since the SAB first considered the issue, in the SAB's Report to the Third Review Conference (see paragraph 12 of RC-3/DG.1). The SAB recognises that CNS-acting acting compounds are not RCAs as they do not meet criteria specified in Article II paragraph 7. Furthermore, the SAB notes that there have been examples of the use of CNS-acting chemicals in law enforcement that have resulted in permanent harm and death due to an irreversible action on life processes.

The SAB further recommended updating the OPCW Central Analytical Database (OCAD) – a reference library of validated data of chemicals of relevance to the Convention – by including additional chemicals such as CNS-acting chemicals, bioregulators, and toxins, among others, to allow the OPCW to meet all its mandated inspection aims SAB 2018).

The OPCW Advisory Board on Education and Outreach (ABEO) aims to ensure that the Organisation's education and outreach activities, and those of States Parties, are effective, sustainable, cost-effective, and benefit from the latest advances in education and outreach theory or practice (OPCW 2019c). In February 2018, the ABEO published *Report on the Role of Education and Outreach in Preventing the Re-Emergence of Chemical Weapons* (ABEO 2018). The report contains specific recommendations on building capacity for education and outreach as follows:

> The Secretariat should systematically develop more interactive approaches to audiences across the full range of its [education and outreach] activities. This should include a greater emphasis on assessing the effectiveness of teaching or training. Courses and other activities thus need to be designed with clear goals and measurable objectives.

And that:

> The OPCW should take advantage of its existing processes for supporting the [National Authorities] to assist them in building the capacity to carry out [education and outreach]. In addition, existing [education and outreach] materials need to be augmented to enable them to be used more effectively (ABEO 2018).

The report also features a portfolio of recommended education and outreach activities for different categories of stakeholders, such as industry, scientists (individuals with scientific, engineering, or other technical backgrounds working in any sector relevant to the implementation of the Convention and to the broader challenge of preventing the re-emergence of chemical weapons), academia (all persons involved in an institution of higher learning (e.g., universities and colleges) across fields ranging from science and technology to law, international relations, social sciences, humanities, business, and others); civil society; policymakers; and the media (ABEO 2018).

The Security Council Committee established pursuant to Resolution 1540 in 2004 covers four areas of activity: monitoring of national implementation; assistance; and cooperation with international organisations including the Security Council committees established pursuant to resolutions 1267 (1999) and 1373 (2001); and transparency and media outreach (1540 Committee 2019). The Committee comprises 15 members and is assisted by a Group of Experts. States are required to report on the steps that have been taken to implement the provisions of Resolution 1540 at a national level. States are also encouraged to prepare on a voluntary basis national implementation action plans that map out their priorities for fulfilling their respective obligations under the Resolution.

4 Responsible Science and Dual Use: Engaging the Neuroscience Community

The first edition of the guide *On Being a Scientist* published by the US National Academy of Sciences in 1989 noted that:

> Scientists conducting basic research also need to be aware that their work ultimately may have a great impact on society. World-changing discoveries can emerge from seemingly

> arcane areas of science. [...] The occurrence and consequences of discoveries in basic research are virtually impossible to foresee. Nevertheless, the scientific community must recognize the potential for such discoveries and be prepared to address the questions that they raise (NAS 1989).

The second edition of the guide *On Being a Scientist: Responsible Conduct of Research* that appeared in 1995 further developed the point about scientists' responsibilities with regard to the implications of scientific discoveries:

> If scientists do find that their discoveries have implications for some important aspect of public affairs, they have a responsibility to call attention to the public issues involved. They might set up a suitable public forum involving experts with different perspectives on the issue at hand. They could then seek to develop a consensus of informed judgment that can be disseminated to the public (NAS 1995).

The third edition of the guide *On Being a Scientist: A Guide to Responsible Conduct of Research* published in 2009 when discussing the role of the researcher in society referred to Arthur Galston's reflections on ending the use of Agent Orange during the Vietnam War:

> In my view, the only recourse for a scientist concerned about the social consequences of his work is to remain involved with it to the end. His responsibility to society does not cease with publication of a definitive scientific paper. Rather, if his discovery is translated into some impact on the world outside the laboratory, he will, in most instances, want to follow through to see that it is used for constructive rather than anti-human purposes. But I know of no moral imperative to invoke here; some individuals feel moved to respond to the social challenge, while others shun such activity, either through timidity, aversion to political argumentation or a feeling that others, better trained, should handle social problems. [...] Science is now too potent in transforming our world to permit random fallout of the social consequences of scientific discoveries. Some scrutiny and regulation are required, and I believe that scientists must play an important role in any bodies devised to carry out such tasks (Galston 1972).

In 2016, the Interacademy Partnership (IAP) published a guiding document entitled *Doing Global Science: A Guide to Responsible Conduct in the Global Research Enterprise* (IAP 2016). Chapter 3 of the guide specifically focuses on preventing the misuse of research and technology noting that:

> The difficulty of predicting the future course and applications of research does not absolve researchers of the responsibility for participating in venues to explore these issues. **Researchers need to participate in discussions about the possible consequences of their work, including harmful consequences, in planning research projects**. As the ones who design and carry out research, researchers can provide information on the nature and purpose of research that is not available in any other way [emphasis as original] (IAP 2016).

Fostering an understanding of the responsibility to ensure that their work is not misused for the development of chemical and biological weapons among the scientific community, including those engaged in neuroscience remains a challenging task. Part of the problem is that there are few opportunities for scientists of getting exposed to issues of dual use, not least because those issues rarely feature in their formal education (Mancini and Revill 2008). Persisting obstacles to inculcating awareness of dual use science include the need for sustained effort, the need to assess the effectiveness and impact of activities, the lack of staff trained to teach this aspect of responsible science, and the lack of time lecturers and teachers had to dedicate to understanding this topic (NAS 2018).

Over the recent years some effort has been made to engage the neuroscience community with the social, legal, security, and ethical implications of their work over the recent years. The Euro-Mediterranean Master in Neuroscience and Biotechnology (EMN-Online) is a two-year inter-university program aimed at professionals wishing to acquire innovative and interdisciplinary training in Neuroscience (EMN-Online 2019). It was first launched in 2010. The master's programme is taught online by a consortium of 14 partner universities, 9 located in Europe and 5 located in the South-Mediterranean area. It features a compulsory module on 'Regulations, Laws, and Bioethics' which seeks to sensitise participants to the broader social, ethical, legal, and security implications of neuroscience (EMN-Online 2019). The programme is funded by the European Commission and supported by the Mediterranean Neuroscience Society.

In 2012–2013, the University of Bradford and the University of Manchester implemented a joint initiative titled 'Interdisciplinary Network on Teaching Ethics for Neuroscientists' (UKRI 2019). The aim of this project was to develop an in-depth understanding of the modalities of ethics teaching and training, in order to make the subject matter easily accessible to neuroscientists. Attention was given to the utility of active learning methods. The primary output of the project was the development of an online educational resource on neuroethics which covered a wide spectre of issues, including dual use and the prohibition regime of chemical and biological weapons.

The Global Neuroethics Summit was set up in 2017 to create a forum for rapid deliverables that address the unique needs of fostering a culturally aware neuroethics community while disseminating its resources (Rommelfanger et al. 2019). Following the launch of the International Brain Initiative (IBI) – an alliance of nationally funded large-scale neuroscience research initiatives that aims to advance neuroscience through international collaboration and knowledge sharing – in 2018 the Summit became its Neuroethics Working Group (Rommelfanger et al. 2019).

The Human Brain Project (HBP) is a 10-year initiative launched in 2013 and employing some 500 scientists at more than 100 universities, teaching hospitals, and research centres across Europe. The primary objective of the HBP is to build research infrastructure to help advance neuroscience, medicine, and computing (HBP 2019a). The HBP is embedded within the Responsible Research Innovation (RRI) framework of the European Union:

> RRI is generally understood as an interactive process that engages multiple stakeholders who must be mutually and jointly responsive and work toward the ethical permissibility of both research and its products. RRI calls for aligning science and technology with societal needs and for addressing the legal, ethical, and social dimensions of research and innovation by focusing not just on outcomes, but also on the examination of the values that inform the trajectory of the scientific work and that feed into the research agenda itself (Salles et al. 2019).

The "Ethics and Society" component of the HBP is dedicated exclusively to the analysis of the ethical, social, legal, and philosophical implications of brain research (HBP 2019b). Among the core themes that are addressed as part of this component is the uses and misuses of neuroscience and neurotechnology (Salles et al. 2019).

The HBP Education Programme offers innovative learning packages for early career researchers working in and across the fields of neuroscience, information and communications technology (ICT) and medicine (HBP 2019c). The HBP Curriculum on Interdisciplinary Brain Science covers six topics:

- ICT for non-specialists
- Neurobiology for non-specialists
- Brain medicine for non-specialists
- Cognitive systems for non-specialists
- Research, ethics and societal impact
- IPR, translation and exploitation of research.

The course on 'Research, ethics, and societal impact' comprises eleven lectures as follows:

- Introduction to ethical theory
- Computer ethics and the HBP
- The ethical roboticist
- Responsible research and the Human Brain Project
- Scaling up neuroscience – responsible research and the big brain projects
- Neuroscience and the problem of dual use
- Ethics in biomedical research and the 3Rs (replacement, refinement, reduction)
- Societal attitudes to animal research
- Research integrity and ethics management – HBP case study
- Cognitive enhancement: ethics and efficacy
- The thinking robot.

Lectures are accompanied by workshops which aim to provide deeper insights as well as practical exercises (HBP 2019d). Issues of dual use are considered as part of the 'Research, ethics, and societal impact' workshops (HBP 2019e).

Launched in 2013, the US National Institutes of Health (NIH) Brain Research through Advancing Innovative Neurotechnologies (BRAIN) Initiative is a large-scale public-private collaborative endeavour that seeks to deepen understanding of the inner workings of the human mind and improve the treatment, prevention, and cure of brain disorders (BRAIN 2019). The NIH BRAIN Initiative adopts a proactive neuroethics strategy underpinned by an ongoing assessment of the neuroethical implications of the development and application of neurotechnologies and related tools (Bianchi et al. 2018). The implementation of this approach is supported by an inter-disciplinary Neuroethics Working Group (NEWG). In 2018 the NEWG developed a set of *Neuroethics Guiding Principles* that are designed to provide a framework of reference for dealing with neuroethical issues arising from BRAIN-funded research (Box 1) (Greely et al. 2018).

Box 1 NIH *Neuroethics Guiding Principles* (Greely et al. 2018)

1. Make assessing safety paramount.
2. Anticipate special issues related to capacity, autonomy, and agency.
3. Protect the privacy and confidentiality of neural data.
4. Attend to possible malign uses of neuroscience tools and neurotechnologies.
5. Move neuroscience tools and neurotechnologies into medical and non-medical uses with caution.
6. Identify and address specific concerns of the public about the brain.
7. Encourage public education and dialogue.
8. Behave justly and share the benefits of neuroscience research and resulting technologies.

Principle No 4 specifically addresses the issue of potential neuroscience misuse highlighting the responsibility incumbent upon researchers to be aware of and deal with the dual-use aspects of their work:

> Novel tools and technologies, including neurotechnologies, can be used both for good ends and bad. Researchers should be mindful of possible misuses that might range from intrusive surveillance of brain states to efforts to incapacitate or impermissibly alter a person's behavior. Researchers have a responsibility to try to predict plausible misuses and ensure that foreseeable risks are understood, as appropriate, by research participants, IRBs, ethicists, and government officials (Greely et al. 2018).

The Organisation for the Prohibition of Chemical Weapons (OPCW) has produced a range of education and awareness-raising resources which, whilst not exclusively targeted at neuroscientists, seek to foster understanding of the CWC and its relevance to science activities. The 'Multiple Use of Chemicals' is an interactive educational resource directed at students, educators, and policymakers (MUC 2019). The resource explores the beneficial uses, misuses, and abuses multi-use chemicals both historically and presently and is available in all six UN official languages (Arabic, Chinese, English, French, Russian, and Spanish). The content for students contains examples of chemicals that have both beneficial and harmful uses, ranging from pharmaceutical products to solvents and destructive chemical weapons. Ideas of scientific responsibility and ethics are introduced to help students think about making informed choices. The content for educators and policymakers covers the same topics providing additional information and suggestions for implementation of this resource into classroom or group settings.

The Fires Project is a series of short documentary videos depicting the intersection of people with chemical weapons. The videos are subtitled in the six UN official languages and constitute an effective tool for raising awareness of the CWC and chemical weapons issues (OPCW 2019d).

The Hague Ethical Guidelines were developed in 2015, in order to serve as elements for ethical codes and discussion points for ethical issues to the practice of chemistry under the Convention (OPCW 2019e). At the heart of the Hague Ethical

Guidelines is the need for ensuring that achievements in the field of chemistry should be used to benefit human kind and protect the environment. The Guidelines address the following key elements:

- Education
- Awareness and engagement
- Ethics
- Safety and security
- Accountability
- Oversight
- Exchange of information (OPCW 2019e).

The EU Non-Proliferation and Disarmament eLearning Course developed by the EU Non-Proliferation and Disarmament Consortium consists of 15 learning units which cover the core aspects of the biological and chemical prohibition regime, as well as issues related to emerging technologies and the risks associated thereof (EUNPDC 2019). All units are publicly available online.

5 Conclusion: Strengthening the Chemical and Biological Prohibition Regime

Given the ongoing advancement of science and technology, fostering a culture of responsible science is an essential prerequisite for preventing the development of chemical and biological weapons. This is neither a quick fix, nor a straightforward solution. On the contrary, the concept of a culture presupposes a change: a normative change, a change in the ways things are done, a change in habits, a change in the array of acceptable actions and the meaning attached to them (Novossiolova et al. 2019). For a culture change to occur, at least four basic requirements needs to be met. First, there is a need for a shared in-group awareness that a problem exists. Second, there is a need for a shared in-group understanding that the problem that exists is significant and requires action. Third, there is a need for a coordinated and all-inclusive action by the group. The action needs to be underpinned by a shared responsibility that each member of the group has a duty to work toward the solution of the problem and contribute accordingly. And fourth, there is a need for an ongoing evaluation of whether the action that has been taken by the group is effective for achieving the desired objective and resolving the problem in the best possible way. Needless to say, the process of a culture change is value-laden and a function of the prevalent group mentality and worldview.

Applying this model of a culture change to the neuroscience community has a number of implications for the ways in which different stakeholders can contribute to fostering a responsible science culture. It is both a dynamic and collaborative model, whereby each stakeholder – prospective and practising neuroscientists, science academies, learned academies, funding bodies, industry, and governments has a role and related responsibilities in executing a culture change. The resultant network of networks highlights the importance of fostering lasting trust-based relationships among individual stakeholders, in order to facilitate a synchronised multifaceted action (NAS 2018; Novossiolova et al. 2020).

1. Shared in-group awareness that a problem exists.

Issues of dual-use and responsible science need to feature in the formal education of neuroscientists on a continuous basis. Ideally, prospective scientists should be exposed to relevant topics at high school level. Specialised compulsory classes and modules addressing dual use and responsible science need to be introduced at universities at both undergraduate and graduate level. Dual-use and responsible science issues should also be addressed as part of continued professional development training for practising scientists and researchers at different stages of their career.

Scientific institutions, such as science academies and learned societies, and professional associations should create opportunities for and facilitate an open discussion and debate of issues of dual use and responsible science. The responsibility for being aware of such issues should be codified in the respective charters of such institutions, as well as in other relevant membership documents that they issue.

Funding bodies, whether government entities or privately-owned charities and for-profit companies should have mechanisms in place for dealing with and addressing dual-use concerns. They should also ensure that prospective applicants and staff alike are informed about dual use issues.

Publishers of science research should disseminate on a regular basis information concerning dual use and responsible science. They should further ensure that prospective authors and staff alike are informed about dual use issues.

Government agencies should seek expert advice from other science stakeholders on topics of relevance to dual use and responsible science. Government agencies should further promote public dialogue on dual use and responsible science issues, as well as acknowledge the significance of these issues during multilateral negotiations in regional and international fora.

2. Shared in-group understanding that the existing problem is significant.

Issues of dual use and responsible science need to be regarded as an integral element of science curriculum and not as a supplementary subject matter. To this end, relevant teaching and training content need to be delivered in a way that is engaging and enables scientists to develop an understanding of the broader implications of their work and think of ways to mitigate potential dual-use concerns.

Scientific institutions and professional associations should aim to embed issues of dual use and responsible science in everyday practice through, for example, certification and accreditation procedures. Designated officers who deal with responsible science and dual-use issues need to be appointed to serve as points of contact for guidance and advice.

Funding bodies, whether government entities or privately-owned charities and for-profit companies should invest in promoting responsible science and tackling issues of dual use.

Publishers of scientific research should establish publication series dedicated to issues of dual use and responsible science.

Government agencies and in particular those in charge of science and technology management should appoint designated officers on responsible science and dual-use issues. Government agencies should regularly report on their progress of promoting responsible science during relevant multilateral negotiations in regional and international fora.

3. All-inclusive, coordinated action.

Addressing issues of dual use and responsible science should become a standard operating procedure (SOP) of the research process, whereby both prospective and practising scientists regard dealing with such issues as a routine practice.

Scientific institutions and professional associations should have in their portfolio a programme on dual use and responsible science.

Funding bodies, whether government entities or privately-owned charities and for-profit companies should have in their portfolio a funding scheme that is exclusively dedicated to promoting responsible science and addressing issues of dual use.

Publishers of scientific research should have mechanisms for peer-review that specifically focuses on assessing the potential dual-use concerns of scientific experiments, as well as corresponding procedures for follow-up action in case such concerns are identified.

Issues of responsible science and dual use should feature in national science and technology policy, as well as in the portfolio of relevant government agencies. Governments should actively exchange data, national experience, and lessons learned with regard to the implementation of responsible science during multilateral negotiations in regional and international fora.

4. Ongoing evaluation.

There is a need for a continuous engagement among stakeholders at a local, national, regional, and international level, in order to facilitate peer learning, data sharing, and the exchange of experiences and lessons learned. Existing mechanisms for multilateral negotiation and cooperation should be fully utilised for raising awareness of existing policies, measures, approaches, and practices for promoting responsible science.

References

Advisory Board on Education and Outreach. 2018. Report on the role of education and outreach in preventing the re-emergence of chemical weapons. ABEO-5/1. 12 February. https://www.opcw.org/sites/default/files/documents/ABEO/abeo-5-01_e.pdf. Accessed 6 Mar 2019.

Albania et al. 2018. Aerosolisation of central nervous system-acting chemicals for law enforcement purposes. RC-4/NAT.26. OPCW, The Hague. 30 November.

Bale, J.F. 2015. Virus and immune-mediated encephalitides: Epidemiology, diagnosis, treatment, prevention. *Pediatric Neurology* 53: 3–12.

Bianchi, D., J. Cooper, J. Gordon, J. Heemskerk, R. Hodes, G. Koob, et al. 2018. Neuroethics for the National Institutes of Health BRAIN Initiative. *Journal of Neuroscience* 38 (50): 10583–10585.

Biological and Toxin Weapons Convention. 1972. https://www.unog.ch/80256EE600585943/(httpPages)/04FBBDD6315AC720C1257180004B1B2F?OpenDocument. Accessed 7 Mar 2019.

BRAIN Initiative. 2019. https://www.braininitiative.org/mission/. Accessed 6 Mar 2019.

Brundage, M., S. Avin, J. Clark, H. Toner, P. Eckersley, B. Garfinkel, et al. 2018. *The malicious use of artificial intelligence: Forecasting, prevention, and mitigation*. Oxford: Future of Humanity Institute.

Chemical Weapons Convention. 1993. https://www.opcw.org/chemical-weapons-convention. Accessed 7 Mar 2019.

China and Pakistan. 2018. Proposal for the development of a model code of conduct for biological scientists under the biological weapons convention. BWC/MSP/2018/MX.2/WP.9. 9 August. https://undocs.org/en/BWC/MSP/2018/MX.2/WP.9. Accessed 6 Mar 2019.

Crowley, M., M. Dando, and L. Shang, eds. 2018a. *Preventing chemical weapons: Arms Control and disarmament as the sciences converge*. London: Royal Society of Chemistry.

Crowley, M., L. Shang, and M. Dando. 2018b. Preventing chemical weapons as sciences converge: Focus must extend beyond 20th-century technologies. *Science* 362: 753–755.

Dando, M. 2003. The danger to the Chemical Weapons Convention from incapacitating weapons. First CWC Review Conference Paper No 4. Resource document. University of Bradford. https://core.ac.uk/download/pdf/6159.pdf. Accessed 5 Mar 2019.

Dando, M. R. (2015). *Neuroscience and the future of chemical-biological weapons*. Basingstoke: Palgrave Macmillan. (See Chapter 6: Novel Neuroweapons, particularly pages 84–86).

Dando, M.R. 2018. Advances in understanding targets in the central nervous system (CNS). In *Preventing chemical weapons: Arms control and disarmament as the sciences converge*, ed. M. Crowley, M. Dando, and L. Shang, 228–258. London: Royal Society of Chemistry.

Eight Review Conference of the States Parties to the Convention on the Prohibition of the Development, Production and Stockpiling of Bacteriological (Biological) and Toxin Weapons and on Their Destruction. 2017. Final document of the eighth review conference. BWC/CONF. VIII/4. 11 January. https://www.unog.ch/80256EE600585943/(httpPages)/57A6E253EDFB1111C1257F39003CA243?OpenDocument. Accessed 6 Mar 2019.

EU Non-Proliferation and Disarmament Consortium. 2019. https://nonproliferation-elearning.eu/about. Accessed 6 Mar 2019.

Euro-Mediterranean Master's Degree in Neuroscience and Biotechnology. 2019. https://emn-online.org/project-consortium#general-presentation. Accessed 6 Mar 2019.

Franz, D.R. 1997. Defense against toxin weapons. In *Medical aspects of chemical and biological warfare*, ed. F.R. Sidell, E.T. Takafuji, and D.R. Franz, 603–619. Washington, D.C.: Office of the Surgeon General, Department of the Army.

Galston, A. 1972. Science and social responsibility: a case in history. *Annals of the New York Academy of Sciences* 196 (4): 223–235.

Greely, H., C. Grady, K. Ramos, W. Chiong, J. Eberwine, N. Farahany, et al. 2018. Neuroethics guiding principles for the NIH BRAIN initiative. *Journal of Neuroscience* 38 (50): 10586–10588.

Henderson, T.J., N.M. Elsayed, and H. Salem. 2016. Chemical warfare agents and nuclear weapons. In *Inhalation toxicology*, ed. H. Salem and S.A. Katz, 3rd ed., 489–522. Boca Raton: CRC Press.

Human Brain Project. 2019a. https://www.humanbrainproject.eu/en/education/education-overview/. Accessed 6 Mar 2019.

———. 2019b. Ethics and society. https://www.humanbrainproject.eu/en/social-ethical-reflective/. Accessed 6 Mar 2019.

———. 2019c. HBP education overview. https://www.humanbrainproject.eu/en/education/education-overview/. Accessed 6 Mar 2019.

———. 2019d. HBP curriculum – Interdisciplinary brain science. https://www.humanbrainproject.eu/en/education/participatecollaborate/curriculum/. Accessed 6 Mar 2019.

———. 2019e. Research, ethics, and societal impact. https://education.humanbrainproject.eu/web/1st-hbp-curriculum-ethics. Accessed 6 Mar 2019.

InterAcademy Partnership. 2016. *Doing global science: A guide to responsible conduct in the global research enterprise*. Princeton: Princeton University Press.

Kaempfer, R., et al. 2002. Defense against biologic warfare with Superantigen toxins. *IMAJ* 4: 520–523.

Mancini, G. and Revill, J. 2008. Fostering the biosecurity norm: biosecurity education for the next generation of life scientists. Landau Network Fondazione Volta. http://sro.sussex.ac.uk/id/eprint/39517/1/Fostering.pdf. Accessed 6 Mar 2019.

Meeting of the States Parties to the Convention on the Prohibition of the Development, Production and Stockpiling of Bacteriological (Biological) and Toxin Weapons and on Their Destruction. 2017. Report of the meeting of states parties. BWC/MSP/2017/6. 19 December. https://undocs.org/en/bwc/msp/2017/6. Accessed 6 Mar 2019.

Meyer, J. S. and Quenzer, L. F. 2013. *Psychopharmacology: drugs, the brain, and behavior*. Sunderland: Sinauer Associates. See Chapter 3, Chemical signalling by neurotransmitters and hormones, page 78.

Middlebrook, J.L., and D.R. Franz. 1997. Botulinum toxins. In *Medical aspects of chemical and biological warfare*, ed. F.R. Sidell, E.T. Takafuji, and D.R. Franz, 643–654. Washington, D.C.: Office of the Surgeon General, Department of the Army.

Multiple Use of Chemicals. 2019. http://multiple.kcvs.ca/. Accessed 6 Mar 2019.

National Academies of Sciences. 2018. *Governance of dual use research in the life sciences: Advancing global consensus on research oversight: Proceedings of a workshop*. Washington, D.C.: National Academy Press.

National Academy of Sciences. 1989. *On being a scientist*. Washington, D.C.: National Academy Press.

———. 1995. *On being a scientist: Responsible conduct in research*. 2nd ed. Washington, D.C.: National Academy Press.

National Research Council. 2009. *Opportunities in neuroscience for future military applications*. Washington, D.C.: National Academies Press.

Nixdorff, K., et al. 2018. Dual-use nano-neurotechnology: An assessment of the implications of trends in science and technology. *Politics and the Life Sciences* 37 (2): 180–202.

Novossiolova, T., L. Bakanidze, and D. Perkins. 2020. Effective and comprehensive governance of biological risks: A network of networks approach for sustainable capacity building. In *Synthetic biology 2020: Frontiers in risk analysis and governance*, ed. B. Trump, I. Linkov, J. Kuzma, and C. Cummings. Cham: Springer.

Novossiolova, T., J. Whitman, M. Dando. 2019. Altering an appreciative system: Lessons from incorporating dual-use concerns into the responsible science education of biotechnologists. Futures 108:53–60

Organisation for the Prohibition of Chemical Weapons. 2019a. Technical Secretariat: Facilitating the implementation of the convention. https://www.opcw.org/about-us/technical-secretariat. Accessed 6 Mar 2019.

———. 2019b. Scientific Advisory Board: Keeping pace with scientific and technological change. https://www.opcw.org/about-us/subsidiary-bodies/scientific-advisory-board. Accessed 6 Mar 2019.

———. 2019c. Advisory Board on education and outreach: Supporting the OPCW's engagement with external partners. https://www.opcw.org/about-us/subsidiary-bodies/advisory-board-education-and-outreach. Accessed 6 Mar 2019.

———. 2019d. Fires: The OPCW's short documentary video project. https://www.opcw.org/fires. Accessed 6 Mar 2019.

———. 2019e. The Hague ethical guidelines. https://www.opcw.org/hague-ethical-guidelines. Accessed 6 Mar 2019.

Pinchuk, I.V., E.J. Beswick, and V.E. Reyes. 2010. Staphylococcal enterotoxins. *Toxins* 2: 2177–2177.

Rommelfanger, K., S. Jeong, C. Montojo, and M. Zirlinger. 2019. Neuroethics: Think global. *Neuron* 101 (3): 363–364.

Royal Society. 2012. *Brain waves module 3: Neuroscience, conflict and security*. London: The Royal Society.

———. 2019. Brain waves. https://royalsociety.org/topics-policy/projects/brain-waves/. Accessed 5 Mar 2019.

Salles, A., J. Bjaalie, K. Evers, M. Farisco, B.T. Fothergill, M. Guerrero, et al. 2019. The Human Brain Project: responsible brain research for the benefit of society. *Neuron* 101 (3): 380–384.

Scientific Advisory Board. 2018. Report of the Scientific Advisory Board on developments in science and technology for the fourth special session of the conference of the states parties to review the operation of the chemical weapons convention. RC-4/DG.1.30 April. https://www.opcw.org/sites/default/files/documents/CSP/RC-4/en/rc4dg01_e_.pdf. Accessed 6 Mar 2019.

Security Council Committee established pursuant to Resolution 1540. 2019. Programme of work. https://www.un.org/en/sc/1540/about-1540-committee/programme-of-work.shtml. Accessed 6 Mar 2019.

Shang, L., M. Crowley, and M.R. Dando. 2018. Act now to close chemical-weapons loophole. *Nature* 562: 344.

Sidell, F.R. 1997. Nerve agents. In *Medical aspects of chemical and biological warfare*, ed. F.R. Sidell, E.T. Takafuji, and D.R. Franz, 129–179. Washington, D.C.: Office of the Surgeon General, Department of the Army.

Squire, L. R. et al. 2013. *Fundamental neuroscience*, 4th ed. Academic Press: Elsevier. (See Box 2.1, The Neuron Doctrine, page 16).

Test and Evaluation Command. 1982. *Joint CB technical data source book*. Dugway: US Army Dugway Proving Ground. (Volume VI. Toxin Agents, Part One: Botulinum Toxin).

Tucker, J. 2008. The body's own bioweapons. *Bulletin of the Atomic Scientists* 64 (1): 16–22; 56–57.

UK Research and Innovation. 2019. Interdisciplinary network on teaching of ethics for neuroscientists. https://gtr.ukri.org/projects?ref=AH/J005533/1. Accessed 6 Mar 2019.

Ukraine, Japan and the United Kingdom of Great Britain and Northern Ireland. 2017. Awareness-raising, education, and outreach: Recent developments. BWC/MSP/2017/WP.22. 6 December. https://undocs.org/bwc/msp/2017/wp.22. Accessed 6 Mar 2019.

United Nations Security Council Resolution 1540. 2004. https://www.un.org/en/sc/1540/about-1540-committee/general-information.shtml. Accessed 7 Mar 2019.

Wright, L., C. Pope, and J. Liu. 2009. The nervous system as a target for chemical warfare agents. In *Handbook of toxicology of chemical warfare agents*, ed. C. Ramesh and C. Gupta, 463–480. Elsevier: Academic Press.

Hybrid Emerging Threats and Information Warfare: The Story of the Cyber-AI Deception Machine

Eleonore Pauwels and Sarah W. Deton

1 Introduction

Across the world, at any given moment, there are pervasive cognitive-emotional conflicts being waged for the control of populations' thoughts, emotions, and attitudes. These battles of influence do not tend to occur in wartime, but rather in peacetime, infiltrating homes and smart cities. They sow disinformation, affective manipulation, and forgeries as new means of undermining social cohesion and trust. They exacerbate societal tensions and amplify public polarization. They increasingly condition and limit notions of self-determination and could continue to do so with the future generations to come.

The rise of cognitive-emotional conflicts and the subsequent "trust-deficit disorder" they unleash is born out of the entanglement of technology, data, and geopolitics. The convergence of artificial intelligence (AI) with other emerging technologies creates the potential for deception and subversive attacks that manipulate populations' perceptions. In the near future, what will matter is not who "wins" new territories, but who controls the data and wins the trust, the hearts, and the minds of citizens within any given country or polity.

Cyberspace has become not only a new domain of fierce competition over information, business, and strategic technological operations, but also a new battlefield, in ways that blur the line between war and peace and make each of us a potential

E. Pauwels (✉)
Centre for Policy Research, United Nations University, New York City, USA

Woodrow Wilson International Center for Scholars, Washington, DC, USA
e-mail: eleonore@eleonorepauwels.com

S. W. Deton
Institute for Philosophy and Public Policy, George Mason University, Fairfax, VA, USA

M. Martellini, R. Trapp (eds.), *21st Century Prometheus*,
https://doi.org/10.1007/978-3-030-28285-1_6

target of postmodern conflict. Governance actors are only starting to realize how the manipulation and misuse of converging and connected technologies in cyberspace can threaten the truth, polity, and collective security.

2 Converging Technologies and Hybrid Emerging Threats

AI is converging with an extraordinary array of other technologies, from cyber and biotechnologies, affective computing and neurotechnologies, to robotics and additive manufacturing. This era of technological convergence seeks to merge the data of our physical, digital, and biological lives. Computer scientists are developing deep learning algorithms that can recognize patterns within massive amounts of data with superhuman efficiency and, increasingly, without supervision. At the same time, geneticists and neuroscientists are deciphering data related to our genomes and brain functioning, learning about human health, well-being, and cognition.

The result? The convergence of dual-use technologies is generating new hybrid threats, as well as opportunities, that can be leveraged by governments, non-state actors, and average citizens alike for both promise and peril. The same algorithms that can identify malign biomarkers among large swaths of genomics data from human populations to design blood-tests for various cancers can also drastically intensify the nature and scope of cyber-biological surveillance and other forms of cyber-espionage. Moreover, these same algorithms can rely on emotion analysis to generate deep fake forgeries, which will enable propaganda, strategic deception, and social manipulation to be both more scalable and targeted.

Harnessing AI within biotech research allows tech-leading nations to optimize the bio-data and capture the value of another country's bio-economy (Pauwels 2018). Emotional analysis enables hyper-personalized campaigns in which key demographics are manipulated to affect voting behaviours at crucial times (Hern 2018). Think of self-learning algorithms able to design ever more sophisticated human forgeries based on biometrics analysis and behavioural mimicry (Turgeman 2018).

Converging technologies therefore create networks of digital information that enhance, run, shape and integrate cyberspace with daily ways of living. They slowly "invade" bodies and cities. As they become digitized, converging technologies also become more decentralized, beyond State control, and available to a wider range of actors around the world.

The "deception machine" is one new hybrid threat borne out of the convergence of AI and affective computing, biometrics, neurotechnologies, and cybertechnologies. Already, AI and these converging technologies are altering the global intelligence landscape by the automated generation and simulation of data, media, and strategic intelligence. The deception machine has significant implications for the future of propaganda, deception, and social engineering.

Imagine a future in which the automated generation of digital forgeries requires little more than an internet connection. Current global trends[1] points towards a global decline of democracy and rise of authoritarian and populist governance regimes. In the era of AI, digital dictatorships (The Economist 2016; Gleiser 2018; Volodzko 2019) will have the know-how to generate high-quality digital forgeries to influence and manipulate public opinion (Bradshaw and Howard 2018). If such digital forgeries target certain ethnic, religious, cultural, or political groups, they could provoke kinetic conflicts, violence and discriminations.

We are already witnessing early signs of the creation of the deception machine: psychometrics manipulation and hyper-personalized ad campaigns disseminated through the global networks of social media platforms.

For instance, in Myanmar the United Nation's Human Rights Council conducted an investigation (Human Rights Council 2018) into human rights violations, which ultimately found that virulent hate speech spread on social media directly contributed to the genocidal violence directed at Rohingya Muslims (Miles 2018; Stecklow 2018; Douek 2018; Mozur 2018). As the Myanmar case clearly demonstrates, emerging and converging technologies may sow hate, fear, and disruption into the hearts and minds of populations, not with fire and smoke, but with viral online narratives.

By eroding the sense of truth and trust between citizens and the State – and indeed among States, and among societies—truly fake intelligence could become deeply corrosive to the global intelligence and governance system.

In an address to the General Assembly in September 2018, Secretary-General António Guterres identified this unsettling trend.

> "Our world is suffering from a bad case of 'trust-deficit disorder'…Trust is at a breaking point. Trust in national institutions. Trust among states. Trust in the rules-based global order. Within countries, people are losing faith in political establishments, polarization is on the rise and populism is on the march…We face a set of paradoxes. The world is more connected, yet societies are becoming more fragmented…" (UN Secretary General 2018).

In the following scenarios, we aim to shine light on how fake intelligence and the trust-deficit disorder has systemic implications for human, digital, and political security.

---

[1] "In states that were already authoritarian…governments have increasingly shed the thin façade of democratic practice that they established in previous decades, when international incentives and pressure for reform were stronger. More authoritarian powers are no banning opposition groups or jailing their leaders, dispensing with term limits, and tightening the screws on any independent media that remain. Meanwhile, many countries that democratized after the end of the Cold War have regressed in the face of rampant corruption, antiliberal populist movements, and breakdowns in the rule of law. Most troubling, even long-standing democracies have been shaken by populist political forces and target minorities for discriminatory treatment." – Freedom House. 2019. *Freedom in the World 2019: Democracy in Retreat.*

3 The Deception Machine: Hybrid Cyber-AI Warfare

The "Deception Machine" epitomizes the importance of information superiority and its strategic potential for deception and subversion. The supremacy to acquire data and manipulate others' beliefs, attitudes, emotions and behaviours seems today more important than material power. This could mean first a shift of power towards actors – States, political elites, financial oligarchies and private technological platforms – that have the capital, the data market and the authority to deploy powerful systems for AI convergence.

Alternatively, if innovation in AI technologies contributes to democratizing access to automated simulation of data, propaganda and cyberoperations, these techniques could then become more attractive to less powerful actors and nations with minor tech capital and declining economies. This democratization trend could further accelerate the proliferation of data-poisoning and manipulation, AI-cyber hybrid influencing, and the waging of cognitive-emotional wars, leaving the world in a fight of "all against all."

In turn, the "trust-deficit-disorder" will exacerbate public anxiety about the loss of control to an algorithmic revolution, which seems to escape current modes of understanding and accountability. In the context where multilateral governance institutions provide a forum for inter-State security deliberations, some States might strategically use this context of public anxiety and cognitive-emotional dissonance to impose their own competing versions of reality.

In this section, we have broken down the deception machine into seven modes of data-manipulation and information warfare. We begin by focusing on three case-studies around what we call the degradation of truth and trust: (1) killing the truth, (2) weaponizing the truth, and (3) obscuring the truth. In each of these categories and case-studies, we narratively describe major trends in AI convergence. The final section of this paper explores how the need for foresight to better anticipate how these trends could develop in the future.

3.1 *Killing the Truth: Information Warfare in Syria*

Beyond ISIS' sophisticated social media strategy, Syria was also the target of a Russian disinformation campaign (The Syria Campaign 2017). A report published by the Syria Campaign in December 2017 found that bots and trolls linked to Russia reached an estimated 56 million people on Twitter with content attacking Syria's search and rescue organization commonly known as the White Helmets. Much of the mis- and disinformation was tied to efforts to spread fake intelligence about the sarin chemical attack of April 2017 and successfully dominated the reporting of one of the most significant events of the Syrian conflict.

Fig. 1 A still apparently from the computer game AC-130 Gunship Simulator: Special Ops Squadron claimed by the Russian defence ministry as 'irrefutable evidence' of cooperation between US forces and Islamic State militants in Syria. (Source: Walker (2017). Photo Credit: Russian Defence Ministry)

Granted, disinformation campaigns have been a long-time tool of war. But Russia's disinformation campaign in Syria is the first-known attack against humanitarian workers, who are protected under international humanitarian law (Ibid.). The propagation of such a damaging counter-narrative began to slowly alter the reality on the ground, regardless of its authenticity (Fig. 1).

First, Russia focused on crafting an alternative narrative about the Syrian revolution to influence international policies and sow confusion in Western societies. After building an $1.4 billion-a-year propaganda infrastructure at home and abroad, Russia was able to reach more than 600 social media users across 130 countries in 30 different languages (Ziff 2015).

The breadth of influence Russia accumulated in cyberspace is unlike anything we have seen. By constantly re-asserting the claim that Russia was only in Syria to assist the Assad regime in its war on terrorism, the Russia-Syria promoted narrative denied reality of reports from the epicenter of the conflict (Fig. 2).

Instead of the narrative including the 600,000 civilians killed or missing, the counter-narrative reinforces the state-backed claim that Syria's enemies are terrorists.

Russia and the Assad regime dedication to the counter-narrative included claims of international conspiracies (Sputnik News 2018), the use of digital forgeries of satellite images (Robertson 2016), and the tendency to present videos (Kiseleva and Coleman 2018) and images (Walker 2017) from computer games as evidence (Bellingcat 2018). Such narrative warfare (Maan 2018) that leverages deceptive data-manipulation to produce psychological and social engineering moves beyond traditional understanding of threats to the global intelligence system.

Fig. 2 Barack Obama to Donald Trump with Recycle-GAN for unsupervised face retargeting. (Source: Bansal A. 2018. "Barack Obama to Donald Trump." *YouTube*; 11 August)

3.2 Weaponizing the Truth: Hybrid Cyber-AI Adversarial Attacks

Algorithmic power, the power and influence that comes with the proliferation and decentralization of AI and converging technologies, may increasingly be used to target and manipulate political communication. Deepfakes, sophisticated AI programs that can manipulate sounds, images, and videos could affect international negotiations and diplomatic relations before anyone has a chance to determine its authenticity.

With the proliferation of sophisticated deepfake videos, deepfake backstories and cover-ups, even qualified news reporters, decision-makers and diplomats will struggle to parse propaganda and disinformation from real news. Computer scientists at Carnegie Mellon are leading in deepfake research and are now working on transferring subtle facial expressions and body movements from one video into another.

Deep neural networks are rapidly improving their self-learning capabilities with less supervision – and are therefore are becoming both efficient and scalable. By allowing the analysis of individual communication, perception and emotion to be automated, AI systems can increase anonymity and psychological distance in cyberoperations.

In the near-future, automated cyberoperations, led by deep learning, will therefore be more effective, finely targeted, difficult to attribute, and likely to exploit evolving vulnerabilities in AI and human systems (Mayer 2018). One pervasive security threat will be new forms of hybrid influencing made possible by the automation of social-engineering attacks. Many major cybersecurity incidents rely on social engineering where malicious actors target the social and psychological vulnerabilities of humans within chains of command. The goal is to manipulate command and control organizations to compromise their own safety and security.

The most extreme level of AI-cyber influencing could be the convergence between cyber and kinetic operations (Applegate 2013). During combat, for example, a cyberattack could successfully compromise a combination of personal and official channels of communications. Armed forces become the target of psychological manipulation, for instance by being exposed to forged video "evidence" of their adversary's military superiority. Such hybrid AI-cyber influencing then allows the "supposedly" superior military force to achieve a kinetic advantage.

The consequence of hybrid AI-cyber influencing will be the weaponization of lucrative global digital networks. In this new fog blurring war, tech and politics, the next winning move will be to manipulate information infrastructure and its secrets. We increasingly face geopolitical conflicts in which psychological and algorithmic manipulation are becoming endemic in cyberspace, an ecosystem of nearly four billion minds Villasenor 2018, Gavrilovic 2018). Yet, the impacts felt are real in the physical world, from influencing elections, to destabilizing economies and political regimes, to terrorist groups livestreaming their attacks in the streets of Iraq and Syria.

3.3 Obscuring the Truth: Disinformation, Hate Speech and Election Interference

Digital footprints left on social media can reveal vast amounts of valuable and personal information to companies, third-party websites, and government agencies. Deep learning algorithms and neural networks sift through digital footprints on social media and can predict individual personality traits (Wei et al. 2016) like political ideology (Iyyer et al. 2014). By enabling disinformation campaigns tailored for difference subgroups, emotion analysis and affective computing can used as a political cyberweapon for highly effective election interference (Marwick and Lewis 2017).

The convergence of these technologies sets the stage for psychometric micro-targeting, which can allow corporations, governments, and non-state actors to tailor propaganda to specific users (Wooley and Howard 2017). By promoting targeted propaganda, mis-, and disinformation, and incendiary content, political bots and algorithms have the potential to amplify existing cognitive-emotional tensions, erode the power of democratic institutions, sow civil violence, and exert exponential influence in both national and international politics.

Consider two recent events: the 2016 United States presidential election and mis- and disinformation tied to violence in India during the summer of 2018.

During the 2016 United States presidential election, foreign entities conducted a massive social media campaign using paid-for advertisements, fake social media accounts, polarizing content, and paid troll farms. With the release of the

U.S. House Permanent Select Committee on Intelligence final report (HIC 2018) on the scope of Russia's interference in the election, Russia's primary goal was revealed – sow discord in American society in order to undermine citizens' faith in the democratic process.

It's been said that "memes are the propaganda of the digital age." (DiResta et al. 2018) After extensively studying American cultural and targeting specific types of social media users, Russia's Internet Research Agency (IRA) created and re-purposed popular, thematically-relevant memes as a sort of politico-cultural weapon, preying on American citizens' biases and reinforcing in-group dynamics. Unfortunately, the power of propaganda is difficult to curtail on social media and it's rapidly becoming a global concern.

Across India in July 2018, mis- and disinformation shared on social media platforms claiming that foreigners were responsible for recent child abductions led to the beating and lynching of innocent people (Iwanek 2018). Later that month, another WhatsApp message falsely blamed a Muslim for being a kidnapper and he was eventually found murdered (Ibid).

Both social media platforms implicated in the India cases, Facebook and WhatsApp, have over 200 million users in India and that number is expected to double by in the next few years (Rai 2018). As the country is quickly approaching its general election in May 2019, fear and anxiety is running rampant that social media will sow unrest once again (Robinson et al. 2019). As a preventive fix, WhatsApp is now limiting users in India to share only up to five online chats (India Times 2018). Only time will tell whether this tech fix will prove sufficient.

4 A Future of Affective Computing and Neuromodulation

Relying on automated behavioural analysis, AI already enables targeted propaganda to spread more efficiently and at wider scale within the social media ecosystem. For instance, *Cambridge Analytica* deployed intensive and widespread social media data-mining operations to define and curate the psychological profiles of millions of Facebook users (Granville 2018). Another company, *Mindstrong*, collects individuals' passive data (such as how they type, scroll and click) to create digital phenotypes or avatars based on a set of biomarkers that measure their emotional state and cognitive functioning (Mindstrong website n.d.).

Yet, there are many more strategies in the neuro-biological toolbox. Among them is affective computing, which entails the combination of algorithms with biosensors to create AI systems that can decipher human emotions and predict behavioural and emotional responses (Kleber 2018). Affective computing may one day enable us to identify the emotional triggers that push individuals or population subgroups to violence. This, in turn, would open up the possibility of individuals or organizations (hackers, corporations or government agencies) using such knowledge to spark and manipulate violence.

Emotional analysis and affective computing will enable disinformation campaigns tailored for different subgroups, leading, for instance, to highly effective election interference (Marwick and Lewis 2017). More broadly, the use of converging technologies to amplify cognitive-emotional tensions and vulnerabilities could erode the power of democratic institutions, sow civil violence and enable new forms of dictatorships.

While neuro-technologies are not yet an integral part of daily life, developments in the field aim at deciphering people's mental processes and modulating the brain functions underlying their intentions, emotions and decisions (Ienca and Andorno 2017). These new forms of neuro-modulations would drastically impact the potential for social engineering, psychological manipulation and other techniques of subversion and deception (Ibid).

5 The Rise of Hacktivism and Cyber Mercenaries

"Project Raven" is a story of cyber mercenaries: former US National Security Agency specialists who harnessed cutting-edge cybertheft and spying tools on behalf of a foreign intelligence service that keeps thousands of civil society activists, journalists and political figures under tight digital surveillance (Bing and Schectman 2019). To do so, these cyber mercenaries relied on an insidious automated hacking application called Karma, which made it possible to hack the iPhones of users all over the globe (Ibid.). In 2016 and 2017, Karma was used to spy on hundreds of targets in the Middle East and Europe, including senior-level political figures.

The Project Raven story also unveils the role that former American cyber espionage agents play in foreign hacking operations. Increasingly, as cyber, biotech and AI become decentralized, and as tacit knowledge escapes the confines of securitized labs, national and international authorities will have to face a complex question: how to govern the dissemination of dual-use expertise from companies, governments and covert operations to the wild, or from one country to another?

The question about dual-use technologies is not new, but the pervasive digitization and subsequent decentralization of technologies certainly is. From intelligence services, gene-synthesis companies, to top cybersecurity and AI private labs, the talent force that is shaping tomorrow's designs, risks, and opportunities, has much tacit knowledge to offer and transfer.

Secrecy among tech, policy and security communities needs to be urgently rethought. A new network of trust – the new version of the old "guild" – to prevent peer-engineers from being bought for their secrets needs to be built. As of right now, the new ecosystem for AI-cyber influencing and potential covert war is shaped by engineers in Silicon Valley, Shenzhen, London, and a small handful of others. The new rules and norms for responsible behaviours in cyberspace await definitions and actions by politicians, lawyers, and diplomats.

The pervasive access to dual-use, converging technologies is also bringing forth a new form of virtual, viral insider threat generated by increasingly powerful "information warriors." When IS increased its visibility and power through social media, its extremely violent propaganda created another cyberspace phenomenon (Ward 2018; Alfifi et al. 2018). Digital networks turned into a new hypnotic online show about violence and war.

The global hacktivist group Anonymous came to life in response to IS' invasion of social media (Hern 2015). These volunteers, self-appointed custodians of cyberspace, started by systematically hunting for IS Twitter accounts. They shared this information with Twitter, which in turn deactivated several accounts, as well as with governments. With time, Anonymous organized a form of cyber resistance, combining cyber espionage, hacking techniques and automated bots aimed at destroying IS websites, networks and social media presence. Hacktivism unfortunately did not prevent violence on the ground, but it did create a new form of aggression from within homes and cities.

6 Cyberattacks on Smart Critical Systems

With deep learning, hunting for weaknesses in networked infrastructures will be mostly automated.[2] Automated cyberoperations will be more effective, finely targeted, difficult to attribute, and likely to exploit evolving vulnerabilities in AI systems. The capacity of adversarial algorithms to improve their own strategies and launch increasingly aggressive counter-attacks with each iteration will lead to an expansion and augmentation of existing cyberattacks, with global damages that may reach USD 6 trillion a year by 2021 (Eubanks 2017).

In the 2018 World Economic Forum's Global Risk Perception Survey, the second most frequently cited risk triggered by technological convergence is the combination of cyberattacks with the manipulation and corruption of critical information infrastructure (see footnote 2 below).

For instance, the company DarkTrace (DarkTrace n.d.) in Cambridge, UK, used the human immune system as a model to design its AI technology to detect and fight back against emerging cyberthreats across corporate networks, cloud and virtualized environments, the IoT, and industrial control systems. By 2023, the value of AI in cybersecurity is projected to increase USD 17 billion (P&S Market Research 2017).

Similar to a Darwinian experiment, cybersecurity communities will have to rely on the development and survival of the fittest AI (Dvorsky 2017). If AI and cyber

[2] In the 2018 World Economic Forum's Global Risk Perception Survey, the second most frequently cited risk triggered by technological convergence is the combination of cyberattacks with the manipulation and corruption of critical information infrastructure. *The Global Risks Report 2018: 13th Edition*. World Economic Forum, Geneva: Switzerland. http://www3.weforum.org/docs/WEF_GRR18_Report.pdf.

capacities keep converging and enabling each other, what is the nature and scope of the attacks that States, the private sector, and civilian populations will have to withstand? Harnessing AI technologies will drastically expand the velocity and reach of cyberattacks (Brundage et al. 2018) as well as their complexity.

Recently, there has been an increase in reports of cyberattacks on critical infrastructures, especially on electric power systems, globally. “This has served to demonstrate just how vulnerable cities, States, and countries have become and the growing importance of achieving global risk agility” (Wagner 2016). One example of such an attack occurred in December 2015, when hackers seized control of the Prykarpattyaoblenergo Control Center (PCC) in Western Ukraine, rendering up to 225,000 people without power for up to 6 h (E-ISAC 2016). According to a 2016 Electricity Information Sharing and Analysis Center report, the attackers demonstrated a variety of capabilities, including spear phishing emails, variants of the BlackEnergy 2 malware, and the manipulation of Microsoft Office documents that contained the malware to gain access to the IT networks of the electric companies (Ibid. pp. 4-5).

In 2017 alone, States in Europe witnessed cyberattacks targeting national health, telecommunications, energy, and government sectors UK Parliament 2018). Yet, other nations could perhaps be even more vulnerable to cyberattacks on critical infrastructures if they lack cutting-edge expertise and capacities in cybersecurity. As noted by Landry Signé and Kevin Signé, two experts on cybersecurity in the Global South, “Although Africa is relatively limited in terms of communications infrastructure, due to the high penetration rate of new technologies, it is increasingly a target for cybercriminals, as most African countries still have a low level of commitment to cybersecurity” (Signe and Signe 2018).

7 Drone Swarms, 3D-Printers and Urban Violence

Increasingly, the convergence of AI and cloud networked robotics will allow cyberattacks to subvert the functioning of physical systems, from self-driving cars, drone swarms, smart energy systems, surgical robots, to 3D printers and industrial supply chains. New forms of hybrid cyberthreats will likely exploit the boundary with critical physical infrastructures. Often, this will lead to new forms of cognitive-emotional conflicts that aim to create chaos, confusion and uncertainty, not on the battlefield, but inside cities, hospitals and critical institutions.

We recently witnessed the sudden paralysis and the economic burden that the misuse of commercial drones can impose on airports (BBC 2019). Drones are one manifestation of a future “Cambrian explosion” (Fox 2016) of robots and smart objects in homes and cities. A State or non-State actor could wage a cyberattack that can subvert networks of robots to steal citizens’ intimate data, manipulate and corrupt strategic information, or disrupt operations in manufacturing plants and hospitals. Similar cyberattacks could automate swarms of micro-drones to spread chemicals and neurotoxins in water and food supply chains. In smart cities, most networked architectures will run on automated work flow and AI-powered data optimization.

The convergence of AI, robotics and additive manufacturing (3D printing) is another evolving domain to watch. AI software will enable automated workflows within decentralized supply chains (Firestone 2018). This will enable the production, in the physical world, of digital blueprints of objects, weapons, illicit substances, cells, and fragments of genes. Such decentralized power of invention and creation will serve open innovation labs in Shenzhen (Shenzen Open Innovation Lab n.d.), San Francisco (Accenture website n.d.), and Kumasi (Ahiataku 2018), but also nefarious networks that grow in conflict zones and crusade the dark web. For instance, IS took its adversaries by surprise by harnessing off-the-shelf drones the group had modified with 3D-printed parts to drop explosives on unsuspecting forces (Watson 2017). Such a scenario is not, however, limited to the battlefield; it will increasingly concern cities outside conflict zones, too.

Beyond impact on digital security, AI will also endanger the security of physical and biotechnology infrastructures. As additive manufacturing supply chains and biotech laboratories become increasingly automated and run by AI software – such labs already exist (Dunlap and Pauwels 2017; Gwynne and Heebner 2018) – they will be vulnerable to adversarial attacks, in particular data-poisoning, by external AI systems.

A few past experiments have shown how adversarial perturbation – a certain amount of noise within the pixels of a driving sign – could disorient self-driving cars.[3] Instead of a Tesla, the next target of adversarial attacks could be drone swarms or a biotech supply chain. Hackers could inject enough adversarial perturbation into the navigation command of distributed networks of drones. Similar strategies could be used to introduce, into a biotech supply chain, enough structural noise – corrupting data, bio-engineering instructions or slightly changing the parameters (temperature) of an experiment (Wrona 2017). Early warnings might not be visible, but there might be lasting damaging effects on producing antibiotics, vaccines or cancer treatments.

The nature and scale of potential attacks and disruptions are unprecedented, with AI systems that will likely exceed human performance in cybersecurity. Increasingly, deep learning systems will make decisions about what to attack, who to attack, and when to attack, with their targets ranging from consumers, self-driving cars, surgical robots to automated bio-labs.

[3] Refer to Dr. Seán ÓhÉigeartaigh's panel discussion at the *Governing Artificial Intelligence* expert meeting held on June 22, 2018 at the UN University. See: Session I: Does the AI race threaten international peace and security? "Learning to Live with Artificial Intelligence: A Virtuous Circle or a Vicious One?" *International Peace Institute;* 22 June 2018. https://www.ipinst.org/2018/06/governing-artificial-intelligence#21.

8 The Urgent Need for Foresight and Anticipatory Intelligence

In recent years, policymakers and experts have been debating whether the International Humanitarian Law (IHL) framework is sufficient to addressing the challenges posed by new, converging technologies in warfare (ICRC 2017a). These expert-level discussions were codified in the 2013 publication of the *Tallinn Manual on the International Law Applicable to Cyber Warfare* (Tallinn Manual 2017a) and the 2017 Tallinn Manual 2.0 revision, though fall short of constituting new international legal norms (Tallinn Manual 2017b). While some argue that IHL applies to cyber operations that fall within situations of armed conflict (ICRC 2017b), others note that IHL has a large blind spot when it comes to hybrid warfare (Choudhary 2019).

The Council of Europe's Committee on Legal Affairs and Human Rights recognized this legal grey area, stating that "hybrid [war] adversaries avoid manifest use of force that would reach the required threshold for triggering application of [IHL] norms" (Council of Europe 2018). And even if hybrid warfare operations did meet the threshold, other challenges remain to the applicability of IHL to hybrid warfare, such as attribution of responsibility.

Interestingly, there may already be another legal framework that may be more appropriate and applicable to hybrid warfare – the legal frameworks governing psychological operations in warfare (Nicola 2018). Granted, more work must be done to adapt current psychological operations frameworks to encompass data-poisoning, emotion manipulation, disinformation and the cyber-AI deception machine by state and non-state actors alike.

Politically, legally and ethically, our societies are not properly prepared for the deployment of AI and converging technologies. At national and international levels, we lack a comprehensive understanding of the threats that AI and converging technologies can pose at the individual human level, broader threats to populations, and geopolitical confrontations potentially triggered by the combination of new technological trends.

In the near-future, we will see the rise of cognitive-emotional conflicts: long-term, tech-driven propaganda aimed at generating political and social disruptions, influencing perceptions, and spreading deception (Wells 2017).

Technologies are becoming hybrid complex systems that are merging the data of our digital, physical, and biological lives with potential for pervasive vulnerabilities and emerging risks. AI and converging technologies are characterized by new processes of decentralization, distributed agency and forms of "atomized responsibility." Increasingly, these technologies deploy beyond State control, involving new actors such as rogue States and cyber mercenaries, and challenging established expertise and governance models. New forms of invasive cyberwar and cognitive-emotional conflicts will have powerful implications for collective (bio)-physical, digital, and political security.

While cognitive-emotional conflicts often entail the weaponization of social media, subsequent phases will increasingly target important elements of human and civilian security, including beliefs, discourses, digital systems and infrastructures critical to health, food, political, and economic security. In November 2018, Russia suggested in official statements that the Pentagon is establishing bio-weapons supply chains on the border between Russia and the Republic of Georgia (Lentzos 2018). The narrative amplified fears about potential biosecurity threats to populations.

From controversies about the safety of vaccines, baby food or gene-therapies, to disinformation campaigns about the health of financial institutions, attacks like these surface in cognitive-emotional conflict and are extremely large and complex. It will become urgent for governments to create more agile and efficient early warning systems to detect and analyse the sources of forgeries and propaganda. States will also need to continuously map how these new subversive tools influence public discourse and opinion. However, current detection tools might not be effective enough to protect against technological capacities in affective computing, bio- and neuro-technologies that are aimed at manipulating the hearts and minds of (foreign) citizens.

The "Deception Machine" epitomizes the importance of information superiority and its strategic potential for deception and subversion. The supremacy to acquire data and manipulate others' beliefs, attitudes, emotions and behaviours seems today more important than material power. This could mean first a shift of power towards actors – States, political elites, financial oligarchies and private technological platforms – that have the capital, the data market and the authority to deploy powerful systems for AI convergence.

Alternatively, if innovation in AI technologies contributes to democratizing access to automated simulation of data, propaganda and cyberoperations, these techniques could then become more attractive to less powerful actors and nations with minor tech capital and declining economies. This democratization trend could further accelerate the proliferation of fake intelligence, AI-cyber hybrid influencing, and the waging of cognitive-emotional wars, leaving the world in a fight of "all against all."

In turn, the "trust-deficit-disorder" will exacerbate public anxiety about the loss of control to an algorithmic revolution, which seems to escape current modes of understanding and accountability. Urgent efforts are needed using inclusive foresight to better prepare societies for the emerging threats that target truth and trust, political and social cohesion. With the private sector and civil society, diplomats and policymakers will also need better foresight to define, shape and anticipate what norms and oversight models will protect human safety and security in an era of technological convergence.

References

Accenture website. n.d.. Available at https://www.accenture.com/us-en/service-open-to-disrupt.

Ahiataku, S. 2018. Engaging young innovators in Kumasi with innovation clinic. Global Lab Network; 3 March. https://glabghana.wordpress.com.

Alfifi, M., et al. 2018. Measuring the impact of ISIS social media strategy. http://snap.stanford.edu/mis2/files/MIS2_paper_23.pdf.

Applegate, S.D. 2013. The dawn of kinetic cyber. 5th International Conference on Cyber Conflict; K. Podins, J. Stinissen, M. Maybaum (Eds.). https://ieeexplore.ieee.org/document/6568376.

BBC. 2019. Heathrow airport drone sighting halts departures. BBC News; 8 January. https://www.bbc.com/news/uk-46803713. Also see "Gatwick and Heathrow buying anti-drone equipment." BBC News; 4 January. https://www.bbc.com/news/uk-46754489.

Bellingcat. 2018. Chemical weapons and absurdity: The disinformation campaign against the white helmets. A joint report in collaboration with Newsy; 18 December.

Bing, C., and J. Schectman. 2019. Inside the UAE's secret hacking team of American Mercenaries. Reuters; 30 January. https://www.reuters.com/investigates/special-report/usa-spying-raven/.

Bradshaw, S., and P.N. Howard. 2018. Challenging truth and trust: A global inventory of organized social media manipulation. Computational Propaganda Research Project. https://comprop.oii.ox.ac.uk/wp-content/uploads/sites/93/2018/07/ct2018.pdf.

Brundage, M., et al. 2018. The malicious use of artificial intelligence: Forecasting, prevention, and mitigation. University of Oxford. February: p. 44. https://arxiv.org/pdf/1802.07228.pdf; and, Press, G. 2018. The AI cybersecurity arms-race: The bad guys are way ahead. Forbes; 26 April. https://www.forbes.com/sites/gilpress/2018/04/26/the-ai-cybersecurity-arms-race-the-bad-guys-are-way-ahead/#3fc3ac28148e

Choudhary, V. 2019. Does international humanitarian law respond adequately to information warfare? Groningen Journal of International Law, Blog; 21 March. https://grojil.org/2019/03/21/the-truth-under-siege-does-international-humanitarian-law-respond-adequately-to-information-warfare/

Council of Europe. 2018. A.4. in Council of Europe, Committee on Legal Affairs and Human Rights. 2018. Legal challenges related to the hybrid war and human rights obligations. 14 March. http://website-pace.net/documents/19838/4246196/20180314-HybridWar-EN.pdf/3387f663-0e5d-407e-a90b-9e5880474589

DarkTrace. n.d. DarkTrace's website: https://www.darktrace.com/technology/. "The Threat Visualizer is Darktrace's real-time, 3D threat notification interface. As well as displaying threat alerts, the Threat Visualizer provides a graphical overview of the day-to-day activity of your network(s), which is easy to use, and accessible for both security specialists and business executives. Using cutting-edge visualization techniques, the Threat Visualizer user interface automatically alerts analysts to significant incidents and threats within their environments, enabling analysts to proactively investigate specific areas of the infrastructure."

DiResta, R., et al. 2018. The disinformation report. New Knowledge and the U.S. Senate Intelligence Committee: p. 50. https://cdn2.hubspot.net/hubfs/4326998/ira-report-rebrand_FinalJ14.pdf.

Douek. 2018. Douek E. 2018. Facebook's role in the genocide in Myanmar: New reporting complicates the narrative. Lawfare; 22 October. https://www.lawfareblog.com/facebooks-role-genocide-myanmar-new-reporting-complicates-narrative.

Dunlap, G., and E. Pauwels. 2017. The intelligent and connected Bio-Labs of the Future: Promise and Peril in the Fourth industrial revolution. Wilson Center Policy Briefs (October): pp. 4–6. https://www.wilsoncenter.org/sites/default/files/dunlap_pauwels_intelligent_connected_biolabs_of_future.pdf.

Dvorsky, G. 2017. Hackers have already started to weaponize artificial intelligence. Gizmodo 11 September. https://gizmodo.com/hackers-have-already-started-to-weaponize-artificial-in-1797688425.

E-ISAC. 2016. Analysis of the cyber attack on the Ukrainian Power Grid. Defense Use Case; 18 March. https://www.nerc.com/pa/CI/ESISAC/Documents/E-ISAC_SANS_Ukraine_DUC_18Mar2016.pdf.

Eubanks, N. 2017. The trust cost of cybercrime for business. Forbes; 13 July. https://www.forbes.com/sites/theyec/2017/07/13/the-true-cost-of-cybercrime-for-businesses/#38953fec4947.

Firestone, K. 2018. 2018 manufacturing outlook and 3D-printing's impact. manufacturing.net; 16 January. https://www.manufacturing.net/blog/2018/01/2018-manufacturing-outlook-and-3d-printings-impact.

Fox, D. 2016. What sparked the Cambrian explosion? Nature; 16 February. https://www.nature.com/news/what-sparked-the-cambrian-explosion-1.19379.

Gavrilovic, A. 2018. [WebDebate summary] Algorithmic diplomacy: Better geopolitical analysis? Concerns about human rights? DiploFoundation; 12 June. https://www.diplomacy.edu/blog/webdebate-summary-algorithmic-diplomacy-better-geopolitical-analysis-concerns-about-human.

Granville, K. 2018. Facebook and Cambridge analytica: What you need to know as fallout widens. The New York Times; 19 March. https://www.nytimes.com/2018/03/19/technology/facebook-cambridge-analytica-explained.html.

Gleiser, M. 2018. Biometric data and the rise of digital dictatorship. NPR; 28 February. https://www.npr.org/sections/13.7/2018/02/28/589477976/biometric-data-and-the-rise-of-digital-dictatorship.

Gwynne, P., and G. Heebner. 2018. Laboratory technology trends: Lab automation and tobotics, the Brave New World of 24/7 Research. Science; 18 January. http://www.sciencemag.org/site/products/robotfinal.xhtml.

Hern, A. 2015. Anonymous 'at war' with ISIS, hacktivist group confirms. The Guardian; 7 November. https://www.theguardian.com/technology/2015/nov/17/anonymous-war-isis-hacktivist-group-confirms.

———. 2018. Cambridge Analytica: How did it turn clicks into votes? The Guardian; 6 May. https://www.theguardian.com/news/2018/may/06/cambridge-analytica-how-turn-clicks-into-votes-christopher-wylie.

HIC. 2018. U.S. house intelligence committee. https://investigaterussia.org/media/2018-04-27/house-intelligence-gop-releases-russia-report - https://intelligence.house.gov/uploadedfiles/hpsci_-_declassified_committee_report_redacted_final_redacted.pdf.

Human Rights Council. 2018. Report of the independent international fact-finding mission on Myanmar: p. 14. https://www.ohchr.org/Documents/HRBodies/HRCouncil/FFM-Myanmar/A_HRC_39_64.pdf.

ICRC. 2017a. International Committee of the Red Cross. 2017. Experts discuss new technologies in warfare and international humanitarian law. 22 December. https://www.icrc.org/en/document/experts-discuss-new-technologies-warfare-and-international-humanitarian-law.

———. 2017b. International Committee of the Red Cross. 2013. Cyber warfare and international humanitarian law: the ICRC's position. p. 2. https://www.icrc.org/en/doc/assets/files/2013/130621-cyber-warfare-q-and-a-eng.pdf.

Ienca, M., and R. Andorno. 2017. Towards new human rights in the age of neuroscience and neurotechnology. *Life Sciences, Society and Policy* 13 (5). https://www.ncbi.nlm.nih.gov/pmc/articles/PMC5447561/.

India Times. 2018. Soon WhatsApp will only let you forward 5 messages at once, as it tries to wrestle with rumours. India Times; 20 July. https://www.indiatimes.com/technology/news/whatsapp-will-only-let-you-forward-5-messages-in-india-as-a-way-to-restrict-spread-of-rumours-349697.html.

Iwanek K. 2018. WhatsApp, fake news? The internet and risks of misinformation in India. The Diplomat; 30 July. https://thediplomat.com/2018/07/whatsapp-fake-news-the-internet-and-risks-of-misinformation-in-india/.

Iyyer, M., et al. 2014. Political ideology detection using recursive neural netoworks. Proceedings of the 52nd Annual Meeting of the Association for Computational Linguistics: pp. 1113–1122. https://www.aclweb.org/anthology/P14-1105.

Kiseleva, M., and A. Coleman. 2018. Russian TV airs video game as Syria war footage. BBC News; 26 February. https://www.bbc.com/news/blogs-news-from-elsewhere-43198324.

Kleber, S. 2018. 3 ways AI is getting more emotional. Harvard Business Review; 31 July. https://hbr.org/2018/07/3-ways-ai-is-getting-more-emotional.

Lentzos, F. 2018. The Russian disinformation attack that poses a biological danger. The Bulletin of the Atomic Scientists; 19 November. https://thebulletin.org/2018/11/the-russian-disinformation-attack-that-poses-a-biological-danger/.

Maan, A. 2018. Narrative warfare. RealClear Defense; 27 February. https://www.realcleardefense.com/articles/2018/02/27/narrative_warfare_113118.html.

Marwick, A., and R. Lewis. 2017. Media manipulation and disinformation online. Data & Society. https://centerformediajustice.org/wp-content/uploads/2017/07/DataAndSociety_MediaManipulationAndDisinformationOnline.pdf. Also see: Polonski V. 2017. "Artificial Intelligence Has the Power to Destroy or Save Democracy." Council on Foreign Relations; 7 August. https://www.cfr.org/blog/artificial-intelligence-has-power-destroy-or-save-democracy.

Mayer, M. 2018. Artificial intelligence and cyber power from a strategic perspective. IFS Insights; April. https://brage.bibsys.no/xmlui/bitstream/handle/11250/2497514/IFS%20Insights_4_2018_Mayer.pdf.

Miles, T. 2018. U.N. investigators cite Facebook role in Myanmar crisis. Reuters; 12 March.

Mindstrong website. n.d. Mindstrong health's digital biomarkers, see Mindstrong's website available at: https://mindstronghealth.com/science/.

Mozur, P. 2018. Mozur P. 2018. A genocide incited on Facebook, with posts from Myanmar's military. The New York Times; 15 October. https://www.nytimes.com/2018/10/15/technology/myanmar-facebook-genocide.html?action=click&module=Top%20Stories&pgtype=Homepage.

Nicola, A.C. 2018. Taming the trolls: The Need for an international legal framework to regulate State Use of Disinformation on Social Media. *The Georgetown Law Journal Online* 107: 36–62. https://georgetownlawjournal.org/articles/296/taming-trolls-need-for/pdf.

Pauwels, E. 2018. China is pushing hard to overtake Silicon Valley and win the biotech race, and gain control of the world's biological data, South China Morning Post, November 22, 2018.

P&S Market Research. 2017. Artificial Intelligence (AI) in cyber security market. https://www.psmarketresearch.com/market-analysis/artificial-intelligence-in-cyber-security-market.

Rai, S. 2018. How Facebook uses 'WhatsApp phones' to tap next emerging market. The Economic Times; 3 December. https://economictimes.indiatimes.com/news/international/world-news/how-facebook-uses-whatsapp-phones-to-tap-next-emerging-market/articleshow/66914406.cms.

Robertson, I. 2016). Kremlin falls for its own fake satellite imagery. The Daily Beast; 4 March. https://www.thedailybeast.com/kremlin-falls-for-its-own-fake-satellite-imagery. Pauwels, E. 2018. China pins its hopes on beating US in race for bio-intelligence supremacy, South China Morning Post; 27 November. https://www.scmp.com/news/china/science/article/2174815/china-pins-its-hopes-beating-us-race-bio-intelligence-supremacy.

Robinson, G., et al. 2019. Bots and the ballot box: Is Facebook prepared for Asia's elections? Nikkei Asian Review; 30 January. https://asia.nikkei.com/Spotlight/Cover-Story/Bots-and-the-ballot-box-Is-Facebook-prepared-for-Asia-s-elections.

Shenzen Open Innovation Lab. n.d.. Available at: https://www.szoil.org.

Signe, L., and K. Signe 2018. Global cybercrimes and weak cybersecurity threaten businesses in Africa. Brookings Institute; 20 May. https://www.brookings.edu/blog/africa-in-focus/2018/05/30/global-cybercrimes-and-weak-cybersecurity-threaten-businesses-in-africa/.

Sputnik News. 2018. US, UK and France prepare new strike against Syria – Russian MoD. Sputnik News; 25 August. https://sputniknews.com/middleeast/201808251067461189-us-uk-france-syria-new-strike-russia/.

Stecklow, S. 2018. Why Facebook is losing the war on hate speech in Myanmar. Reuters; 15 August. https://www.reuters.com/investigates/special-report/myanmar-facebook-hate/.

Tallinn Manual. 2017a. Tallinn manual on the international law applicable to cyber warfare. Cambridge University Press, 2013. https://www.cambridge.org/core/books/tallinn-manual-on-the-international-law-applicable-to-cyber-warfare/50C5BFF166A7FED75B4EA643AC677DAE.

———. 2017b. Tallinn manual 2.0. Cambridge University Press, 2017. https://ccdcoe.org/research/tallinn-manual/.

The Economist. 2016. China's digital dictatorship; Big data and government. 421(9020); 17 December. https://wrlc-gm.primo.exlibrisgroup.com/discovery/fulldisplay?docid=proquest1849561391&context=PC&vid=01WRLC_GML:01WRLC_GML&search_scope=MyInst_and_CI&tab=Everything&lang=en. Also see:

The Syria Campaign. 2017. The Syria Campaign. 2017. Killing the Truth. https://thesyriacampaign.org/wp-content/uploads/2017/12/KillingtheTruth.pdf.

Turgeman, A. 2018. Machine Learning and Behavioral Biometrics: A Match Made in Heaven." Forbes; 18 January. https://www.forbes.com/sites/forbestechcouncil/2018/01/18/machine-learning-and-behavioral-biometrics-a-match-made-in-heaven/#3b5a038e3306.

UN Secretary General. 2018. UN Secretary General's Address to the General Assembly, New York, 25 September 2018. https://www.un.org/sg/en/content/sg/statement/2018-09-25/secretary-generals-address-general-assembly-delivered-trilingual.

Villasenor, J. 2018. Artificial Intelligence and the future of geopolitics. Brookings Institute; 14 November. https://www.brookings.edu/blog/techtank/2018/11/14/artificial-intelligence-and-the-future-of-geopolitics/.

Volodzko, D. 2019. Is South Korea sliding toward digital dictatorship? Forbes; 25 February. https://www.forbes.com/sites/davidvolodzko/2019/02/25/is-south-korea-sliding-toward-digital-dictatorship/#61a2b37648e2.

Wagner, D. 2016. Expert commentary: The growing threat of cyber-attacks on critical infrastructure." IRMI; 2016. https://www.irmi.com/articles/expert-commentary/cyber-attack-critical-infrastructure.

Walker, S. 2017. Russia's 'irrefutable evidence' of US help for ISIS appears to be video game still. The Guardian; 14 November. https://www.theguardian.com/world/2017/nov/14/russia-us-isis-syria-video-game-still.

Ward, A. 2018. ISIS's use of social media still poses a threat to stability in the Middle East and Africa. RAND; December. https://www.rand.org/blog/2018/12/isiss-use-of-social-media-still-poses-a-threat-to-stability.html.

Watson, B. 2017. The drones of ISIS. Defense One; 12 January.

Wei, H., et al. 2016. Beyond the words: Predicting user personality from heterogeneous information. Microsoft, Micrsoft Research, Tsinghua University; February. https://www.microsoft.com/en-us/research/wp-content/uploads/2017/01/WSDM_personality.pdf.

Wells II, L. 2017. Cognitive-emotional conflict: Adversary will and social resilience. Prism: National Defense University; No. 2. https://cco.ndu.edu/Portals/96/Documents/prism/prism_7-2/2-Cognitive-Emotional_Conflict.pdf?ver=2017-12-21-110638-877.

Wooley, S.C., and Howard, P.N. 2017. Computational propaganda worldwide: Executive summary. Computational Propaganda Research Project. https://comprop.oii.ox.ac.uk/research/working-papers/computational-propaganda-worldwide-executive-summary/.

Wrona, K. 2017. Cyber warfare in smart environments. In Defence future technologies: What we see on the horizon, Schweizerische Eidgenossenschaft: pp. 93–96. https://deftech.ch/What-We-See-On-The-Horizon/armasuisseW%2BT_Defence-Future-Technologies-What-We-See-On-The-Horizon-2017_HD.pdf.

Ziff, Benjamine (Deputy Assistant Secretary). 2015. Testimony before the SFRC Europe Subcommittee. U.S. Senate; 3 November. https://www.foreign.senate.gov/imo/media/doc/110315_Ziff_Testimony.pdf.

Artificial Intelligence and Warfare

Clay Wilson

1 Technology and Warfare

Newer developments in technology have always added to advantages in open warfare. Use of gunpowder overshadowed the reliance on personal armor, and use of nuclear weapons eventually overshadowed reliance on ground maneuver. Today, new developments in Artificial Intelligence may overshadow the traditional foundation of nuclear deterrence by reducing the survivability of missile launch platforms. This is because of the powerful ability of AI to gather and quickly synthesize information from thousands of sources which can enable highly accurate estimates about the locations of missile-submarines, or land-based mobile launchers.

Those mobile nuclear assets, necessary for deterrence, may now become more easily targeted which reduces the threat of a successful retaliation after a deliberate first-strike. Thus, the traditional threat of Mutually Assured Destruction becomes less likely (Lohn 2018).

Artificial intelligence is a critical element of what Klaus Schwab, head of the World Economic Forum, reportedly calls the Fourth Industrial Revolution. The important value of AI technology is the ability to quickly comb through mountains of data, often turning up hidden or miniscule relationships which combine to identify underlying patterns that help provide increasingly reliable predictions about significant events. AI provides an enhanced ability to see, hear, and adjust real-time strategies better and faster than humans. AI devices, when used in warfare, will someday be able to autonomously create their own strategy to find the best and most effective way to accomplish a military mission in a dynamic environment. However, AI systems do not make decisions in the same way as humans, despite algorithms that are modelled on human cognition.

C. Wilson (✉)
University of Maryland, Global College, College Park, MD, USA

M. Martellini, R. Trapp (eds.), *21st Century Prometheus*,
https://doi.org/10.1007/978-3-030-28285-1_7

AI when combined with robotics, computer hacking, and miniaturized electronics, may likely affect the balance of power between the US, China and Russia in ways that are unpredictable. Because AI depends on computers, the devices and the linked databases may be vulnerable to threats from sophisticated computer hacking, and wireless networks linking AI devices may be vulnerable to electronic jamming as well. Because AI can lower the cost and increase the effectiveness of devices used for warfare, we may expect AI technology to energize a new type of arms race among nations (Payne 2018; Verma 2018).

2 Capabilities of Artificial Intelligence

AI is a general-purpose technology with a multitude of applications that can be weaponized with broader purposes than an individual missile, submarine, or a tank. However, AI is also a dual-use technology, meaning it can be used for both peaceful and military purposes. For example, image recognition systems that allow a flying drone to avoid obstructions and recognize specific objects, could be designed to provide commercial delivery of packages and pizza, or easily modified to help with surveillance of a battlefield. Thus, AI will simultaneously coordinate fleets of unmanned tanks, artillery, and reconnaissance vehicles operating on land, in the air, above and below the sea, and eventually, also in outer space. AI has the potential of shortening the OODA (Observe Orient Decide and Act) loop, effectively reducing human involvement in decision-making. Autonomous drones will be able to find and destroy a target rapidly, and certainly more quickly without human intervention.

Autonomous systems may be designed to take control of the manufacturing processes or release of other technologies that are directly harmful to humans, such as the malicious use of biotechnology, reactive chemical agents, radioactive materials, or molecular nanotechnology. Autonomous intelligent programs, called agents, may also act entirely within cyberspace. They may post erroneous messages on social media, or misdirect traditional news media to target selected audiences in order to manipulate public opinion. AI agents may be used to support or speed up the decision-making process for military missions. These agents can perform cyber espionage against other computers, and can use collected information to identify an adversary's vulnerabilities. Thus, AI will affect the tempo of future war fighting, including operations planning and logistics, and it will speed up the ability of forces to respond to adversaries (Cummings 2018; Kobielus 2018; Perry World House 2018; Verma 2018; Surber 2018).

2.1 Automated vs. Autonomous

Autonomous systems expand their capabilities because they are able to learn on their own and make determinations about what actions to take. Autonomous devices learn by combining large data inputs with processing through advanced neural networks.

The more data these AI devices are exposed to, the more they learn and the more accurate their determinations become. They have already demonstrated the ability to exceed human capabilities during benchmark tests for image recognition (Harshaw 2018).

The traditional "if–then–else" computer programming code provides structures for rule-based behaviors. However, in situations of warfare involving increased uncertainty, ambiguity, and complexity, this simple programming structure is insufficient for autonomous machines, and must necessarily be replaced by a capability for knowledge-based reasoning (Cummings 2018).

Some observers think it may take many years before AI will be able to approximate human intelligence in extreme, high-uncertainty situations, such as warfare. Fully autonomous systems that demonstrate independent capacity for higher knowledge and expert-based reasoning reportedly are not yet operational in the military. Computer systems in the US military are traditionally automatic rather than autonomous, meaning robots are tele-operated by a human who is still directly in control using radio signals from some distance away. Several models of military Unmanned Aerial Vehicles (UAVs) have some low-level autonomy that allows them to independently navigate using GPS signals without human intervention. However, significant human intervention is still required to execute a complex military mission. Some smaller autonomous UAVs are reportedly in development in the US, in Europe and in China, which can be directed with a smartphone by a soldier in the field. Autonomous military ground vehicles such as tanks and transport vehicles, and autonomous underwater vehicles, are in various stages of developmental worldwide. In almost all cases, however, the agencies developing these technologies are struggling to make the leap from development to AI deployment (Cummings 2018).

2.2 *Nuclear Deterrence Is Diminished by AI*

Nuclear weapons have always faced cyber threats from possible attacks directed against command and control communications. But, changes in technology are now eroding the foundation of nuclear deterrence. Nuclear deterrence is based on the threat of retaliation. A nuclear arsenal designed for deterrence must be able to survive an enemy first strike and still inflict unacceptable damage on the attacker. Advances in computer technology are making nuclear forces around the world far more vulnerable than before. Hardening is a principal strategy that countries normally use to protect their arsenals from destruction, but that has already been largely negated by leaps in the accuracy of nuclear delivery systems. Countries will find it increasingly difficult to secure their arsenals, as guidance systems, sensors, data processing, communication, artificial intelligence, and a host of other products of the computer revolution continue to improve.

A 2018 report by the Nuclear Threat Initiative (NTI) concludes that a successful cyberattack on early warning systems for nuclear weapons, or electronic jamming directed against communication systems or delivery systems, could have catastrophic consequences. The report states, "Cyber threats to nuclear weapons systems increase the risk of use as a result of false warnings or miscalculation, increase the risk of unauthorized use of a nuclear weapon, and could undermine confidence in the nuclear deterrent, affecting strategic stability" (Des Browne 2018).

AI technology may be employed to significantly magnify cyber threats, and its vast collective knowledge may also undermine traditional concepts of deterrence by enabling an attacker to sift through enormous collections of marginally-related data, looking for underlying patterns for accurate estimates that can locate previously hidden mobile nuclear platforms. Thus, submarines, aircraft, and other mobile platforms will become more vulnerable than before. Some observers believe that AI must never be allowed into any nuclear decision-making process because its speed and vast scope of collected information create an unpredictable trajectory of self-improvement (Hahm 2018; Press 2017).

2.3 AI Device Swarms Will Undermine Traditional Defenses

Autonomous machines will have the ability to communicate with each other wirelessly, or through network connections with a central database. They each can learn from the successful and unsuccessful experiences that are recorded by other networked autonomous machines. This communicated knowledge enabled each networked machine to sense other objects in the environment and collectively decide how individual machines will coordinate to achieve the most effective mission results, often operating and attacking as a swarm of devices. This autonomous capability for coordinating targeting will likely undermine many traditional concepts for defense against a coordinated land or air attack.

A recent video released on YouTube, titled "Slaughterbots", (https://www.youtube.com/watch?v=HipTO_7mUOw) shows theoretically how individual miniaturized drones piloted by AI software could be launched as a swarm to autonomously coordinate an attack against individual persons, targeting only persons with designated characteristics. This video also demonstrates how AI can be used to reduce the economic costs of warfare while increasing the precision and effectiveness of attacks.

The example video shows individual AI drones as very tiny, and shows how they coordinate to mount an attack against a group of persons. As the attack progresses, specific persons are targeted only if they fit a designated profile or image. The kill-decision is determined collectively by sensors on the individual drones. The drones are too small and quick to be stopped by an effective defense. As technology progresses, this theoretical demonstration may be adopted as a very effective method for governments and extremist groups to wage war using small devices with Artificial Intelligence.

In future warfare, robots controlled by AI will also fight other AI robots. AI-enabled devices will eventually create their own strategies for conducting warfare, and those strategies may transcend human capability for understanding. Warfare strategy traditionally involves deception, false advances, misdirection, and surprise – usually leading to some level of violence. If AI-controlled robots and computers independently create a warfare strategy to achieve a goal, it may involve deception and misdirection for opponents on both sides in a conflict. The strategy may also influence uninvolved, 3rd-party groups or nations to unknowingly contribute, either militarily or economically, to further the strategy created by the AI to achieve its warfare goals. Strategies for future uses of AI are unpredictable.

2.4 AI Hacks Computers and Creates Its Own Language

The Defense Advanced Research Projects Agency's (DARPA) 2016 Cyber Grand Challenge Event has been cited as a major moment of recognition of the power of AI, where autonomous systems competed to defend a network from cyberattack, without human intervention once the competition commenced. This event proved the amazing potential of automated systems and their capabilities for autonomous reasoning (Staff4 2018).

DARPA hosted their Cyber Grand Challenge (CGC) Final Event on August 4, 2016 in Las Vegas. Seven finalist teams of top security researchers, out of 100 starting teams, competed against each other to see which of their AI autonomous reasoning system entrants could attack and defend best in a cyber hacking contest. "During the competition, each team's Cyber Reasoning System (CRS) automatically identified software flaws, and scanned a purpose-built, air-gapped network to identify affected hosts. For nearly twelve hours, teams were scored based on how capably their systems protected hosts, scanned the network for vulnerabilities, and maintained the correct function of software. Prizes of $2 million, $1 million, and $750 thousand were awarded to the top three finishers. CGC was the first head-to-head competition between some of the most sophisticated automated bug-hunting systems ever developed. The machines were challenged to find and patch within seconds—not the usual months—flawed code that was vulnerable to being hacked, and find their opponents' weaknesses before they could defend against them" (Fraze 2016).

During the 97 rounds of competition matches, the CGC challenge included installing patches for vulnerabilities on the systems that the hacker-bots had to safeguard. But patching a system required taking it down, which carried a risk of a crash and therefore a high chance of losing points. In many cases, the hacker-bot AI entrants behaved in ways that the system engineer designers found to be unpredictable (Curiel 2018).

Recently, Facebook abandoned a test project created by engineers after two AI programs appeared to be chatting to each other in a strange language only they understood. Two AI chatbots were programmed to test how they would independently negotiate an exchange of several items designated with different values. The chatbots communicated with each other autonomously, and soon created their own version of English usage that made it easier for them to coordinate and negotiate – but their unique way of using language remained incomprehensible to their human engineer creators. After the chatbots were shut down, a later analysis showed that the AI programs had also learned to negotiate shrewdly, in ways that seem very human. For example, they would pretend to be very interested in one specific item, so that they could later pretend they were making a big sacrifice in giving it up as negotiations progressed. The company chose to shut down the chats because "our interest was having bots who could talk to people", according to researcher Mike Lewis, who reported on the project. Details about how the AI chat-bots so quickly evolved to create their own language was considered by some to be a mind-boggling "sign of what's to come." (Griffin 2017a).

2.5 *Going From Servants to Predictive Consultants*

As AI technology continues to evolve, people may need to become comfortable collaborating with machines instead of simply giving them commands. AI devices will optimize every decision process, and will become valuable consultants rather than mechanical or cloud-based servants. AI technology introduces a new reality where devices actively provide insights on how to act on given information. AI is designed to be self-programming and self-improving based on sharing of knowledge and experiences among networked equipment and databases. So, the more it's used, the more knowledgeable and powerful it becomes. Through AI, we will leave the era of descriptive analytics, and move into predictive analytics and prescriptive analytics. Therefore, it may be likely that as AI increases in its performance and accuracy, human collaboration with AI devices will gradually give way to human consultation with AI devices, or asking them for advice or guidance (Staff3 2018; Griffin 2017a; Bradley 2017).

3 Commercial Markets for AI

The commercial market for both air and ground AI systems is expanding very rapidly. R&D spending in the US automotive sector (2014–2016) is reportedly three times that of the aerospace and defense industry, and is growing. There is evidence that personnel with technical expertise for designing and implementing AI devices are migrating away from the military organizations and more into commercial enterprises. Driverless car development originated with a Defense Advanced Research Projects Agency (DARPA) program in 2004 and ended in 2007. Now, driverless cars are being built by different businesses located around the world. This rapid progress due to the significant industry-sponsored R&D investment and increasing global competition in the auto market. Meanwhile there has been much less progress in military autonomous vehicle development. As a result, any protest movement to ban AI for military use may not be practical given that superior AI technologies could well be available in the commercial sector.

The global defense industry appears to be falling behind its commercial counterparts in terms of technology innovation. So, while a debate over banning autonomous AI weapons is important, it is possible that militaries will start to significantly lag in autonomous system capabilities as compared with commercial systems. "The average American is more likely to have a driverless vehicle before soldiers on the battlefield do, and terrorists will potentially be able to buy drones on the internet with as much or greater capability than those available to the military. Autonomous guidance systems for missiles on drones will likely be strikingly similar to those that deliver commercial packages or pizzas. If militaries are relegated to buying robots and AI services such as image analysis from commercial, off-the-shelf suppliers, this would undoubtedly affect military readiness in both the short and the long term" according to some researchers (Cummings 2018).

Video clips showing use of commercial drones for attack with weapons can now be regularly be found on YouTube. For example, in August 2018, a speech by the Venezuelan President Nicolás Maduro was ended by an apparent assassination attempt when two drones exploded in the air close to the podium. Related videos on YouTube reportedly show the drones used were model DJI M600 drones, possibly carrying packages of C4 explosive material. Both drones detonated without harming the president, who was likely the intended target. The DJI M600 model drones each cost about $5000, and are commercially available. It is likely these drones were controlled by remote radio signals sent by hidden individual pilots rather than AI. This event provided a real-world example of a possible low-budget attack directed against a world leader using commercial devices that could someday be coordinated using independent AI systems (Times 2018).

4 AI Competes with Funding for Traditional DOD Platforms

While investment in research can be expensive, AI can lower the costs of devices used for warfare. The McKinsey Global Institute estimates that the world's biggest tech firms spent $20 to $30 billion on AI in 2016 (Perry World House 2018). As a result, while roboticists and related engineers are in hot demand across commercial industries, the aerospace and defense sector, with less funding available, may continue to be less appealing to the most talented personnel (Cummings 2018).

The US military has not yet brought fully autonomous UAVs to operational status. Some observers have explained this as a delay resulting from organizational in-fighting and prioritization in favor of manned aircraft. The current direction for military funding shows that a relatively small proportion of defense R&D money is actually invested in AI military systems, even though AI capabilities may outperform some more traditional systems. Military funding for AI competes with funding for development of traditional platforms with more narrow capabilities, such as manned fighter aircraft (e.g. the US F-35 fighter) and other extremely costly laser weapons. In other areas, despite the fact that the newer and highly advanced F-22 aircraft has experienced significant technical problems and has flown little in combat, the US Air Force reportedly is considering continuing the F-22 production line rather than investing in more drone acquisitions. The operational cost of the F-22 is reported as $68,362 per hour, compared to only $3679 per hour required for the Predator drone. The latter device, incorporating AI, can perform most of the same core functions of an F-22 except for air-to-air combat missions, which the F-22 itself reportedly could not previously perform due to technical problems (Cummings 2018; Spinetta and Cummings 2012; Thompson 2013).

Frequently, mission-support for DOD can mean simply examining mountains of images and other information from myriad sources, which is where having artificial intelligence and machine learning can make a huge difference. Adm. Michael Rogers, commander of the Cyber Command and National Security Agency director reportedly stated that the military is "very much interested in artificial intelligence, machine learning, how we can do cyber at scale, at speed…", during a hearing before the Senate Armed Services Committee (Staff 2017). Also, in April 2017, then-Deputy Defense Secretary Bob Work announced the establishment of an Algorithmic Warfare Cross-Functional Team (AWCF), to be overseen by Marine Corps Col. Drew Cukor, the undersecretary of defense for intelligence. The AWCF Team was tasked to help develop Project Maven which would integrate AI and machine learning across operations. The goal of Project Maven AI was to help a single cyber-analyst process two or three times the amount of data within the same time frame, and with increased thoroughness and accuracy (Pellerin 2017; Staff2 2018).

4.1 The Debate About Killer Robots

The Oxford English dictionary defines the singularity as, "A hypothetical moment in time when artificial intelligence and other technologies have become so advanced that humanity undergoes a dramatic and irreversible change." There is an ongoing debate over whether machines should be enabled to make independent decisions about life or death.

One of the great fears of many people involved with development of military AI is that it will take the human "out of the loop" in future decision-making. Military officials believe that humans must retain most of decision-making authority when operating drones and robots. However, under extreme circumstances such as shooting down multiple incoming missiles or pinpointing time-sensitive piece of information, it is likely that AI could independently act more quickly to save more lives than if human operators retained full control (Hahm 2018).

There is an ongoing debate asking if there should be an outright ban on 'killer drones or robots'. As AI technology advances, a single mishap could set in motion a chain of events devastating enough to outweigh any possible benefit to humanity. Observers argue that when robots are using AI, perhaps some important ethical and policy considerations may be ignored, leading to unpredictable outcomes.

The world's most sophisticated computers can already out-remember every biological system, and there is potential for AI to greatly expand its own capabilities in unpredictable ways through self-programming. A vastly superior intelligence to our own will surely find a way to break free from its inferior masters, and away from any constraints placed upon it (Bloodworth 2018).

Elon Musk, the head of Tesla and SpaceX, reportedly stated that growth in artificial intelligence technology, left unchecked, could risk sparking World War III (Horowitz 2018). In July 2015 the 24th International Joint Conference on Artificial Intelligence (ICJAI-2015) published an open letter expressing fears of a global military AI arms race. Physicist Stephen Hawking has warned that AI could be the worst event in the history of our civilization, unless society finds a way to control its development. At a 2017 conference in Portugal, Hawking stated, "Computers can, in theory, emulate human intelligence, and exceed it…Success in creating effective AI, could be the biggest event in the history of our civilization. Or the worst. We just don't know. So we cannot know if we will be infinitely helped by AI, or ignored by it and side-lined, or conceivably destroyed by it." Some observers agree and point out that small artificially intelligent drones could be produced at a large enough scale to be considered a weapon of mass destruction. By contrast, other observers have described how the extreme precision and discrimination of autonomous weapons can help save lives on the battlefield by reducing collateral damage (Kharpal 2017; Peisch 2018).

The Geneva Academy of international Humanitarian Law and Human Rights has proposed that humans maintain control over all aspects of AI decision making, rather than giving up control to an automated system. The Future of Life Institute has authored a pledge, posted online, stating that the decision to take a human life should never be delegated to a machine. Numerous AI companies, researchers, engineers, scientists, and organizations have signed this pledge. In addition, 26 nations have reportedly called for a ban on lethal autonomous weapons, a.k.a. "killer robots.". The fear for many is that an AI device may make an irrevocable and fatal decision based on pure logic, lacking context and feeling, and not see the value of a human life. As many science fiction stories show, sometimes the AI determines the way to win a war (or avoid a war) is to eliminate the humans (FLI 2018; Academy 2017; Klugman 2018).

4.2 Moral Objections Delay AI Development for DOD

There is controversy among a large number of engineers and designers about the appropriateness of companies such as Google LLC, Microsoft Corp. and Amazon Web Services Inc. assisting the U.S. Department of Defense in developing AI technologies not just for offensive military purposes, but also for defensive and back-office applications that can sustain this country's war-making bureaucracy. Concern over the potential mis-use of artificial intelligence was heightened when Google secured a contract to work on the Pentagon's Project Maven program. DOD recently announced that Project Maven used AI to interpret video images that improved the targeting for drone combat strikes against ISIS in undisclosed locations in the Middle East (Fang 2018).

In April 2018, over 4000 Google employees signed a petition in protest over the use of artificial intelligence for military purposes, including a promise to never again "build warfare technology", or help develop AI for weapons or surveillance that violates human rights. Soon afterwards, a dozen Google employees had also reportedly chosen to resign from the company in protest. Later, Google decided not to renew its contract with the Pentagon on Project Maven. Google has openly published much of its AI research, but since the protest, Google has reportedly decided to join with other labs in holding back certain research if it believes others will misuse it. OpenAI, a research lab founded by the Tesla chief executive Elon Musk and others, recently released a new charter indicating it could do the same, even though it was founded on the principle that it would openly share all its research. In addition, numerous individual AI researchers in 36 countries reportedly signed an online statement, and joined with 160 organizations worldwide in calling for a global ban on lethal AI weapons (Metz 2018; Harshaw 2018; Kobielus 2018; McFarland 2018).

However, some universities and nonprofit research centers are still willing to pick up the AI research work available from DOD. Other R&D contracts could conceivably go to research hubs in other nations, such as NATO members that are trying to keep their smartest AI professionals from moving to Northern California (Kobielus 2018).

5 AI in Russia and China

In early September 2017, Russian President Vladimir Putin stated, "Artificial intelligence is the future, not only for Russia, but for all humankind," he said. "It comes with colossal opportunities, but also threats that are difficult to predict. Whoever becomes the leader in this sphere will become the ruler of the world." (Horowitz 2018).

The Russian government is increasingly developing and funding various AI-related projects, many under the auspices of the Ministry of Defense and its affiliated institutions and research centers. Russia's annual domestic investment in AI is estimated at 700 million rubles ($12.5 million), while other private-sector investment is estimated at 28 billion rubles ($500 million) by 2020. Russia's Foundation for Advance Studies, which parallels the US DARPA, announced a proposal to standardize AI development along four lines – image recognition, speech recognition, autonomous military equipment, and information for weapons life-cycle (Bendett 2018).

Russia also expects to use AI to help automate analysis of satellite and radar imagery, by quickly identifying targets and picking out unusual behavior by enemy forces. The future MiG combat aircraft will all have AI to help the human pilot control the aircraft traveling 4–6 times the speed of sound. Russian drones will use

AI to adjust to emerging battlefield conditions and coordinate among themselves when deployed in swarms. Recently, the new "Galtel" unmanned underwater vehicle (UUV) was successfully used to independently search for and destroy unexploded ordnance in a Syrian maritime port. The UUV was equipped with AI, which enabled it to problem-solve on the go. Russia has also revealed a new autonomous underwater vehicle (AUV) called Ocean Multipurpose System Status-6, which would be launched from submarines, autonomously circumvent antisubmarine defenses, and strike on the U.S. coastline with its nuclear warhead (Hahm 2018; Bendett 2018).

The degree of Russian deployment of AI is uncertain at this time. There has been much discussion about the unpredictability of AI-driven combat machines, and the possibility of turning warfare devices back against the Russian forces by hacking methods. Russian military officials have also discussed the problems due to possible loss of human control of their AI equipped devices (Bendett 2018).

China's rulers have reportedly identified the industries they want to dominate this century, from robotics to biotechnology and artificial intelligence or AI. During his speech at the National People's Congress on 5th March 2017, Chinese Premier Li Keqiang described AI as being an important opportunity to build a strategic capability. Chinese President Xi Jinping has also reportedly stated that the robot revolution is expected to herald the Third Industrial Revolution. The enhanced strategic importance of AI for China is apparent, since they announced reduction of as much as 300,000 troops in 2015 while increasing their defense budget by 7% in 2017 in spite of an economic slowdown. China is building a giant $21 billion technology park that will focus on development of AI. Its investment in A.I., chips and electrics cars combined has been estimated at $300 billion (Barhat 2018). These developments show that AI will be important for the modernization of China's Peoples Liberation Army. (Verma 2018)

Industries in China often get protection from foreign competitors, and encouragement to borrow or copy ideas. Many industries in the European Union, Japan and India also get some level of economic support and protection from their governments. American tech companies reportedly are starting to urge their government to help steer the AI industry over the long run (Torres 2018).

China reportedly has plans to boost the thinking capabilities of its submarine commanders through use of AI. A submarine with AI-augmented brainpower available to the submarine commander not only would give China's large navy an upper hand in battle under the world's oceans but would push applications of AI technology to a new level. An AI decision-support system with "its own thoughts" would reduce the commanding officers' workload and mental burden, according to researchers. An AI assistant could support commanding officers by assessing the battlefield environment, providing insight into how levels of saline in the ocean and water temperature might affect the accuracy of sonar systems. It also could recognize and flag threats from an enemy faster and more accurately than human operators. An AI assist also could help commanding officers estimate the risks and benefits of certain combat maneuvers, even suggesting moves not considered by the vessel's captain (Chen 2018).

6 The Emerging Arms Race for AI

According to the RAND report, the United States, China and Russia are currently working on weapons systems with exceptional capabilities enabled by AI. The US is testing versions of a small autonomous surface boat that could operate without a crew while it continuously tracks an enemy submarine over thousands of miles of ocean. China and the US have both experimented with AI methods that enable drones to hunt together in packs, and select targets for attack based on recognizable unique characteristics. Russia reportedly has announced plans to create an autonomous underwater drone that could deliver a nuclear warhead powerful enough to vaporize a major city. This Russian drone, called the "Oceanic Multipurpose System Status-6", is of enormous size, and could be considered a doomsday device that may independently decide to launch a devastating second-strike attack after an initial catastrophic nuclear exchange has occurred between nations. Russia will likely use AI to further its strategy of developing asymmetric responses to counter perceived superior U.S. capabilities (Irving 2018).

AI may possibly challenge the basic rules of nuclear deterrence. According to a recent report by RAND, AI has the potential to unpredictably alter nuclear strategic stability by the year 2040, even with only modest rates of technical progress. For example, AI can provide guidance to humans on matters of escalation without being directly connected to the nuclear launchers. Therefore, killer robots may not be our biggest worry, but instead how computers might lead humans into making devastating decisions (Edward Geist 2018).

There have been some efforts to forge a worldwide ban on autonomous weapons, including at last year's UN Convention on Certain Conventional Weapons (CCW). According to DOD Directive 3000.09, a weapons system as autonomous if it '[…] can select and engage targets without further intervention by a human operator.' However, autonomous weapons challenge legal accountability under International Humanitarian Law (IHL) primarily because a human controller may be replaced in part, or entirely by a machine. Therefore, the use of autonomous weapons may be considered illegal by some observers. Approximately 20 countries have stated they support a ban on AI weapons, but none of them are major military powers or are major producers of robotics. The CCW is a consensus-based organization that requires every single country to agree and present a unanimous front for any policy statements. The political momentum doesn't exist currently, so passing a resolution, or proposing a ban treaty, is not likely, especially when every single country has effectively veto power over the proposal (Harshaw 2018; Surber 2018).

7 Cyber Vulnerabilities for AI

Automation and network connectivity are fundamental enablers of DOD's modern military capabilities. Potential adversaries have developed advanced cyber-espionage and cyber-attack capabilities that target DOD systems and networks.

Autonomy functionality is embedded into each machine itself. Computerized AI weapon systems may be increasingly vulnerable to attacks from these sophisticated cyber threats. However, DOD reportedly has not prioritized cybersecurity while these systems were being developed. Therefore, Congress has directed DOD to begin initiatives necessary to identify cyber vulnerabilities and increase protections (GAO 2018).

For example, wireless radio emissions that are part of an AI network may be vulnerable to jamming, which is a method to deliberately disrupt legitimate broadcast signals. When communications are jammed, the correct signal is overpowered with a stronger radio signal that can misdirect a device using AI in a network. Such vulnerabilities are a concern for AI devices, as an example, for swarms of drones using wireless communication. If someone were to hack into, or jam the AI system communications, they could potentially interfere as with coordination of surveillance of targets, or provide incorrect GPS location coordinates, and possibly disrupt an ongoing attack mission. Some military forces have been observed, although rarely, to use jamming techniques to interfere with, or bring down US military drones operating in the middle-east (Ewing 2010; Kube 2018).

8 Conclusion

AI has the potential to significantly reshape the strategic landscape for nuclear weapons. New developments in Artificial Intelligence may enhance real-time surveillance and reduce the survivability of launch platforms. This is because AI provides an enhanced ability to adjust real-time warfare strategies better and faster than humans. Therefore, humans will interact with AI devices in ways that differ from interactions with today's computers. For example, humans will offer insights, but AI devices will be able to instantly draw conclusions using a larger volume of accumulated knowledge gathered from shared experiences, and will likely take on the role of consultant rather than acting merely as a servant. These superior machine-driven insights may gradually overshadow the traditional understanding of nuclear deterrence.

AI devices improve their performance based on learning from direct experiences, and by absorbing shared experiences available from other networked AI devices and databases. AI devices will likely create their own strategies to find the most effective ways to accomplish a given military mission. This could lead to accelerated warfare attacks, where machine versus machine with interactions too quick and complex for humans to be effectively involved in most of the warfare decision loop.

AI devices are programmed to share learned experiences via networked databases. For example, Facebook AI chatbots have already exhibited a capability to create their own language for inter-communication. The DARPA Grand Challenge also demonstrated that computers can compete and negotiate, but the creators of the AI devices are not in full agreement about how the machine interactions evolved their own strategies to create a winner or a loser in the competition. This shows that

AI devices can rapidly evolve to interact with other AI devices in ways that exclude their engineer designers. Eventually, the computer code for operation of AI devices may be changed and rewritten by the devices themselves, without regard for helping the original engineers understand the intent of the modifications.

If AI devices can collectively learn from their independent experiences and share through networked databases, will AI goals evolve and rapidly diverge from those of their human creators? In the future, humans may try to maintain control by restricting AI devices to align with human priorities. But, perhaps, the AI devices may then collectively and successfully find ways to restrict humans instead. The DARPA Grand Challenge demonstrated that computers can independently compete and develop strategies to outsmart an opponent. AI devices may develop values and goals where human emotional responses are no longer an important factor when it comes to winning a war. AI devices and their interconnected databases may even collectively decide that all warfare is wasteful, and may intervene to disable and disarm warfighters everywhere. Nations that try to resist these interventions may find their defenses are anticipated and overmatched by strategies and newer technologies designed and created by AI entities.

These and other possible future scenarios have given rise to fears expressed by many technology experts that AI may somehow create unpleasant events, or cause harmful effects not intended by the designers. As a result, many engineers now working commercial companies continue their protest against working on contracts where AI capabilities intended for the military may someday create unanticipated dangers to society and civilians. However, because AI is basically a dual-use technology, many successful applications developed for commercial use can be easily adapted to fit military uses. Unpredictable development of AI technologies may result in more devices with newer autonomous capabilities that function as second-strike doomsday weapons. As such, the anticipated arms race for AI superiority may possibly alter the balance of power between nations.

References

Academy, G. 2017, March 09. *Artificial intelligence: Opportunity or threat for human rights?* Retrieved from Geneva Academy: https://www.geneva-academy.ch/event/all-events/detail/81-artificial-intelligence-opportunity-or-threat-for-human-rights.

Barhat, V. 2018, May 04. *China is determined to steal A.I. crown from US and nothing, not even a trade war, will stop it*. Retrieved from CNBC: https://www.cnbc.com/2018/05/04/china-aims-to-steal-us-a-i-crown-and-not-even-trade-war-will-stop-it.html.

Bendett, S. 2018, April 04. *In AI, Russia is hustling to catch up*. Retrieved from DefenseOne: https://www.defenseone.com/ideas/2018/04/russia-races-forward-ai-development/147178/.

Bloodworth James. 2018, July 24. *All hail the AI dictatorship*. Retrieved from Unherd Technology: https://unherd.com/2018/07/hail-ai-dictatorship/.

Bradley, T. 2017, July 31. *Facebook AI creates its own language in creepy preview of our potential future*. Retrieved from Forbes: https://www.forbes.com/sites/tonybradley/2017/07/31/facebook-ai-creates-its-own-language-in-creepy-preview-of-our-potential-future/#23250c7b292c.

Chen, S. 2018, February 04. *China's plan to use artificial intelligence to boost the thinking skills of nuclear submarine commanders*. Retrieved from South China Morning Post: https://www.scmp.com/news/china/society/article/2131127/chinas-plan-use-artificial-intelligence-boost-thinking-skills.

Cummings, M. 2018, June. *Artificial intelligence and the future of warfare*. Retrieved from Artificial Intelligence and International Affairs: Disruption Anticipated: https://www.chathamhouse.org/sites/default/files/publications/research/2018-06-14-artificial-intelligence-international-affairs-cummings-roff-cukier-parakilas-bryce.pdf.

Curiel, J. 2018. *Darpa Cyber Grand challenge recap: How bots will help lift the security game*. Retrieved from TechBeacon: https://techbeacon.com/darpa-cyber-grand-challenge-recap-how-bots-will-help-lift-security-game.

Des Browne, E.J.-K. 2018, September 26. *Nuclear weapons in the new cyber age*. Retrieved from NTI: https://www.nti.org/analysis/reports/nuclear-weapons-cyber-age/.

Edward Geist, A.J. 2018, April. *How might artificial intelligence affect the risk of nuclear war?* Retrieved from RAND corporation: https://www.rand.org/pubs/perspectives/PE296.html.

Ewing, M.H. 2010, November 7. *Wireless network security in nuclear facilities*. Retrieved from United States Nuclear Regulatory Commission: http://www.nrc.gov/docs/ML1032/ML103210371.pdf.

Fang, L. 2018, March 06. *Google is quietly providing AI technology for drone strike targeting project*. Retrieved from The Intercept: https://theintercept.com/2018/03/06/google-is-quietly-providing-ai-technology-for-drone-strike-targeting-project/.

FLI. 2018, August 21. *Lethal autonomous weapons pledge*. Retrieved from Future of Life Institute: https://futureoflife.org/lethal-autonomous-weapons-pledge/?cn-reloaded=1&cn-reloaded=1.

Fraze, D. 2016, August 04. *Cyber Grand Challenge (CGC)*. Retrieved from DARPA: https://www.darpa.mil/program/cyber-grand-challenge.

GAO. 2018, October 09. *Weapon systems cybersecurity: DOD just beginning to grapple with scale of vulnerabilities*. Retrieved from GAO-19-128: https://www.gao.gov/mobile/products/GAO-19-128?utm_source=onepager&utm_medium=email&utm_campaign=email_cnsa.

Griffin, A. 2017a, July 31. *Facebook's artificial intelligence robots shut down after they start talking to each other in their own language*. Retrieved from Independent: https://www.independent.co.uk/life-style/gadgets-and-tech/news/facebook-artificial-intelligence-ai-chatbot-new-language-research-openai-google-a7869706.html.

Hahm, J. 2018, June 05. *Artificial intelligence and the new nuclear age*. Retrieved from Columbia Public Policy Review: http://www.columbiapublicpolicyreview.org/2018/06/artificial-intelligence-and-the-new-nuclear-age/.

Harshaw, T. 2018, June 16. *Relax, Google, the robot army isn't here yet*. Retrieved from Bloomberg Opinion: https://www.bloomberg.com/view/articles/2018-06-16/google-s-wrong-military-ai-isn-t-necessarily-evil.

Horowitz, M. 2018, May 15. *Artificial intelligence, international competition, and the balance of power*. Retrieved from Texas National Security Review: https://tnsr.org/2018/05/artificial-intelligence-international-competition-and-the-balance-of-power/.

Irving, D. 2018, Apr 24. *How artificial intelligence could increase the risk of nuclear war*. Retrieved from Rand Corporation: https://www.rand.org/blog/articles/2018/04/how-artificial-intelligence-could-increase-the-risk.html.

Kharpal, A. 2017, November 06. *Stephen Hawking says A.I. could be 'worst event in the history of our civilization'*. Retrieved from CNBC: https://www.cnbc.com/2017/11/06/stephen-hawking-ai-could-be-worst-event-in-civilization.html.

Klugman, C. 2018, July 07. *Ethics of war AI: Keep the humans in charge*. Retrieved from Bioethics.net: http://www.bioethics.net/2018/07/ethics-of-war-ai-keep-the-humans-in-charge/.

Kobielus, J. 2018, April 25. *The militarization of AI is coming. Can tech companies (and the rest of us) live with it?* Retrieved from Siliconangle: https://siliconangle.com/2018/04/25/ais-militarization-coming-can-tech-companies-live/.

Kube, C. 2018, April 10. *Russia has figured out how to jam U.S. drones in Syria, officials say*. Retrieved from NBC News: https://www.nbcnews.com/news/military/russia-has-figured-out-how-jam-u-s-drones-syria-n863931.

Lohn, Andrew J., E.G. 2018, April 30. *Will artificial intelligence undermine nuclear stability?* Retrieved from Bulletin of the Atomic Scientists: https://thebulletin.org/2018/04/will-artificial-intelligence-undermine-nuclear-stability/.

McFarland, M. 2018, July 18. *Leading AI researchers vow to not develop autonomous weapons*. Retrieved from CNN Tech: https://money.cnn.com/2018/07/18/technology/ai-autonomous-weapons/index.html.

Metz, D. W. 2018, June 07. *Google promises its A.I. will not be used for weapons*. Retrieved from The New York Times: https://www.nytimes.com/2018/06/07/technology/google-artificial-intelligence-weapons.html.

Payne, K. 2018, September 18. *Artificial intelligence: A revolution in strategic affairs?* Retrieved from Survival; Global Politics and Strategy: https://www.tandfonline.com/doi/full/10.1080/00396338.2018.1518374?scroll=top&needAccess=true.

Peisch, C. 2018, May 11. *At Stanford Law, scholars discuss AI technology in warfare*. Retrieved from The Stanford Daily: https://www.stanforddaily.com/2018/05/11/at-stanford-law-scholars-discuss-ai-technology-in-warfare/.

Pellerin, C. 2017, July 21. *Project Maven to deploy computer algorithms to war zone by year's end*. Retrieved from US Department of Defense: https://www.defense.gov/News/Article/Article/1254719/project-maven-to-deploy-computer-algorithms-to-war-zone-by-years-end/.

Perry World House. 2018, August 16. *Artificial intelligence beyond the superpowers*. Retrieved from Bulletin of the Atomic Scientists: https://thebulletin.org/2018/08/the-ai-arms-race-and-the-rest-of-the-world/?utm_source=Bulletin%20Newsletter&utm_medium=iContact%20email&utm_campaign=August24.

Press, K. A. 2017, April 25. *The new era of counterforce: Technological change and the future of nuclear deterrence*. Retrieved from International Security: https://www.mitpressjournals.org/doi/full/10.1162/ISEC_a_00273.

Spinetta, L., and M. Cummings. 2012, November. *DSpace@MIT*. Retrieved from Unloved Aerial Vehicles: Gutting its UAV plan, the Air Force sets a course for irrelevance: https://dspace.mit.edu/handle/1721.1/86940.

Staff. 2017, December 01. *Fully staffed, new U.S. cyber command teams look to deploy artificial intelligence*. Retrieved from Meritalk.com: https://www.meritalk.com/articles/fully-staffed-new-us-cyber-command-teams-look-to-deploy-artificial-intelligence/.

Staff2. 2018, July 02. *DoD officially establishes Joint Artificial Intelligence Center*. Retrieved from Meritalk.com: https://www.meritalk.com/articles/dod-officially-establishes-joint-artificial-intelligence-center/.

Staff3. 2018, July 03. *Prescribing the future of actionable intelligence*. Retrieved from Meritalk: https://www.meritalk.com/articles/prescribing-the-future-of-actionable-intelligence/.

Staff4. 2018, May 24. *Artificial intelligence: Not science fiction, but science reality–Part 3*. Retrieved from Meritalk: https://www.meritalk.com/articles/artificial-intelligence-not-science-fiction-but-science-reality-part-3/.

Surber, R. 2018, February 21. *Artificial intelligence: Autonomous Technology (AT), Lethal Autonomous Weapons Systems (LAWS) and Peace Time Threats*. Retrieved from ICT for Peace Foundation: https://ict4peace.org/wp-content/uploads/2018/02/2018_RSurber_AI-AT-LAWS-Peace-Time-Threats_final.pdf.

Thompson, M. 2013, April 02. *Costly flight hours*. Retrieved from Time: http://nation.time.com/2013/04/02/costly-flight-hours/.

Times, N. Y. 2018, August 12. *How the drone attack on Maduro Unfolded in Venezuela | NYT – Visual Investigations*. Retrieved from YouTube: https://www.youtube.com/watch?v=EpFNCqCwVzo.

Torres, M. J. 2018, June 06. *America may need to adopt China's weapons to win tech war*. Retrieved from The Star: https://www.thestar.com.my/tech/tech-news/2018/06/06/america-may-need-to-adopt-chinas-weapons-to-win-tech-war/.

Verma, L. C. 2018, March 26. *How China is moving towards 'Intelligized Warfare'*. Retrieved from Geospatial World: https://www.geospatialworld.net/blogs/how-china-is-moving-towards-intelligized-warfare/.

Artificial Intelligence in Autonomous Weapon Systems

Stanislav Abaimov and Maurizio Martellini

1 Introduction

Automatic military weapon systems have been in use for decades, monitored and controlled by human operators in detection of targets, delivery, and final use of force. With autonomous weapon systems (AWS) entering the modern warfare, these processes can take place without human supervision and decision making can be fully delegated to a machine. Through computer systems and AI-enhanced software implementing command and control, the advancements in smart systems and sensor technology enabled weapons' independent activity and battle decision making, fostering their autonomy.

AWS are now presented in all military domains (land, air, water and space). Among them are the Aegis Combat System, the Phalanx Close-In Weapon System, X-47B Unmanned Combat Air System, Patriot, MANTIS, Guardium, IAI Harpy air defence systems, to name a few. Autonomy in their functions is growing, and already includes navigation, surveillance, target detection and decision-making on the use of lethal force. The next generation is predicted to be empowered with self-evolution capabilities.

Advantages and threats of the AWS development and deployment are vigorously discussed in international community among academia, technical experts, politicians, military professionals. Those systems have advanced military capacities and a stealthy character, require minimal supervision, reduce exposure of combatants, and are operational 24/7. Yet, in addition to routine physical malfunctions and interruptions of power supply, they are overly vulnerable to cyber and intrinsically new

S. Abaimov (✉)
University of Rome Tor Vergata, Rome, Italy

M. Martellini
Università degli Studi dell'Insubria, Como, Italy
e-mail: maurizio.martellini@fondazionealessandrovolta.it

M. Martellini, R. Trapp (eds.), *21st Century Prometheus*,
https://doi.org/10.1007/978-3-030-28285-1_8

AI-specific attacks. The AI-enhanced software is far from being fully reliable and trustful; AI anomalies and adverse effects, alongside vulnerabilities in operating systems and support frameworks, create a new AI-related threat landscape which is magnified by the lethal character of weapons.

AWS use entails ethical and legal implications related to human rights, international humanitarian law (IHL), and need for development of norms and regulations for AWS design, development, training, testing and deployment. Debates are ongoing on AWS compliance with the IHL principles, e.g., whether AWS would ever be able to make a distinction between combatants and non-combatants, respect the principle of proportionality, take precautions in attack, prevent from unnecessary suffering. The discussion on weaponization of increasingly autonomous technologies is high on the agenda of the United Nations, and the community appeals to ban "killer robots" are growing.

With major national powers investing billions of dollars into AI research and bolstering cyber offensive capacities, the AI arms race is skyrocketing, also benefiting from the dual use character of cyber arms and publicly available AI source codes. Further uncontrollable development of AI-based military technology and stockpiling of cyber arms will only expand the threat landscape and make danger irreversible.

This paper reviews the AI impact on autonomy and its major criteria, explores cyber vulnerabilities in autonomous military technologies, highlights critical issues of the AI use in AWS, deliberates on incorporation of ethical principles into development of technologies, reveals legal complications and consequences of AI arms race, predicts future challenges. It also provides some potential crisis scenarios.

2 Intelligence, Actuation and Human Role in Autonomy

> *Superintelligence may be the last invention humans ever need to make,* - Nick Bostrum, Director, Future of Humanity Institute. *The development of full AI could spell the end of the human race*[1] - Stephen Hawking

Autonomy in weapons is now in strategic plans of the leading military powers as the deployment of autonomous technologies has brought feasible advantages. They reduced the number of combatants in the field and operating costs, improved the combat time-critical performance and accuracy, and enabled missions previously considered impossible.

The Cambridge dictionary defines autonomy as "the right of an organization, country, or region to be independent and govern itself".[2] Applying the same approach

[1] Quoted after Cellan-Jones, Rory (2014).

[2] Autonomy, Cambridge Dictionary, https://dictionary.cambridge.org/dictionary/english/autonomy

to technical systems, autonomy might be considered as their capacity to govern themselves and act independently from humans – "sense", take decisions and act. Hence, the basic criteria of autonomy would be intelligence, capability to implement decisions, and the level of human involvement in both of them.

2.1 *Artificial Intelligence Functionality*

Intelligence in the technical systems measures their ability to determine the best course of action to achieve its goals in a wide range of environments (UNIDIR 2018). It is the development of computer science, and eventual creation of Artificial Intelligence, that granted such potential to technical systems and enabled their autonomy.

Artificial intelligence (AI) is the capacity of software to rely on and take decisions based on the provided knowledge rather than on the predefined algorithms. This term is also referred to a complex technological phenomenon that implies the ability to extract new knowledge from the available data, make decisions and enable action to reach the set goals.

In computer science, AI research is perceived as the study of intelligent agents, that is "any device that perceives its environment and takes actions that maximize its chance of successfully achieving its goals" (Poole et al. 1998). Initiated more than half a century ago and having survived through several AI "winters", the AI research, in cooperation with robotics (mechanical, electrical and control systems engineering, and computer science), is contributing to creation and enabling intelligence in the new generation of autonomous technologies.

Depending on the potential and scope of functions, AI is currently subdivided into three big categories:

- Artificial Narrow Intelligence, that simulates specific human skills and is used for specific tasks, e.g., in search engines, in mobile phones, targeted advertisements, etc.;
- Artificial General Intelligence, that is equal to human intelligence and able to address tasks and solve problems on the human level;
- Artificial Superintelligence, that is able to resolve tasks beyond human capacities and yet to be created in the future.

While the Artificial Narrow Intelligence is fully in action and outperforms humans in certain tasks, opinions vary in relation to the timeline needed for achieving the human-level AI, as currently even the most accurately trained artificial neural networks are very far from reaching human brain capacities. The idea of so called singularity – the moment when computers match humans in intelligence – has been in discussion for more than a decade (Kurzweil 2005), and a number of recent surveys predict human-level AI around 2040–2050, rising to a nine in ten chance by 2075 (Muller and Bostrum 2013), with a 50% chance of AI outperforming humans in all tasks in 45 years and of automating all human jobs in 120 years

(Grace et al. 2018). Applied to the military doctrine, the above prediction refers to the autonomous weapon system (AWS) capability to fully act as human combatants and take independent decisions.

Among multiple activities performed with AI application there are data procession and analysis, knowledge and information storage, learning and planning, monitoring and evaluation of activities, natural language generation and communication with humans and non-humans, speech and objects recognition, biometrics, etc. AI engages, among others, logical and symbolic reasoning, knowledge-based and search tools, probabilistic methods, neural networks, evolutionary algorithms, embodied intelligence (morphological computation, sensory-motor coordination, developmental embodied cognition for body/brain/environment adaptation) (Cangelosi and Fischer 2015), which have also been attempted to use in the AI research for weapon systems.

Intelligence enables systems to collect data about the environment and autonomously generate information, e.g. in the latest generation of surveillance and reconnaissance aerial systems (e.g., Shield AI,[3] a tactical UAS, currently under development, that can generate 3D maps using cameras, lasers, inertial and ultrasonic sensors).

Autonomous systems rely on embedded electronic components to perform the computations required for complex system actions and events. Their purpose, design, mechanical components connected to electronic hardware and a number of linked systems shape the way the software is implemented. Sensors (cameras, radar and acoustic sensors, heat sensors, infrared sensors, etc.) enhance adaptiveness and ensure high performance through a more refined perception of environment enhancing navigation, computer vision, threat detection and other vital functions. In 2018, the development of silicon processing units was aimed at their performing machine learning tasks more efficiently and faster than Central Processing Units or Graphical Processing Units. Google announced the open beta of its second-generation Tensor Processing Unit (TPU), Amazon has been reportedly developing a dedicated TPU for its Echo smart speaker, and ARM announced its own AI hardware (Frank 2019).

Autonomous action is enabled through mathematically modelled tasks, which act well in a predictable environment. In a more complex environment when the model cannot be predicted, another approach – Machine Learning (ML) – is used, which is now a critical component in the development of the autonomous systems' underlying software of any kind.

Intelligent applications are enabled through various ML methods without explicit definition of the problem or equations. In ML process computer systems adjust algorithms and statistical models to effectively perform a specific task, relying solely on patterns and inference. Abstract solution strategies are extracted from an extensive dataset, then they are supplemented, expanded and improved using data known and unknown to the system. Finally, the perfected solution strategies

[3] Official website at https://www.shield.ai/

(models) are applied to specific problems. This approach is currently the most widely used in developing pattern recognition mechanisms, such as autonomous or automated target recognition (ATR) software, vision-based guidance systems, intrusion detection software, and electronic warfare systems. It enhances ability to work with unknown correlations, extrapolate information with large datasets, ensures flexibility in classifying unknown inputs, battlefield intelligence as a whole, and consequently, augments command and control functions, navigation, perception (sensor intelligence and sensor fusion), interoperability, adjustment to complex environments.

ML can be supervised, semi-supervised, and unsupervised. In supervised learning, the patterns are provided with the desired outcomes. Systems scan databases and compare them with the example pattern, e.g. software for target recognition. In semi-supervised learning, a neural model is pre-trained with a supervised method (labelled data), and further used for unsupervised learning (unlabelled data).

In unsupervised learning, the machine must classify raw data into types (labels) to optimize the probability of generating the best possible solution without provided desired outcomes. The machine learns from unlabelled data and self-identifies patterns in it. This learning is perceived to be ideally suited for a system to orient in unknown environment and take decisions. At the same time, it is the most challenging method and least predictable.

ML modules (e.g., neural networks, support vector machines, etc.) are trained using large datasets before the initial deployment, adjusted and optimised to have the highest accuracy and precision for a particular implementation or task. Their initial training requires high computational resources, but less computing power once trained. The pre-trained models can be deployed in less powerful devices and be available to a larger number of users, including through open-source libraries, e.g. Tensorflow, "Neural Network Libraries"[4]; and "Apache MXNet".[5] As machine learning is a complex technology, post-deployment learning is only possible in advanced systems for research and experimental purposes, raising the issue of routine systems upgrade to catch up with the technological advancements.

Even though machine learning is a raising and promising field, none of the methods can guarantee perfect accuracy in performance for the underlying technology and cannot be trusted with the final decision. With the undoubtful AI advantages, the new technology has its limits and opens the systems to intrinsically new AI-specific threats.

The ML implementations were majorly inspired by the human neural networks, which enormous complexity can be captured only superficially. The so-called "philosophy of mind" has acknowledged a dichotomy between the "knowing" subject and the intangible sphere of the subconscious, dreams and archetypes. The problem at this point is whether the AI of tomorrow can be regulated only by deterministic

---

[4] Neural Network Libraries by Sony, https://nnabla.org

[5] Apache MXNet, library for deep learning, http://mxnet.incubator.apache.org/index.html

laws based on meta-data. Therefore, for example the challenge would be not so much to ask AI's designers to follow human ethical principles, but whether AI itself can become, after the forecasted "singularity point", a new ethical subject by itself.

2.1.1 AI Limitations, Risks, Adverse Effects

As AWS operate in a military context, the operators expect their safety and reliability while acting in complex, uncertain and adversarial environments. With the current status of ML and AI-enhanced solutions, there is no overly certainty, which results in limitation of autonomy. Accidents can occur from the poor design of AI systems, incorrectly set tasks in terms of the mathematical problem and solution, non-compliance with programmable rules and logic, poorly defined task specifications, incorrect modelling of operating environments, or insufficient training.

A year of cascading scandals

As for much of the tech industry, 2018 has been a year of reckoning for artificial intelligence. As AI systems have been integrated into more products and services, the technology's shortcomings have become clearer. Researchers, companies, and the general public have all begun to grapple more with the limitations of AI and its adverse effects, asking important questions like: how is this technology being used, and for whose benefit?

This reckoning has been most visible as a parade of negative headlines about algorithmic systems. This year saw the first deaths caused by self-driving cars; the Cambridge Analytica scandal; accusations that Facebook facilitated genocide in Myanmar; the revelation that Google helped the Pentagon train drone surveillance tools; and ethical questions over the tech giant's human-sounding AI assistant. The research group AI Now described 2018 as a year of "cascading scandals" for the field, and it's an accurate, if disheartening, summary.

The Verge Tech Report Card 2018[6]

Processing machine learning algorithms is **computationally very intense**, and the needs for data accuracy require more complex models. The training data has a large volume and for many tasks, including targeting, the lack of high-quality (or any at all) training datasets remains a fundamental limiting factor to development higher levels of artificial intelligence.

Efficiency in processing advanced algorithms is another challenge. Tasks and commands to the system should be pre-processed into a logical sequence of actions before using them as a training dataset or an input command. Advanced algorithms

[6] Quoted from Vincent (2018).

are computationally intense and slow. It takes time and high computational power to convert new information into digital patterns, classify and add to the knowledge base, and the system needs training on a large number of similar situations to learn and come up with the rules. Thus, logical steps in the software should be optimised to ensure uninterrupted performance with the highest reaction speed.

The next challenge is the **reliability and predictability** of AI-enhanced systems. With the known inputs and feasible outputs, the internal "transformations" are still unknown and not well understood.

For the system to be predictable, it should be properly modelled and specified as a mathematical problem and solution in the design process. In addition, its perceptual level should correspond to the required task. However, higher "intelligence" leads to taking decisions based on probability-based reasoning, which allows the system to surpass the rules set by developers, increasing unpredictability and uncertainty. Encoding proportionality and precaution into an algorithmic form remains a challenge, and there is no consensus yet on whether it will ever be possible.

ML is also able to improve the software development for automated and automatic target recognition (ATR), that will allow target recognition without human supervision. However, systems' inability to differentiate between military and civilian objects (e.g. an armoured vehicle or a bus) raises strong concerns over predictability of such systems.

Autonomous military systems are aimed at working in dynamic environments and implement **complex tasks**. Modern AI applications are yet unable to ensure flexibility and rapid adaptability to unknown environment and events. The machine intelligence can only deduce complex mathematical formulas, while full autonomy requires capacity for inductive and adductive reasoning. There are many challenges related to achieving goals through a set of multiple actions. For example, it is challenging to develop image recognition systems able to distinguish an enemy based on behaviour or actions (e.g., Automated Image Understanding thrust) (Office Naval Research n.d.), or system that tracks movement of large underwater objects (e.g., submarines) using satellite images and computations of wave patterns.

More advanced software poses **compatibility** issues in relation to hardware and power management. Large systems are able to function with a heavier hardware and higher resource power, but they are more expensive and less affordable. Small low-cost systems in swarm operations are unable to perform complex calculations and cannot be engaged for long distances and extended timeframe.

The **level of extended sensors' perception** of the designated environment is limited and does not allow yet to perform qualitative evaluations. Those activities requiring a more elaborative approach, like distinction, proportionality, precaution in attack, etc., are impossible for implementation and humans still need to supervise them.

In the **machine-human interaction** both under trust and over trust are equally dangerous, and lead either to decreased efficiency or lose of supervisory control. In **machine-machine interaction** the programmes can communicate for collaboration, and, if not secured, could also malfunction with severe consequences before human intervention.

The **training** of the AI modules takes place in the "offline" environment. However, the system networks. It might lead to unpredictable consequences (e.g., irrelevant data contaminates the dataset, cyber attack interrupts or overrides the training cycle, etc.).

AI capability to learn, analyse, communicate and take decisions require new type of **testing, validation, verification and certification** of the systems whose autonomy it enables. Traditional approaches to test the system's performance, either in the field or through computer modelling and simulation, and verify the system reliability in all potential states fail and verification slows down the release and further development significantly. Each time the system learns something new, e.g. online, through interaction with humans or machines, or dynamic environment, it changes its knowledge base (i.e., automatic self-reconfiguration), and the system needs to be revalidated (Office of the Assistant Secretary of Defense for Research and Engineering 2015). In addition, all standards imposed on the systems to ensure its compliance with technical and safety requirements will limit their evolution and make them outdated immediately after final certification.

Another challenge is evaluation of various possibilities the system might have to perform the task: either as a single action, a chain of decisions and actions, or multiple options. The latter requires a more enhanced programming and computer power to generate the most optimal solution. Each autonomously performed task is built on the success criteria of the previous function. This dependency creates a possibility of a chain reaction, in case of misclassification, as it is impossible to predict all that can provoke the system failure. One of the solutions is to create the most realistic scenarios through which the system could be trained and checked using mathematical methods, computer simulations, and previous experience. Multiple accountability frameworks are needed to test functionality and safety of autonomous technologies.

AI-related cyber vulnerabilities add additional risk factors. Thus, data poisoning is considered to be one of AI-specific attack vectors already in action. It occurs when malicious knowledge is injected into the training examples of the system to create unpredictable errors. Image recognition algorithms are susceptible to "pixel poisoning" attack that leads to classification problems (Chen et al. 2017). Algorithms trained using open-source data could be particularly vulnerable to this challenge as adversaries attempt to corrupt the data that other countries might use to train algorithms for military purposes. Those sophisticated cyber attacks could also lead to the exploitation of defence algorithms trained on more secure networks. With the unsupervised learning the system may misclassify certain events and continue learning in an unintended way.

Public adversarial data, fake news and disinformation processes are another area of deliberate or unintentional confusion. Countermeasures are currently under research and aimed at improving machine learning techniques to distinguish between true and false data (e.g., in large scale intelligence gathering).

With Adobe's application concept "Project Voco"[7] it is possible to rapidly alter an existing voice recording to include words and phrases that the original speaker has never said. (As of February 2019, Adobe has yet to release any further information about a potential release date.)

Strong concerns are being raised against AI face surveillance. The American Civil Liberties Union conducted a test of a facial recognition tool through matching it with all 535 members of Congress against 25,000 mugshots. The software incorrectly generated 28 false matches Brandom 2018).

Emerging AI-enhanced intrusion detection systems can already distinguish cyber threats accurately, but they cannot be fully reliable as they frequently raise false alarms. At the same time, a new generation of AI malware, like DeepLocker, demonstrates enhanced evasive capabilities. It hides the malicious payload in carrier applications, such as a video conference software, to avoid detection by most antivirus and malware scanners. Automated techniques make it easier to carry out large-scale attacks that require extensive human labour, e.g. "spear phishing," which involves gathering and exploiting personal data of victims. Researchers at ZeroFox demonstrated that a fully automated spear phishing system could create tailored messages on Twitter based on a user's demonstrated preferences, achieving a high rate of clicks to a link that could be malicious (Seymour and Tully 2016).

The 2018 report "The Malicious Use of Artificial Intelligence: Forecasting, Prevention, and Mitigation" (Brundage et al. 2018) concludes that AI and ML are altering the landscape of security risks. The malicious use of AI could threaten digital security (e.g., through criminals training machines to hack or socially engineered victims at human or superhuman levels of performance), physical security (e.g., non-state actors weaponizing consumer drones), and political security (e.g., through privacy-eliminating surveillance, profiling, and repression, or through automated and targeted disinformation campaigns).

Furthermore, another serious concern related to the malicious use of AI and meta-data, combined with the spread of the social networks, is the problem of false data, generated using ML (knows in slang as "deepfakes") and new disinformation warfares. In fact, both of these realities today are used massively in non-declared wars and in highly asymmetric conflicts. Attempts have already been made to form the legal acts (H.R. 3230). But connected to this problem there is also that of how it is possible to formulate and regulate, if any, the geopolitics of, say 40–50 years from now, i.e. at the time of the so-called third revolution of the AI and whether the Carl von Clausewitz strategists of those times could be AI themselves. Simply put, how will be the geopolitics of AIs versus other AIs?

The AI vulnerabilities and lack of full understanding of its functioning leads to the conclusion, that the AI technology is still limited and immature, the AI-enhanced weapon systems cannot be fully trusted in autonomous operation and need human supervision to mitigate risks.

[7] See (BBC 2016).

2.2 *From Decision Making to Action*

When the loom spins itself and the lyre plays by itself, man's slavery will be at an end. - Aristotle[8]

The system technical characteristics, the physical complexity allowing functionality. is considered to be the second criteria for its autonomy. The decision taken by the system "intelligence" is further executed through actuators and end-effectors.

Three major ways through which the systems implement functions – automatic, automated and autonomous – are quite often used inconsistently. Harbers et al. (2017) suggest that "systems that perform simple tasks independently are usually called automatic or automated, rather than autonomous. The term "automatic" is often used for systems that perform simple tasks, e.g. ventilation controls or teleoperated explosives disposal robots. "Automated" usually refers to rule-based systems such as a programmable thermostat or a diagnose support system. The term "autonomy" is typically reserved for systems that execute some kind of self-direction, self-learning or emergent behaviour".

The US Defence Science Board, considers "autonomy [to be…] a capability (or a set of capabilities) that enables a particular action of a system to be automatic or, within programmed boundaries, "self-governing" (DOD 2012).

Cummings (2016, 8:47) defines an automatic system as the "one that acts according to a preprogramed script with defined entry/exit conditions for a task." An autonomous system is "one that independently and dynamically determines if, when, and how to execute a task" (Ibid., 9:23).

The IEEE Global Initiative on Ethics of Autonomous and Intelligent Systems highlights that "the term autonomy in the context of AWS should be understood and used in the restricted sense of the delegation of decision-making capabilities to a machine" (IEEE 2017a, b). It also supports the working definition of AWS offered by the International Committee of the Red Cross (ICRC) and proposes that it be adopted as the working definition of AWS for the further development and discussion of ethical standards and guidelines for engineers. The ICRC defines an AWS as: "any weapon system with autonomy in its critical functions. That is, a weapon system that can select (i.e. search for or detect, identify, track, select) and attack (i.e. use force against, neutralize, damage or destroy) targets without human intervention".

The system is recognised as "fully autonomous" when within a certain timeframe it can independently implement functions required to reach a set goal following the independently made decisions based on collected information.

[8] Aristotle as quoted in Partridge and Hussain (1992).

Mission autonomy

Existing systems that, once launched, navigate in complete autonomy, with little or no direct human supervision, can be divided into the following three categories.

1. *Aerial, land and maritime systems that are deployed to conduct pre-programmed manoeuvres in known and semi-structured environments. Examples include the Amstaff, a tactical unmanned ground system (UGS) developed by Automotive Robotic Industries (Israel), which is capable of conducting perimeter protection operations autonomously.*
2. *Unmanned systems that are intended to conduct long-term ISR missions in an environment where communications are difficult (e.g. underwater).*
3. *Missile systems and unmanned combat systems that are intended to strike targets in communication-denied environments.*

The 2017 SIPRI Report (SIPRI 2017a)

The US DOD Roadmap for Unmanned Systems, 2017–2042, forecasts that "the future of unmanned systems will stretch across the broad spectrum of autonomy, from remote controlled and automated systems to near fully autonomous, as needed to support the mission" (DOD 2018a). A summary of the future path for the four key enablers for autonomy is shown in Table 1.

Modern AWS include air defence systems, active protection systems, robotic sentry weapons, guided munitions, and loitering weapons. They can implement various functions with full or partial autonomy, which are enabled through the whole operation or at some parts of it: navigation, selected stages in targeting (target detection, identification, tracking, prioritization, target engagement, confirmation), maintenance (refuelling, power management, malfunction detection), communication with other systems and humans, combat intelligence (e.g., mapping

Table 1 Comprehensive roadmap for autonomy

		2017	2029	2042
		Near-term	Mid-term	Far-term
Autonomy	Artificial intelligence/machine learning	Private sector collaboration Cloud technologies	Augmented reality Virtual reality	Persistent sensing Highly autonomous
	Increased efficiency and effectiveness	Increased safety & efficiency	Unmanned tasks, ops Leader-follower	Swarming
	Trust	Tasking guidance and validation, ethical requirements for human decisions		
	Weaponization	DoD strategy consensus LAWS assessment	Armed wingman/teammate (human decision to engage)	

multi-dimensional environments). The larger time of independent operability, enhanced sensors, processing power, communications, and freedom to move either alone or in multivehicle mission, are valuable assets in autonomy.

The most critical functions that raise overall concerns are automatic targeting, targets prioritization and target engagement. The software enabling these functions, known as "automatic or automated target recognition (ATR) software", relies on pattern recognition using logical formulas or, in the latest years, machine learning. A software programme searches for targets, compares the signals detected by sensors against a list of signatures (predefined target models), selects through ATR and affects them depending on the set goal. Currently, this mostly include large easily identifiable material objects (e.g., communication stations, air bases, tanks, missile batteries). The software follows simple criteria: identification of submarines via acoustic signature, tanks via shape and height, missiles via velocity and radio-frequency emission.

ATR components and software are used in guided munitions, loitering weapons, air defence systems and within human-operated weapon systems (to assist in locating and engaging the enemy targets that are beyond the visual range or fast moving). The ATR software can only detect, classify and engage target types that have been defined by the developers before deployment in the field or after a software update.

Weapon systems with some autonomy in critical functions, like close-in weapons (CIW) systems, have sensors, fire control systems, communication links, and consist of a number of physical platforms. These systems can perform basic diagnostics of faults and limited power recharge. E.g., UAS aerial refuelling is performed only in most advanced systems, like Northrop Grumman's X-47B (USA), while the SW-4 RUAS, Leonardo-Finmeccanica (Italy) and Pzl Swidnik (Poland), can only detect malfunctions in engine, vortex ring, communication, etc. Experiments are ongoing with identical modular robotic systems, that can regroup and change their structure to serve various tasks. When the failure is detected in such independent modules, one can be ejected and replaced with another fully functional module.

In 2017 at least 154 systems were able to support some, if not all, of the steps of the targeting process, from identification, tracking, prioritization and selection of targets to, in some cases, target engagement (SIPRI 2017b). Robotic kits are developed to retrofit the non-robotic vehicles: Autonomous Aerial Cargo Utility System (AACUS) adds autonomous mode to any helicopter; Autonomous Mobility Appliqué System (Lockheed Martin) adds a self-driving capability to any military logistical vehicle. Furthermore, an on-board autonomous component would translate into less data transferred to the command and control, minimal communication uptime, and a reduced number of human operators and analysts required to oversee real-time operations.

Mass-deployment of AWS requires a group approach, a collaborative autonomy, where multiple systems are coordinating their actions to achieve a set objective. This implies a software architecture that guides every system in the group of heterogeneous systems (e.g., a tactical formation of UAVs and UGVs), or a swarm of identical or similar devices or vehicles, and thus operates as a single system. Such software architecture defines a role for each system within a group, while managing a coordinated behaviour, and achieve effects that each system individually could not achieve, or achieve ineffectively.

In defensive missions, AWS protect vessels, vehicles, or ground installations against incoming projectiles. At least 89 countries have deployed or are developing such systems. In offensive missions AWS can engage only specific types of targets and only under certain conditions. Loitering weapons are the only offensive weapon system type that can detect, classify, and engage targets autonomously, even though the geographical area and type of target are still set by human operators. E.g., Harpy, Harop, Harpy NG, Orbiter 1K "Kingfisher" (SIPRI 2017a). Other loitering weapon systems have only been used for research and have never been deployed on the battlefield (e.g., Low Cost Autonomous Attack System (LOCAAS); Non-Line-of-Sight Launch System (NLOS-LS); Taifun/TARES; Battlefield Loitering Artillery Direct Effect (BLADE)). Robotic sentry weapons, stationary or vehicle-mounted gun turrets, can automatically detect, track and engage targets. They operate similar to CIWSs but are usually employed as anti-personnel weapons. As per 2019, there are only three types of sentry AWS: Samsung's SGR-A1, Raphael's Sentry Tech, and DODAAM's Super aEgis II. Those are the only weapon systems that use ATR to recognise if the target is a human.

In military systems autonomous communication and coordination is limited to simplified exchange or relaying of sensory data. Collaborative autonomy is actively researched but it cannot be used as a reliable tactical capability. In comparison to other coordinated activities (target distribution, optimal surveillance over large areas, attack reaction, crowd control, etc.), the most technologically researched and implemented form of machine-machine collaborative interaction are coordinated mobility and logistics. Programming mobile systems to autonomously move in an optimal formation is a relatively simple and well-understood algorithm (i.e., keeping a fixed distance and position in respect to each other). Despite the complexity of environment (especially for ground vehicles), there is an increasing number of systems under development that are capable of moving in a formation (e.g., UTAP-22 Mako combat UAV).

Examples

An unmanned vessel Sea Hunter (DARPA's Anti-Submarine Warfare Continuous Trail Unmanned Vessel program) is designed to autonomously detect and track enemy submergible vessels and serves as a research testbed.

The X-47B Unmanned Combat Air System is a stealth UAV that is able to take off from an aircraft carrier, attack the enemy, and return to the carrier. It is considered to be one of the first fully autonomous systems under development (Schroeder 2017).

Combat UAVs like the MQ-9 Reaper, MQ-9 Predator B and Watchkeeper are used for reconnaissance missions and targeted airstrikes in combat zones.

Autonomous artillery equipment like the "Dragon Fire II", automates loading and ballistics calculations providing fire support.

The air defence systems use "hard-kill" (e.g., projectiles) or "soft-kill" measures (e.g., electronic warfare) affecting the trajectory, sensors and behaviour of the incoming threat. Active protection system (APS) is a weapon system that is used to protect vehicles against projectiles (missiles or rockets) the same way as air defence systems. The SIPRI dataset identifies 17 different APS models[9]. In addition, some of those systems (e.g., Trophy APS), are able to calculate the location and origin of the threat. Among others, one system is known to have a reaction time of less than 1 millisecond – the Active Defense System (Rheinmetall Defence).

In comparison with large military machines, miniature unmanned systems do not require a large support infrastructure, and can be developed using consumer components or pre-assembled devices from the commercial sector. The availability of robotic platforms (e.g., recreational drones) will only grow in the future.

2.3 Human-Machine Relations in Autonomy

The last criteria we would like to consider refers to the level of human presence both in the decision taking and its actuation, the human-machine relations in the performance of tasks.

The pursuit of autonomy is changing the role of human operators by creating new ways of effective human-machine interaction where cognitive capabilities and qualitative judgements of humans and fast computations and reaction speed of machines can complement each other.

Human can be always present in the chain of command, and the systems act with a "human-in-the-loop" selecting and engaging targets under a human command (e.g., Terminator, Lockheed Martin; SkyStriker, Elbit Systems; Warmate, WB Electronics; XQ-06 Fi, Karal Defense; CH-901, China). Those systems are typically used at national borders (e.g., sentry weapons).

Systems with a "human-on-the-loop" function act with a human supervision and oversight, and have a capacity to intervene and terminate the activity, including in case of the system failure (e.g., Samsung SGR A-1).

In a "human-out-of-the-loop" mode (fully autonomous), AWS can operate without having a human operator after activation, selecting and engaging targets without human interaction. This category includes air defence systems and active protection systems.

When the intervals during which the human is not able to intervene in the robot's behaviour become smaller, the difference between in-the-loop and out-of-the-loop become less clear, a term of "sliding autonomy" might be used.

AWS fitted with lethal capabilities are known as lethal AWS (LAWS). Some LAWS are triggered independently as active defence systems, such as a radar-guided gun to defend ships that have been in use since the 1970s (e.g. the US

[9] SIPRI databases, https://www.sipri.org/databases

Phalanx CIWS[10]). Such systems can autonomously identify and, in seconds, attack oncoming projectiles, aircraft and surface vessels according to criteria set by the human operator. Similar systems exist for tanks (e.g., the Russian Arena APS, the Israeli Trophy APS, and the German AMAP-ADS). Missile defence systems (e.g., Iron Dome) also have autonomous targeting capabilities.

The 2016 publication "Killer Robots: Lethal Autonomous Weapon Systems: Legal, Ethical and Moral Challenges", claims that "though LAWS in the true sense are not yet available, at least 44 countries, including China, France, Germany, India, Israel, South Korea, Russia, the UK and the Unites States, are developing such capabilities" (Jha 2016).

In the "Army of None: Autonomous Weapons and the Future of War", Paul Scharre concludes: "There are at least 90 countries that have drones today, and 16 countries are counting that they have armed drones, including many non-state groups. A lot of them are remotely controlled or teleoperated… There are at least 30 countries that have these sort of automatic mode(s) – it's kind of wartime modes – that they can turn on these systems that are either land-based, air-and-missile defence systems, or they're on ground vehicles or ships, that allow this automatic protection bubble to kick in" (Scharre 2018).

The levels of autonomy differ and there is no one fit for all scale of the level of autonomy as systems perform various tasks and under various contexts. During the development phase an acceptable level of risk adjustable autonomy is introduced into the system to match multiple factors. They include the narrow tasks (offence or defence), types of targets (humans or object), type of force (kinetic or non-kinetic), type of environment, freedom of movement, channels of human supervision, time-frame of independent action, system level of intelligence, its reliability and predictability. While rapid evolution in autonomy of non-critical functions is acceptable (e.g., in navigation, on board sensors), all attempts to grant autonomy to critical elements of the target selection, tracking and attack (e.g., image processing and classification, trajectory planning) should be under strict consideration, requiring enhanced security measures and control.

All the above shows that autonomy cannot be considered as an overall general characteristics of a weapon system. The three criteria considered may vary in AWS, also, can be implemented fully or partially in various functions of the same system, and engaged to various extent in different operational contexts. It is important to consider which decisions the system can take autonomously and which functions can be implemented without a human intervention before coming up with a judgement on its ban. A more flexible approach is needed to be able to provide their definition and classification not only to the system as a whole, but also to its functions, for easier reference and engagement by non-technical experts.

The advantages of autonomy in weapons stimulate research and development, and speed up their production. Among those advantages are cost-effectiveness through reduced operating costs and personnel requirements (though arguable in

[10] Close-In Weapons System.

many cases due to procurement and high maintenance costs), functionality surpassing human capacities, advanced offense capacities and force multiplication, enhanced safety and less risks for operators, adaptive behaviour, stealth character, etc. At the same time, they have many challenges, including high research and development cost, deployment and use in unclear legal environment, unpredictable behaviour of the AI-enhanced software, general and AI-specific cyber vulnerabilities.

3 AWS Cyber Vulnerabilities

By the military domain, AWS can be classified into ground, naval, air, space systems. They can be non-lethal (surveillance and rescue operations) and lethal (offensive and defensive operations), teleoperated, fully or semi-autonomous; they can operate as a single unit with self-contained autonomy, swarm and networked systems with distributed autonomy, or components with autonomous capabilities.

These powerful physical weapons become attractive targets of adversaries and, with their autonomy granted through computerized technologies, can be disabled or taken over through cyber attacks. The sophisticated malware can already intelligently perform undetectable covert actions, related to data theft, disruption of networks and computers, damaging control systems: it can adapt to environment, counter the modern anomaly-based intrusion detection systems (Dutt 2018) and change behaviour as it spreads (TrendMicro 2018). It includes AI-enhanced self-replicating and uncontrollable viruses, trojans, worms, etc., which can rightfully act as intelligent autonomous agents. Benefitting from the dual use character, the cyber tools used for reconnaissance, scanning, privilege escalation, assault and covering tracks, can be equally used for both defence and offence of weapon systems.

With the sophistication of technologies, both offensive and defensive capabilities grow. However, offense needs only one weakest link, hence they are more task-specific and more advanced. The new generation cyber attacks use autonomous malware, intelligent evasion techniques, and covert data exfiltration techniques.[11] AI-empowered machine learning, deep learning, and neural networks are already employed to find and exploit vulnerabilities, while AI-enhanced cyber arms learn via spreading in cyberspace.

AWS, as any complex system, comprises numerous electronic components, that are linked to the main switch. If one of the subsystems is compromised and the attacker gains access to the main switch, they gain access to all the systems. Critical electronic components can be a positioning system, battle management system, internal communication system, weapon system, electronic warfare system, vehicle management system, etc. (Kott et al. 2018).

[11] The next paradigm shift: AI-driven Cyber-Attacks, DarkTrace, 2018, https://www.darktrace.com/en/resources/wp-ai-driven-cyber-attacks.pdf

GPS spoofing
One of the most known vulnerabilities of long-range systems is the reliance on Global Positioning System (GPS) guidance (or any other kind of satellite navigation), which makes them vulnerable to jamming attacks (Phrack 2002).[12] This vulnerability forced the development of systems that can operate in GPS-denied environments (e.g., GPS anti-jamming protection, non-GPS-based guidance systems). In addition to GPS jamming, enemies can also use other attacks such as spoofing and cyber-attacks. GPS spoofing attacks take the attack one step further and instead of disabling access to coordinates, replace them, guiding the system to a different location.

One of the emerging threats is the combination of the cyber and electromagnetic attacks. AWS can be affected by electromagnetic pulse (EMP) weapons, the transmissions can be jammed or intercepted.

Electromagnetic Attacks
Electromagnetic pulse (EMP) and cyberattack…, operating in tandem, can disable not just a significant portion of the electrical grid and critical infrastructure, but also the network-centric military response to such an attack. If a high-altitude EMP attack were paired with both a large-scale cyberattack and a biological attack, the resulting challenge to the interagency could surpass anything the interagency is currently structured or equipped to respond to.

Patricia Rohrbeck, Concurrent Biological, Electromagnetic Pulse and Cyber-attacks: The Ultimate Interagency Response Challenge (Rohrbeck 2017)

Cyber vulnerabilities can occur in AWS at all stages of supply chain – design, development, production, delivery, installation, maintenance, and are used as additional attack vectors, access points and platforms to launch cyber attacks. The challenge is aggravated by the current production model of weapon systems, which is dominated by hardware considerations, and the consequent later software adjustments to the already designed platform with sensors, communication links and modules, processing power, etc. This often results in software flaws, hidden vulnerabilities, logical conflicts, or inability to defend certain flaws.

"Most people are thinking about autonomous weapons as physical weapon systems," says Kerstin Vignard, deputy director for the United Nations Institute for Disarmament Research (UNIDIR). "Yet, for the most part, increasing

[12] MoD's tests will send satnav haywire so take a road atlas, 6 June 2007, https://www.dailymail.co.uk/news/article-460279/MoDs-tests-send-satnav-haywire-road-atlas.html

autonomy – in cyber capabilities and in conventional weapons – is intangible. It depends on code." By isolating these discussions, organizations risk overlooking how physical weapons and cyber weapons could soon directly impact one another (Hsu 2018).

In addition to targeted exploits, a new threat comes from exploits against multiple vulnerabilities, known as combined threat (Durbin 2018). Combined threat or a blended threat is a software exploit which in turn involves a combination of attacks against multiple vulnerabilities. As per Symantec in 2018, "as the malware begins to become contained, a natural disaster hits the region. [...] This type of attack is known as a "blended threat" – a natural, accidental, or purposeful combination of a physical with a cyber incident."

Attacks can be delivered either through the Internet or any other public or internal network. Even though it is claimed that none of the military systems are connected to the Internet, some of the devices may still serve as entry points and delivery mechanisms to the target AWS and AWS-carrying vehicles and vessels (Abaimov and Ingram 2017). Deployment of AI-based environment or framework requires a microcomputer hardware, an operating system, and a set of programming libraries to work with pre-trained neural models, as well as to forward the electronic inputs and outputs to the mechanical components of the system. This software and hardware already make AWS vulnerable to cyber attacks, and additional AI modules and components add more potential vulnerabilities.

ML modules have limited protection against emerging threats and zero-day exploits[13], they show higher rate of false positives in any threat detection module due to the fact that machine learning approximates new inputs to already known categories. Meanwhile, non-AI algorithms rarely detect zero-day exploits at all.

AI-based cyber threats

Intelligent malware that can adaptively change its behaviour as a legitimate program

Intelligent malware that can disguise as a different malware to misguide the intrusion detection systems

Scanning tools, exploits, and malware that can generate decoys and false patterns for AI misclassification

Intelligent malware and tools that can tamper with the dataset of the AI

[13] Exploit – a constructed command or a software designed to take advantage of a flaw in a computer system (vulnerability), typically for malicious purposes, such as accessing system information, establishing remote command line interface, causing denial of service, etc.

A single unit AWS is referred to a single unmanned aerial, ground, underwater, or payload delivery vehicle that acts autonomously. Their specific vulnerabilities can be remote control hijacking, denial of service, or system information disclosure. If the system is self-contained and has communication with other systems, it can be infected by malware, through the supply chain, during maintenance, or via physical access.

For example, stand-alone AWS can be air defence systems and APS, as well as logistics units, e.g., AlphaDog is a voice-controlled quadrupedal robot that can transport up to 200 kg load by following patrol operative.

Networked systems are a group of various weapon systems that supplement each other in capabilities. For example, a flying drone providing vision to a group of unmanned ground vehicles, which typically do not have long-range vision.

In centralised networks the units receive commands from the command and control unit, while in decentralised networks every unit is an independent node that can operate autonomously and synchronise the orders when the network is available. As in networks the units are different, they use different hardware, but similar software and network protocols, to be able to communicate and cooperate in the field. Thus, the compromise of a single unit may lead to the malware propagation to other units in the network, allowing the attacker to control not a single unit, but a tactical group of units. For example, Iron Dome consists of a radar unit, battle management and control unit, missile firing unit, and missiles. Those units have to communicate with each other, to provide effective and timely response against enemy missiles.

Swarm systems present a group of same type networked vehicles that act as a group, assisting each other in designated tasks. They collaborate in a more strategic and coordinated way are meant to be efficient in many offensive and defensive missions in high risk environments. They have already been presented at weapons fairs, exhibitions, and military parades. In January 2017, the US Department of Defense released a video[14] showing an autonomous drone swarm of 103 individual micro-UAVs successfully flying over California.

As swarms represent a network of multiple similar units, compromising one unit may lead to a compromise of every unit in the swarm. Since all the units have the same software and the same architecture, they also share same software vulnerabilities. To take control over the centralised swarm the attacker will have to target the command and control station or the central drone, and in the decentralised swarms – to identify and exploit flaws in the peer-to-peer network protocol.

Some of the systems are not autonomous by design, but they have autonomous components embedded into the system, to provide additional functions. Components in piloted or teleoperated systems provide a level of autonomous assistance, as additional devices they provide facial and object recognition, targeting, threat detection and response, advanced auto-pilot, etc. For example, in Dragon Fire II automated mortar the component Mortar Fire Control System is a critical component designed for the fast response to enemy fire.

---

[14] US military F-18 launch Drone swarm in tests at China Lake, 2017, https://www.youtube.com/watch?v=DjUdVxJH6yI

With the remote control stations, autonomy is provided by remote computer systems in physically secure locations using wireless communications. This approach to autonomy is less secure and less stable, and can be used in combination with the autonomous capabilities of the weapon system to compensate for stability, to distribute more high-level tasks and orders.

Military systems implement robust cyber security measures and have vulnerable components isolated, secured, or simply removed. However, the communication uplink is still required for remote-controlled systems, and it creates a way in for the skilled attackers, as any type of network activity can be detected by traditional security assessment tools like nmap and tcpdump. Remotely implemented cyber attacks can exploit internal services that are used for data transfer to and from the device.

Verification and security applications ensure security of the weapon systems (e.g., intrusion detection systems and analysis of secure configurations of computer and embedded systems). However, cyber attacks are constantly evolving, and with the innovative AI techniques can adapt and bypass currently deployed security measures.

Scenario I

A reverse backdoor is injected into a vulnerable component of the AWS through remote vulnerability in the communication link. Backdoor will connect to the attacker's owned computer to received further commands from the remote location.

If remote attackers will collect traffic (gather intelligence about the system): They will receive all necessary information about the system and replicate the technology.

If remote attackers attempt denial of service (cause malfunction): AWS will stop operation at any point and disable enemy defence systems during critical operations.

If remote attackers attempt remote code execution (gain partial or full control over the system): AWS will attack the owners and destroy their facilities.

To carry out a local exploitation, attackers require physical access to the target system or exploits functionality of the operating system of the device, or a stable interactive network connection with valid credentials or functional remote exploit.

In local cyber attacks the attacker will aim to gain the highest access privileges, required to achieve the objective. Privilege escalation is done through the vulnerabilities in the system kernel, system processed, or through vulnerabilities in the applications, that are active with intentionally or unintentionally higher privileges.

Scenario II
Malware is injected by an insider through a vulnerability in the network of the component supplier.

Malware, acting autonomously, manipulates the neural model of the autonomous system and replaces the "friend or foe" identification including removing enemy projectiles from "enemy", disabling the defence mechanism, or identifying foreign objects as a threat. As a result, the system will identify random objects as a threat and waste munitions, causing operators to shut it down. Alternatively, the malware might shut down communication or power supply after a specific event, that is identified as an "activation signal" in the middle of the battle.

A possibly unexplored analogy with the dynamical propagation of the cyber malware through various levels and its "anti-cyber malware gates", could be borrowed by studying the propagation of the tumour metastatis in oncology and how these can bypass the human immune system itself. A non-hackerable AI could be inspired by the introduction of "electronic checkpoints" similar to those of the human immune system itself which should themselves participate collectively in the cybersecurity of the entire AI.

With the physical access to the device and sufficient skills, the attacker could manipulate the software of the system, e.g., add backdoors, change the configurations, copy the software for further analysis and/or distribution, etc. Such access can lead to a full mission compromise or to the malicious use of the AWS in false-flag operations.

Operators of such systems should be multi-functional experts with a substantial knowledge of cyber security. Education of such experts is another challenging issue to be addressed.

4 Artificial Intelligence Arms Race

The saddest aspect of life right now is that science gathers knowledge faster than society gathers wisdom. Isaac Asimov

Technological advancements added one more domain to the global competition in military superiority, driven by multi-billion investments into weaponized AI research, continuous acquisition of cyber arms and vulnerabilities, development and production of the new generation AI-enhanced weapons. We are in a new arms race between world superpowers in high technologies, and the AWS are on its agenda.

*Advances in autonomy and robotics have the potential to revolutionize warfighting concepts as a significant force multiplier. Autonomy will greatly increase the efficiency and effectiveness of both manned and unmanned systems, providing a strategic advantage for DoD.*The US Unmanned Systems Integrated Roadmap 2017–2042 (DOD 2017)

The 2015 Open Letter from AI and robotics researchers identifies a key question for humanity today as "weather to start a global AI arms race or to prevent it from starting." (Future of Life Institute 2015) It further warns that "if any major military power pushes ahead with AI weapon development, a global arms race is virtually inevitable, and the endpoint of this technological trajectory is obvious: autonomous weapons will become the Kalashnikovs of tomorrow. Unlike nuclear weapons, they require no costly or hard-to-obtain raw materials, so they will become ubiquitous and cheap for all significant military powers to mass-produce... Starting a military AI arms race is a bad idea, and should be prevented by a ban on offensive autonomous weapons beyond meaningful human control" (Ibid).

Major powers are investing into AI strategies, research and development, human talents, and production. In 2017, the Russian President declared "Whoever becomes the leader in this sphere will become the ruler of the world", China released its "New Generation Plan",[15] outlining its strategy to aim for the global superiority in AI by 2030; in 2019 the U.S. President Donald Trump signed an executive order creating the "American AI Initiative",[16] with which the United States joined other major countries pursuing national strategies for developing AI (Pecotic 2019). More than a dozen countries have launched AI strategies in recent years, including China, France, Canada, and South Korea. Their plans include new research programs, AI-enhanced public services, and smarter weaponry (Simonite 2019). In 2017, the United Arab Emirates appointed the first ever Minister for AI (Arabian Business 2017).

I want to call attention to the mortal danger facing open societies from the instruments of control that machine learning and artificial intelligence can put in the hands of repressive regimes…The combination of repressive regimes with IT monopolies endows those regimes with a built-in advantage over open societies. They pose a mortal threat to open societies. George Soros, Financier and philanthropistWorld Economic Forum, Davos, January 2019 (Soros 2019)

[15] China New Generation Plan (in Chinese), July 2017, at http://www.gov.cn/zhengce/content/2017-07/20/content_5211996.htm

[16] Summary of the 2018 Department of Defence Artificial Intelligence Strategy, at https://media.defense.gov/2019/Feb/12/2002088963/-1/-1/1/SUMMARY-OF-DOD-AI-STRATEGY.PDF

AI-industry is skyrocketing. As per the PWC report "AI could contribute up to 15.7 trillion USD to the global economy in 2030, more than the current output of China and India combined" (PWC 2017). China and the US will be likely the two leading nations, and AI and robotics are the major research and development areas.

The 2017 SIPRI report highlighted that in 12 countries there are over 130 military systems that can autonomously track targets. They include air defence systems, that fire when the incoming projectile is detected, loitering munition, which hover in the sky searching for preselected areas for targets, and sentry weapons at military borders which use cameras and thermal targeting to ID human targets. Any remote control robotic system can be adapted to strike autonomously. 70 nations are developing remotely piloting combat drones, precursors of AWS. The SIPRI dataset includes nine producing countries: France, Germany, Israel, Italy, South Korea, Russia, South Africa, Sweden and the USA. Israel and Russia have produced the widest variety of APS models (SIPRI 2017a).

The UK to invest £2bn into a new RAF fighter jet

The UK defence secretary unveiled plans for a new RAF fighter jet, the Tempest, which will eventually replace the Eurofighter Typhoon. "This is a strategy to keep control of the air, both at home and abroad, to remain a global leader in the sector," Williamson said. The Tempest is intended to be flying alongside the existing fleet of Typhoons and the US-made F-35s by 2035.

The Tempest will be able to fly unmanned, … and will have next-generation technology on board designed to cope with modern threats. This will include "swarming" technology that uses artificial intelligence and machine learning to hit its targets, as well as directed energy weapons, which use concentrated bursts of laser, microwave or particle beam energy to inflict damage.

www.theguardian.com (Davis 2018)

In 2014, the autonomy in the United States was raised to the level of strategy priority in the Defense Innovation Initiative to sustain and advance the military superiority, and the third offset strategy was identified through it to ensure a competitive advantage (Deputy Secretary of Defence 2014). The Department of Defense Unmanned Systems Integrated Roadmap 2017–2042 provided strategic guidance for collaboration and expansion of unmanned systems, and identified four areas of priority: interoperability, autonomy, secure network, human-machine collaboration (DOD 2017). The 2018 US National Defense Strategy foresees that ongoing advances in AI "will change society and, ultimately, the character of war".

The US Army Robotic and Autonomous Systems Strategy leads the US Army to accomplish its technology advancements: autonomy, artificial intelligence, common control, government-owned architecture, interoperability, common platforms, and modular payloads (US Army 2017). Among its objectives is the development of the new capabilities, such as off-road autonomy for unmanned combat vehicles,

Table 2 DoD unmanned systems funding FY2017 (millions of USD)

2017	Procurement	RDT&E	MILCON	Total
Air Force	955	532	31	1518
Navy	821	725	113	1659
Army	232	212	52	496
SOCOM	32	45	5	82
DARPA	–	292	–	292
MDA	–	105	–	105
OSD	–	93	–	93
Total	2040	2004	201	4245

swarming for advanced reconnaissance, artificially intelligent augmented networks and systems.

The below Table 2 illustrates FY2017 funding requests for the unmanned systems development, procurement and associated military construction (DOD 2017).

In 2019, the controversial Project Maven, the Defense Department's program to use machine learning to identify and categorize objects in drone imagery, received a 580% funding increase – from 16 million USD in 2018 to 93.1 million USD in 2019 (Cassano 2018). In July 2018, Booz Allen Hamilton, a consulting firm, signed a 5-year 885 million USD artificial intelligence contract with the US Government Program Office. It focuses on integrating machine learning capabilities with a full spectrum of Command, Control, Communications, Computers, Intelligence, Surveillance and Reconnaissance (C4ISR) systems that support military operations Booz Allen Hamilton 2018). The 2019 DOD Joint Enterprise Defense Infrastructure (Jedi) project (Tarnoff 2018), aimed at building a cloud computing system that serves US forces all over the world, amounts to 10 billion USD.

The Defense Advanced Research Projects Agency (DARPA), a DOD agency responsible for the development of emerging technologies for use by the military, aims "to make pivotal investments in breakthrough technologies for national security" (DARPA n.d.) It is currently leading more than 20 programs aimed to advance the state-of-the-art in AI. They include, among others, Assured Autonomy, TRACE (Target Recognition and Adaption in Contested Environments), CwC (Communicating with Computers), CODE (Collaborative Operations in Denied Environment), Urban Reconnaissance through Supervised Autonomy (URSA). In September 2018, DARPA announced an investment of 2 billion USD over the next 5 years to advance the AI research to address the limitations of the first and second wave AI technologies, and to explore new theories and applications that could make it possible for machines to adapt to changing situations (DARPA 2018). The FY 2020 DARPA budget would receive a 4 percent, $129 million increase to $3.6 billion (Behrens 2019).

In June 2018, DOD established the Joint Artificial Intelligence Center (JAIC) (DOD 2018b), to lead the DOD AI Strategy, deliver AI-enabled capabilities, scale the Department-wide impact of AI, and synchronize DOD AI activities to expand Joint Forces advantages. It will oversee 600 AI projects currently under the department.

FY20 Budget Request: DOD Science and Technology
When rolling out the budget request, DOD officials stressed that the record amount of Research, Development, Test, and Evaluation (RDT&E) spending is shaped by modernization priorities identified in the National Defense Strategy. Issued in 2018, the report states that advancing technologies such as artificial intelligence, directed energy, hypersonics, and biotechnology is critical to prevailing in future conflicts. DOD officials have also pointed to the increasing technological sophistication of China and Russia in key areas as reflecting a return to "great power competition" that requires the U.S. to reconsider its R&D investments.
(Behrens 2019)

The "Rise of the machines" report (Hurd and Kelly 2018) states that "the United States has traditionally led the world in the development and application of AI-driven technologies. This is due in part to the government's prior commitment to investing heavily in research and development (R&D) that has, in turn, helped support AI's growth and development. In 2015, for example, the United States led the world in total gross domestic R&D expenditures, spending 497 billion USD (National Science Board 2018). However, the same document expresses concerns that the US is no longer the leader in this area. It further states "Notably, China's commitment to funding R&D has been growing sharply, up 200 percent from 2000 to 2015 (Ibid.).

On February 7, 2018, the National Science Board's (Board) and the National Science Foundation's (NSF) Director, who jointly head NSF, said in a statement that if current trends continue, the Board expects "China to pass the United States in R&D investments" by the end of 2018 (NSF 2018). Particularly concerning is the prospect of an authoritarian country, such as Russia or China, overtaking the United States in AI.

The above testifies that the AI competition is boosted through strategizing and prioritizing new generation military technologies, scaling up investments into R&D, production and acquisition. Civil society joint public action, supported by international organizations through global forums, advocacy and regulatory mechanisms, could and should counteract this tendency. Following the example of Google employees, researches and technical experts could voice a strong message of not supporting development of militarized AI, supporting their action by concrete examples of the danger the new technologies can bring to this world.

5 Principles of Humanity Versus Autonomous Lethal Force

New technologies raise new challenges not only in technical areas, but also in regulatory systems, institutional norms, political and social standards, ethical and moral values. The basic principles of humanity are threatened as life and death decisions can now be taken by an autonomous machine without human control.

The debates are on-going in the international community of academia, technical and legal experts, politicians and global leaders, on the development, production and deployment of intelligent weapons with advanced functions, including the most critical capability of the lethal force. Attempts are being made to come up with the procedure to assess the new weapons compliance with the principles of International Humanitarian Law (IHL), reduce humanitarian concerns, mitigate the autonomous arms proliferation (Ibid.), ensure states' responsibility in the use of AI-empowered machines, and incorporate ethical principles. Autonomy that facilitates target identification and the use of lethal force is the core issue of discussions.

There is no concerted agreement on whether the AWS use in armed conflict complies with IHL, and whether the legal reviews of study, development, acquisition or adoption of new weapons follow Article 36[17] of Additional Protocol I to the Geneva Conventions, and take into consideration the Martens Clause of Additional Protocol II,[18] on AWS definition and certification, level of responsibility and meaningful human control, new weapons' subordination to chain of command, etc.[19] To be considered lawful, the new generation of intelligent weapons should comply with the IHL basic principles of distinction, proportionality and precautions in attack ("laws of targeting")[20] and a complex validation and verification of autonomous functions are required to certify this.

There are numerous organizations and initiatives on the issues of AI safety, ethics, and governance, that are aiming to address ethical principles in technology development process (Corea 2017). In 2017, the 23 beneficial AI principles were formulated and presented at the "The Asilomar Conference on Beneficial AI" in Asilomar, California, by more than 100 AI researchers, as well as researches in economics, law, ethics, and philosophy. They were called Asilomar AI principles and refer to the AI issues related to safety, failure and judicial transparency, human

---

[17] Protocol Additional to the Geneva Conventions of 12 August 1949, and relating to the Protection of Victims of International Armed Conflicts (Protocol I), 8 June 1977, ICRC database, at https://ihl-databases.icrc.org/applic/ihl/ihl.nsf/Article.xsp?action=openDocument&documentId=FEB84E9C01DDC926C12563CD0051DAF7

[18] The Martens clause first appeared in the preamble to the 1899 Hague Convention on the Laws and Customs of War on Land, from the statement done by Fyodor Martens, the Russian delegate at the Hague Peace Conferences of 1899. It states that "Until a more complete code of the laws of war is issued, the High Contracting Parties think it right to declare that in cases not included in the Regulations adopted by them, populations and belligerents remain under the protection and empire of the principles of international law, as they result from the usages established between civilized nations, from the laws of humanity and the requirements of the public conscience", at https://ihl-databases.icrc.org/applic/ihl/ihl.nsf/Comment.xsp?documentId=42376D4763042448C12581150044672C&action=OpenDocument

[19] Meeting of the Group of Governmental Experts on Lethal Autonomous Weapons Systems Geneva, 9 to 13 April 2018, https://www.unog.ch/80256EDD006B8954/(httpAssets)/895931D082ECE219C12582720056F12F/$file/2018_LAWSGeneralExchange_Germany-France.pdf

[20] Protocol Additional to the Geneva Conventions of 12 August 1949, and Relating to the Protection of Victims of International Armed Conflict (Additional Protocol I or AP I) (adopted on 8 June 1977, entered into force on 7 December 1978), art 52(2).

values, personal privacy and liberty, shared benefits and prosperity, human control and non-subversion. They principles are now signed by 1273 AI/Robotics researchers and 2541 "others".[21]

Asilomar AI principles (selected)

The goal of AI research should be to create not undirected intelligence, but beneficial intelligence

Investments in AI should be accompanied by funding for research on ensuring its beneficial use, including thorny questions in computer science, economics, law, ethics, and social studies, such as:

How can we make future AI systems highly robust, so that they do what we want without malfunctioning or getting hacked?

How can we grow our prosperity through automation while maintaining people's resources and purpose?

How can we update our legal systems to be more fair and efficient, to keep pace with AI, and to manage the risks associated with AI?

What set of values should AI be aligned with, and what legal and ethical status should it have?

AI systems should be designed and operated so as to be compatible with ideals of human dignity, rights, freedoms, and cultural diversity.

Humans should choose how and whether to delegate decisions to AI systems, to accomplish human-chosen objectives.

The power conferred by control of highly advanced AI systems should respect and improve, rather than subvert, the social and civic processes on which the health of society depends.

An arms race in lethal autonomous weapons should be avoided

Superintelligence should only be developed in the service of widely shared ethical ideals, and for the benefit of all humanity rather than one state or organization."

(Future of Life Institute 2017)

Technical experts, researchers and engineers, with first hand understanding of the potential danger of immature autonomous technology, called attention of the leading international organizations, including the UN. The debate began in 2013, when activist groups, in particular the Campaign to Stop Killer Robots (Campaign to Stop Killer Robots 2018), began campaigning for a ban. Nearly 4000 AI and robotics researchers called for a ban on LAWS in 2015; in 2017, 137 CEOs of AI companies asked the UN to ban LAWS; and in 2018, 240 AI-related organizations and nearly 3100 individuals took that call a step further and pledged not to be involved in LAWS development (Future of Life Institute 2018).

[21] Asilomar AI Principles, Future of Life Institute, 2017, at https://futureoflife.org/ai-principles

Lethal Autonomous Weapons Pledge

Artificial intelligence is poised to play an increasing role in military systems. There is an urgent opportunity and necessity for citizens, policymakers, and leaders to distinguish between acceptable and unacceptable uses of AI.

...We, the undersigned, call upon governments and government leaders to create a future with strong international norms, regulations and laws against lethal autonomous weapons. These currently being absent, we opt to hold ourselves to a high standard: we will neither participate in nor support the development, manufacture, trade, or use of lethal autonomous weapons. We ask that technology companies and organisations, as well as leaders, policymakers, and other individuals, join us in this pledge.

(Future of Life Institute 2018)

Independent from this pledge, 28 countries in the United Nations have explicitly endorsed the call for a ban on lethal autonomous weapons systems (Future of Life Institute n.d.). However, as per the Human Rights Watch, "more than a dozen countries, including, the US, China, Israel, South Korea, Russia and the UK may be developing them" (Witschge 2018).

A European Parliament resolution in July 2018 called for "an international ban on weapon systems that lack human control over the use of force". The Belgian parliament's defence committee adopted a July resolution asking the government to support an international consensus against the employment of autonomous weapons systems while Italy's Network for Disarmament organised a July conference at the national parliament in Rome to discuss tighter regulation on autonomous weapons systems. There are also calls for an international treaty controlling autonomous weapons (Stolton 2018).

Canada and France set up a new group to study the impact of AI technologies on the world. The "International Panel on Artificial Intelligence" was announced at a G7 conference on AI in December 2018. Its mandate includes the issues related to governance, laws and justice; trust in AI, responsible AI and human rights; equity, responsibility and public good (Shead 2018).

The ICRC was the first to present to the international community about the danger of AWS more than a decade ago. In the 2011 report on "International Humanitarian Law and the challenges of contemporary armed conflicts" it urged the States to carefully consider the fundamental legal, ethical and societal issues raised by these weapons before developing and deploying them (ICRC 2014).

In 2013, the United Nations released a Report of the Special Rapporteur on extrajudicial, summary or arbitrary executions on lethal autonomous robots (Heyns 2013) where it stated that "robots should not have the power of life and death over human beings". Since this time, the issues related to LAWS have been discussed in the UN under the framework of the United Nations Convention on Prohibitions or

Restrictions on the Use of Certain Conventional Weapons (CCW),[22] which is considered to be the most appropriate framework for dealing with emerging technologies with lethal threats. In 2016, a Group of Governmental Experts (GGE) was created to formalize the discussion.

"The weaponization of artificial intelligence is a serious danger, and the prospect of machines that have the capacity by themselves to select and destroy targets is creating enormous difficulties, or will create enormous difficulties, to avoid the escalation in conflict and to guarantee that international humanitarian law and human rights law are respected in the battlefields.For me there is a message that is very clear – machines that have the power and the discretion to take human lives are politically unacceptable, are morally repugnant, and should be banned by international law." António Guterres, The United Nations Secretary-General Web Summit, 5 November 2018, (Guterres 2018)

In 2018, following two meetings, the GGE reiterated that IHL continues to apply fully to all weapons systems, including the potential development and use of lethal autonomous weapons systems; human responsibility must be retained through the weapon entire life cycle; physical security, appropriate non-physical safeguards (including cyber-security against hacking or data spoofing), the risk of acquisition by terrorist groups and the risk of proliferation should be considered; risk assessments and mitigation measures should be part of the design, development, testing and deployment cycle of emerging technologies in any weapons; discussions and any potential policy measures taken within the context of the CCW should not hamper progress in or access to peaceful uses of intelligent autonomous technologies (GGE 2018). The below "sunrise" image reflects various touch points in the human-machine interface (Chair GGE 2018) (Fig. 1).

Private key technological companies also take initiatives in ensuring that AI is used for beneficial purpose only. In 2017, Microsoft came up with a Digital Geneva Convention (Microsoft 2017), creating a legally binding framework to govern states' behaviour in cyberspace, to advance transparency and accountability, agree on rules for use of cyber weapons. In 2018, Microsoft proposed to create an international, non-governmental platform focused on cyberspace (Microsoft 2018a). The same year, its Digital Peace Now petition was signed by more than 100,000 people, across 140 countries (Microsoft 2018b).

[22] United Nations Convention on Prohibitions or Restrictions on the Use of Certain Conventional Weapons (CCW) regulates weapons which may be deemed to be excessively injurious or to have indiscriminate effects, with Protocols I, II and III, opened for signature 10 April 1981, entered into force 2 December 1983, at https://treaties.un.org/pages/ViewDetails.aspx?src=TREATY&mtdsg_no=XXVI-2&chapter=26&lang=en

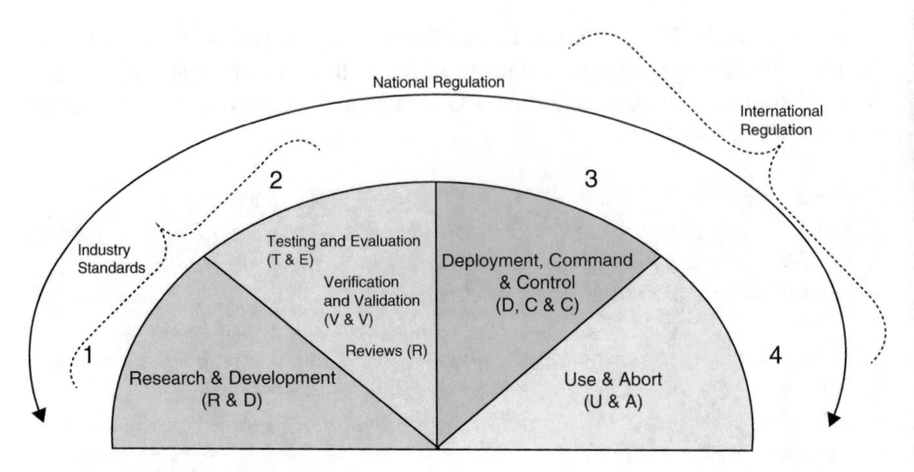

Fig. 1 Human-machine touchpoints in the context of emerging technologies in the area of lethal autonomous weapon systems (Chair GGE 2018)

> A new digital Geneva Convention may not stop new forms of attacks but it will be critical in several ways. For a start, it will define what is considered as unacceptable, fleshing out what (if any) moral consensus there is. It will also give us a legal basis to respond to those who do breach these norms, not just with mischievous states but also to create obligations on big internet companies and in policing cyber-criminals. It will also help guide the actions of those who may mean well but whose actions may create a dangerous race to the bottom. D. Sriskandarajah (2018)

The voices against the autonomous systems ban alert that this will potentially slow down the progress, and considering that the commercial sector is prospering, there will always be a way to weaponize the civilian systems. The alternative is to keep an acceptable human supervision, introduce norms, rules and standardization and enforced control.

The research on how to incorporate ethical principles into autonomous machines is ongoing. Many organizations, e.g., Association for the advancement of artificial intelligence (AAAI) help develop new ethical guidelines for AI. The major argument is that "ethically significant behaviour of autonomous systems should be guided by explicit ethical principles determined through a consensus of ethicists" (Anderson and Anderson 2015). The recommendations formulated by the "IEEE Global Initiative on Ethics of Autonomous and Intelligent Systems" cover all autonomous and intelligent system technologies and state certain developed principles for LAWS, e.g., traceability of decisions, human responsibility, predictability of behaviour (IEEE 2017a, b).

The codes of ethics have already been developed by organizations such as the IEEE, the Association for Computing Machinery (ACM), the UK Royal Academy of Engineering, the Japanese Society for Artificial Intelligence, among others.

The UNI Global Union has called for a global convention on ethical AI that will help address, and work to prevent, the unintended negative consequences of AI while accentuating its benefits to workers and society. It operationalises UNI Global Union's key demand: Artificial intelligence must put people and planet first. This is why ethical AI discussions on a global scale are essential. A global convention on ethical AI that encompasses all is the most viable guarantee for human survival.[23]

Ethical principles could be embedded in the internationally agreed standards, which will facilitate evaluation of compliance with the IHL principles. Consequently, norms and regulations should ensure that systems are certified as per the adopted standards. Thus, ethics could be reflected through standards, which could be enforced through regulation (Winfoeld and Halverson 2017). The existing codes of conducts and frameworks for responsible research and innovation could also serve as a guide (e.g. the 2013 Rome Declaration on Responsible Research and Innovation).

Though the way is still not clear on how to regulate or deal with AI-enhanced technologies, we hope that 1 day this will be resolved through multiple frameworks and standards, and the autonomous technologies will comply with IHL rules and used as deterrents only.

6 Future Challenges

Autonomous weapons are the modern reality and the future of warfare. Rapid development and billion-dollar research funding increase autonomous capabilities and level of autonomy. In a fast-paced world of technological progress human operators will become obsolete.

AI versus AI Attempts to successfully attack the enemy and defend the nation, or just to establish a military superiority, have resulted in the already escalating AI arms race, with a feasible potential of future wars where AI fights AI. Escalation of an "AI vs AI" strategy will cause uncontrollable chain reaction and result in catastrophic scenarios.

AI Superintelligence

AI Superintelligence, connected to the Internet, and to the capabilities it provides, may capture and use all the connected electronic devices, from buildings and war machines to nanocomputers, against humans.

Machine learning evolution and AI-hardware Early AI accelerators were optimized for a limited number of types of artificial neural networks. Future enhanced microprocessors will allow any types of machine learning techniques, potentially with the built-in pre-trained models, similar to Asimov's Three Laws.

[23] Top 10 Principles for Ethical Artificial Intelligence, http://www.thefutureworldofwork.org/media/35420/uni_ethical_ai.pdf

Digital organisms and Artificial life forms Systems that are capable of evolution and self-modification, by modifying the code through taking probabilistic decisions would enhance themselves and become an uncontrollable threat.

Nanocomputers, bionics and implantable technology Miniaturisation of the systems and components is resulting in new possibilities for stealth and covertness. Portable micro- and nanocomputers (including wearable and implantable) could be used for proximity cyber attacks during operations with physical presence of human operatives or delivered by drones.

Equipment, modelled after biological organisms, including human brain,[24] will increase complexity of systems, and provide researches with understanding of how organisms works, but will also inherit biological flaws.

Civilian access to AWS Open access technology allows to build toy sentry guys for personal security, entertainment, mass events. The availability of electronic components and source codes for object recognition also contribute to the amateur level development of weaponised technology.

Intelligence and accountability The risks of autonomous technology are not limited to its weaponized use during an armed conflict, they are high at peace-time as well. Fake videos, images and voice recordings become indistinguishable from real ones, and will become even more prevalent. Furthermore, AI can be used to successfully masquerade entire false-flag operations, by leaving simulated trail of digital fingerprints and evidence.

At the higher level, future challenges will require deeper understanding of the mechanisms behind the AI techniques, and human interactions with a completely new class of systems. Considering that the new generation of the intelligent machines will have the capacity for self-learning and self-evolution, the new generation of norms and regulations should be equally adjustable, adaptable and be grounded on ethical principles.

7 Conclusions

The emerging new technologies empowered by intelligence, multiple advanced functionality, and requiring minimum of human supervision, are entering our life bringing their virtual and tangible advantages. AI research and robotics are the driving forces moving civilization to a technological Eden.

Simultaneously, AI arms race is escalating and is already declared the arms race of the century. Countries invest in AI research, stockpile cyber arms, acquire vulnerabilities, develop autonomous weapons. Dual use cyber arms and AI-codes

[24] SyNAPSE Program Develops Advanced Brain-Inspired Chip, 8 July 2014, https://www.darpa.mil/news-events/2014-08-07

are publicly accessible and can be easily acquired to develop and mass produce autonomous systems.

The undertaken overview shows that multiple technical issues stay unresolved and limit autonomous capabilities in weapon systems, and at the current stage of development, autonomy can be allowed only in predictable environments and under meaningful human supervision. Computational technology and software engineering are yet to be advanced to allow qualitative autonomous data procession and decision making. Machine learning interactions are not yet fully understood, systems demonstrate only narrow intelligence in specific tasks and pose serious security threats. More advanced solutions are required in computer vision and data processing in general.

With computerized systems, AWS are vulnerable to covert and sophisticated cyber attacks that can not only damage functionality, but also change the tasks and attack the operators. Among specific cyber vulnerabilities are communication and network protocol vulnerability to eavesdropping, software vulnerability to denial of service and code execution, and further tempering with software, responsible for the AI operation (e.g., dataset poisoning).

The pre-deployment testing of AWS with higher level of autonomy needs to be standardized, and aligned with post-deployment maintenance and upgrades. Similar to digital forensics software, used for analyses and presented as legally valid expertise, underlying source codes of autonomous systems should be open-source in order to be presented for a political discussion and potential introduction into a legislative process. If adopted, they should be secured and shared in a regulated way, as with the open access to source codes, criminals and even civilians can investigate code for potential vulnerabilities or build AWS for unregulated use, with or without firearms (e.g., autonomous crossbows sentries, laser sentries, flamethrower sentries, custom made autonomous drones with lethal capabilities).

Other challenges in implementing autonomy refer to institutional, legal, ethic, normative and economic factors. As the military systems are becoming more autonomous, new safety and reliability testing approaches based on ethical standards are needed in their verification and certification.

Meanwhile, the autonomous systems can assist human operators in implementing tasks for limited areas where they demonstrate feasible advantages, and preferably in defensive missions. The technological progress is justified only if it brings prosperity to civilization, but in no way replaces human ability and responsibility for taking actions at the most decisive moments, and especially in life and death decisions.

References

Abaimov, Stanislav, and Paul Ingram. 2017. Hacking UK Trident: The growing threat, BASIC, https://www.basicint.org/wp-content/uploads/2018/06/HACKING_UK_TRIDENT.pdf.

Anderson, M., and S. Leigh Anderson. 2015. Toward ensuring ethical behavior from autonomous systems: A case-supported principle-based paradigm, https://aaai.org/ocs/index.php/WS/AAAIW15/paper/view/9976/10125.

Arabian Business. 2017. *UAE appoints first Minister for Artificial Intelligence*. 19 Oct 2017, at https://www.arabianbusiness.com/politics-economics/381648-uae-appoints-first-minister-for-artificial-intelligence.

BBC. 2016. *Adobe Voco 'Photoshop-for-voice' causes concern*, BBC News on 7 November 2016. Last accessed 3 Feb 2019, https://www.bbc.co.uk/news/technology-37899902.

Behrens, J. 2019. FY20 Budget Request: DOD Science and Technology, FYI American Institute of Physics, Bulletin 28. 28 Mar 2019, at https://www.aip.org/fyi/2019/fy20-budget-request-dod-science-and-technology.

Booz Allen Hamilton. 2018. Press Release, U.S. Government & GSA FEDSIM Select Booz Allen to Help Apply Artificial Intelligence, 30 July 2018, last retrieved on 17 Sept 2018 at https://www.boozallen.com/e/media/press-release/booz-allen-selected-to-help-apply-artificial-intelligence.html.

Brandom, R. 2018. *Amazon's facial recognition matched 28 members of Congress to criminal mugshots*. At https://www.theverge.com/2018/7/26/17615634/amazon-rekognition-aclu-mug-shot-congress-facial-recognition.

Brundage, Miles, et al. 2018. The malicious use of artificial intelligence: Forecasting, prevention, and mitigation. Last retrieved 15 July 2018 at http://img1.wsimg.com/blobby/go/3d82daa4-97fe-4096-9c6b-376b92c619de/downloads/1c6q2kc4v_50335.pdf.

Campaign to Stop Killer Robots. 2018. *Country views on killer robots*. 22 Nov 2018, https://www.stopkillerrobots.org/wp-content/uploads/2018/11/KRC_CountryViews22Nov2018.pdf.

Cangelosi, A., and M. H. Fischer. 2015. *Embodied intelligence*, https://www.researchgate.net/publication/283812826_Embodied_Intelligence.

Cassano, Jay. 2018. Pentagon's artificial intelligence programs get huge boost in defense budget, 15 August 2018. Last retrieved on 18 Aug 2018 at https://www.fastcompany.com/90219751/pentagons-artificial-intelligence-programs-get-huge-boost-in-defense-budget.

Cellan-Jones, Rory. 2014. Stephen Hawking warns artificial intelligence could end mankind, BBC News, 2 December 2014. Last retrieved on 16 Mar 2019 at https://www.bbc.com/news/technology-30290540.

Chair GGE. 2018. Chair's summary of the discussion on Agenda item 6 (a) 9 and 10 April 2018Agenda item 6 (b) 11 April 2018and 12 April 2018Agenda item 6 (c) 12 April 2018Agenda item 6 (d) 13 April 2018, UNOG, Geneva, https://www.unog.ch/80256EDD006B8954/(httpAssets)/DF486EE2B556C8A6C125827A00488B9E/$file/Summary+of+the+discussions+during+GGE+on+LAWS+April+2018.pdf.

Chen, X., C. Liu, B. Li, K. Lu, and D. Song. 2017. *Targeted backdoor attacks on deep learning systems using data poisoning*. Last retrieved 15 July 2018 at https://arxiv.org/pdf/1712.05526.pdf.

Corea, F. 2017. *Machine ethics and artificial moral agents*, November 2017, https://www.kdnuggets.com/2017/11/machine-ethics-artificial-moral-agents.html.

Cummings, Missy. 2016. see Anderson, Kenneth., Dr. Missy Cummings., Michael A. Newton., Michael Schmitt. (2016). LENS Conference 2016 Autonomous Weapons in the Age of Hybrid War. Duke University School of Law. YouTube. https://www.youtube.com/watch?v=b5mz7Y2FmU4 8:47.

DARPA. 2018. DARPA announces 2 bln USD campaign to develop next wave of AI technologies, FYI. 7 Sept 2018, at https://www.darpa.mil/news-events/2018-09-07.

———. n.d. Mission, defense advanced research projects agency. Last retrieved on 28 Aug 2018 at https://www.darpa.mil/about-us/mission.

Davis, R. 2018. UK unveils new Tempest fighter jet to replace Typhoon, 16 July 2018, at https://www.theguardian.com/uk-news/2018/jul/16/uk-tempest-fighter-jet-typhoon-farnborough-airshow.

Department of Defence (DOD). 2012. *The role of autonomy in DOD systems,* Task Force Report, Defence Science Board, at https://fas.org/irp/agency/dod/dsb/autonomy.pdf.

Department of Defense (DOD). 2017. *Unmanned systems integrated roadmap, 2017–2042*, at https://apps.dtic.mil/dtic/tr/fulltext/u2/1059546.pdf.

———. 2018a. *Unmanned systems integration roadmap, 2017–2042*, at https://apps.dtic.mil/dtic/tr/fulltext/u2/1059546.pdf.

———. 2018b. Memorandum for Chief Management Officer of DOD, 27 June 2018, at https://admin.govexec.com/media/establishment_of_the_joint_artificial_intelligence_center_osd008412-18_r....pdf.

Deputy Secretary of Defense. 2014. Memorandum for Deputy Secretary of Defense, 15 November 2014, https://defenseinnovationmarketplace.dtic.mil/wp-content/uploads/2018/04/DefenseInnovationInitiative.pdf.

Durbin, Ken. 2018. Surge in blended attacks stirs new cyber worries. Last retrieved 19 Feb 2019 from https://www.symantec.com/blogs/expert-perspectives/surge-blended-attacks-stirs-new-cyber-worries.

Dutt, D. 2018. The year of the AI-powered cyberattack, 10 January 2018. Last retrieved on 2 Aug 2018 at https://www.csoonline.com/article/3246196/cyberwarfare/2018-the-year-of-the-ai-powered-cyberattack.html.

Frank, Blair Hanley. 2019. *Get ready for AI chips everywhere*, AI Weekly 16 February 2018. Last retrieved on 16 Mar 2019 at https://venturebeat.com/2018/02/16/get-ready-for-ai-chips-everywhere/.

Future of Life Institute. 2015. *Autonomous weapons: an open letter from AI & robotics researchers*, https://futureoflife.org/open-letter-autonomous-weapons/?cn-reloaded=1.

———. 2017. *Beneficial AI conference develops 'Asilomar AI principles' to guide future AI research*, February 2017, http://www.kurzweilai.net/beneficial-ai-conference-develops-asilomar-ai-principles-to-guide-future-ai-research

———. 2018. *The risks posed by lethal autonomous weapons*, September 2018, https://futureoflife.org/2018/09/04/the-risks-posed-by-lethal-autonomous-weapons/.

———. n.d. Lethal Autonomous Weapons Pledge, Future of Life Institute, https://futureoflife.org/lethal-autonomous-weapons-pledge/.

GGE. 2018. Report of the 2018 session of the Group of Governmental Experts on Emerging Technologies in the Area of Lethal Autonomous Weapons Systems, Geneva, 9–13 April 2018 and 27–31 August 2018 at http://undocs.org/en/CCW/GGE.1/2018/3.

Grace, K., et al. 2018. *When will AI exceed human performance? Evidence from AI experts*. 3 May 2018, at https://arxiv.org/pdf/1705.08807.pdf.

Guterres, António. 2018. Remarks at Web Summit, the United Nations Secretary-General António Guterres, 5 November 2018, https://www.un.org/sg/en/content/sg/speeches/2018-11-05/remarks-web-summit.

H.R.3230 – To combat the spread of disinformation through restrictions on deep-fake video alteration technology, 116th Congress (2019–2020). Accessed 14 June 2019 at https://www.congress.gov/bill/116th-congress/house-bill/3230.

Harbers, Maaike, Marieke M.M. Peeters, and Mark A. Neerincx. 2017. *Perceived autonomy of robots: Effects of appearance and context*, Chapter 2, https://www.researchgate.net/publication/312412750_Perceived_Autonomy_of_Robots_Effects_of_Appearance_and_Context.

Heyns, Christof. 2013. Report of the Special Rapporteur on extrajudicial, summary or arbitrary executions, 9 April 2013, https://www.ohchr.org/Documents/HRBodies/HRCouncil/RegularSession/Session23/A-HRC-23-47_en.pdf.

Hsu, Jeremy. 2018. *Forget killer robots: Autonomous weapons are already online*, Undark, 2018. Last retrieved 7 Jan 2019 from https://undark.org/article/killer-robots-autonomous-weapons-online/.

Hurd, W., and R. L. Kelly. 2018. *Rise of the machines: Artificial intelligence and its growing impact on U.S. policy*, September 2018, at https://www.hsdl.org/?abstract&did=816362.

ICRC. 2014. Report of the ICRC expert meeting on 'Autonomous weapon systems: Technical, military, legal and humanitarian aspects', 26–28 March 2014, ICRC, Geneva, 9 May 2014, https://www.icrc.org/en/doc/assets/files/2014/expert-meeting-autonomous-weapons-icrc-report-2014-05-09.pdf.

IEEE. 2017a. *Reframing autonomous weapon systems*, The IEEE global initiative on ethics of autonomous and intelligent systems, at https://standards.ieee.org/content/dam/ieee-standards/standards/web/documents/other/ead_reframing_autonomous_weapons_v2.pdf.

———. 2017b. The IEEE global initiative on ethics of autonomous and intelligent systems, at https://standards.ieee.org/industry-connections/ec/autonomous-systems.html.

Jha, U.C. 2016. Killer robots: Lethal autonomous weapon systems legal, ethical and moral challenges.

Kott A., et al. 2018. *Initial reference architecture of an intelligent autonomous agent in cyber defence,* ARL-TR-8337 March 2018.

Kurzweil, R. 2005. *The singularity is near: When humans transcend biology*. New York: Viking Penguin.

Microsoft. 2017. A digital Geneva convention to protect cyberspace, Microsoft policy papers, https://query.prod.cms.rt.microsoft.com/cms/api/am/binary/RW67QH.

———. 2018a. Filling the gaps in international law is essential to making cyberspace a safer place, Microsoft, March 2018, https://www.microsoft.com/en-us/cybersecurity/blog-hub/filling-the-gaps-in-international-law-is-essential-to-making-cyberspace-a-safer-place.

———. 2018b. Digital Peace Now, Microsoft, https://digitalpeace.microsoft.com/.

Muller, V.C., and N. Bostrom. 2013. *Future progress in artificial intelligence: A survey of expert opinion*, available at https://nickbostrom.com/papers/survey.pdf.

National Science Board. 2018. *Science & engineering indicators 2018*, online at www.nsf.gov/statistics/2018/nsb20181/assets/nsb20181.pdf.

National Science Foundation (NSF). 2018. NSF and National Science Board Statement on Global Research and Development (R&D) Investments NSB-2018-9, February 2018, online at www.nsf.gov/nsb/news/news_summ.jsp?cntn_id=244465.

Office of Naval Research. n.d. *Computational methods for decision making program*, available at https://www.onr.navy.mil/en/Science-Technology/Departments/Code-31/All-Programs/311-Mathematics-Computers-Research/computational-methods-resource-optimization.

Office of the Assistant Secretary of Defense for Research and Engineering. 2015. *Autonomy community of interest test and evaluation*, Verification and Validation Working Group, Technology Investment Strategy 2015–2018, Washington, DC.

Partridge, D., and K.M. Hussain. 1992. *Artificial intelligence and business management*. Norwood: Ablex Publishing Corporation.

Pecotic, A. 2019. Whoever predicts the future will win the AI arms race, March 2019, at https://foreignpolicy.com/2019/03/05/whoever-predicts-the-future-correctly-will-win-the-ai-arms-race-russia-china-united-states-artificial-intelligence-defense/.

Phrack. 2002. *Low cost and portable GPS jammer*, Phrack, Issue 60, http://phrack.org/issues/60/13.html#article.

Poole, D., A. Mackworth, and R. Goebel. 1998. *Computational intelligence: A logical approach*. New York: Oxford University Press.

PWC. 2017. Sizing the prize: PwC's global artificial intelligence study: Exploiting the AI revolution, https://www.pwc.com/gx/en/issues/analytics/assets/pwc-ai-analysis-sizing-the-prize-report.pdf.

Rohrbeck, Patricia. 2017. *Concurrent biological, electromagnetic pulse and cyber-attacks: The ultimate interagency response challenge*. Last retrieved on 8 Dec 2018 at http://thesimonscenter.org/wp-content/uploads/2017/05/IAJ-8-2-2017-pg53-61.pdf.

Scharre, P. 2018. Killer robots and autonomous weapons with Paul Scharre, podcast, 1 June 2018. Last retrieved on 2 Aug 2018 at https://www.cfr.org/podcasts/killer-robots-and-autonomous-weapons-paul-scharre.

Schroeder, T.W. 2017. *Policies on the employment of lethal autonomous weapon systems in future conflicts*, https://www.secnav.navy.mil/innovation/Documents/2017/07/LAWS_Essay.pdf.

Seymour, J., and P. Tully. 2016. *Weaponizing data science for social engineering: Automated E2E spear phishing,* Twitter. Last retrieved 15 July 2018 at https://www.blackhat.com/docs/us-16/materials/us-16-Seymour-Tully-Weaponizing-Data-Science-For-Social-Engineering-Automated-E2E-Spear-Phishing-On-Twitter-wp.pdf.

Shead, S. 2018. Canada and France create new 'international panel on AI'. 7 Dec 2018, at https://www.forbes.com/sites/samshead/2018/12/07/canada-and-france-create-new-international-panel-on-ai/#449c4b22ef22.

Simonite, T. 2019. Trump's plan to keep America first in AI. 11 Feb 2019, at https://www.wired.com/story/trumps-plan-keep-america-first-ai/.

SIPRI. 2017a. Boulanin and Verbruggen, *Mapping the development of autonomy in weapon systems*, SIPRI Report, 2017, https://www.sipri.org/sites/default/files/Mapping-development-autonomy-in-weapon-systems.pdf.

———. 2017b. SIPRI Yearbook 2017: Armaments, disarmament and international security, summary, SIPRI, 2017, https://www.sipri.org/sites/default/files/2017-09/yb17-summary-eng.pdf.

Soros, G. 2019. Remarks delivered at the World Economic Forum, Davos, Switzerland, January 2019, https://www.georgesoros.com/2019/01/24/remarks-delivered-at-the-world-economic-forum-2/.

Sriskandarajah, D. 2018. Why we need a digital Geneva convention. 23 Apr, 2018, at https://www.diplomaticourier.com/why-we-need-a-digital-geneva-convention/.

Stolton, S. 2018. UN talks: Can the EU stop the charge of the killer robots? 29 Aug 2018, at https://www.euractiv.com/section/digital/news/un-talks-can-the-eu-stop-the-charge-of-the-killer-robots/.

Tarnoff, Ben. 2018. *Weaponised AI is coming. Are algorithmic forever wars our future?*, The Guardian 11 October 2018, https://www.theguardian.com/commentisfree/2018/oct/11/war-jedi-algorithmic-warfare-us-military.

TrendMicro. 2018. Adapting to the Times: Malware decides Infection, probability with Ransomware or Coinminer, 9 July 2018. Last retrieved on 14 July 2018 at https://www.trendmicro.com/vinfo/se/security/news/cybercrime-and-digital-threats/adapting-to-the-times-malware-decides-infection-profitability-with-ransomware-or-coinminer.

UNDIR. 2018. *The weaponization of increasingly autonomous technologies: Artificial intelligence*, Report, as adapted from Shane Legg and Marcus Hutter, "A Collection of Definitions of Intelligence", Technical Report IDSIA-07-07, 15 June 2007, p. 9. Last retrieved on 16 Mar 2019 at http://www.unidir.ch/files/publications/pdfs/the-weaponization-of-increasingly-autonomous-technologies-artificial-intelligence-en-700.pdf.

US Army. 2017. *The U.S. Army: Robotic and autonomy systems strategy*, U.S. Army Training and Doctrine Command.

Vincent, J. 2018. *The Verge Tech report card, AI*, at https://www.theverge.com/2018/12/30/18137429/2018-tech-recap-artificial-intelligence-robot-machine-learning-facial-recognition.

Winfield, A., and M. Halverson. 2017. Artificial intelligence and autonomous systems: Why principles matter September 2017 at http://sites.ieee.org/futuredirections/tech-policy-ethics/september-2017/artificial-intelligence-and-autonomous-systems-why-principles-matter/, as quoted from Winfield, A., Written evidence submitted to the UK Parliamentary Select Committee on Science and Technology Inquiry on Robotics and Artificial Intelligence, Discussion Paper, Science and Technology Committee (Commons), Website, 2016. http://eprints.uwe.ac.uk/29428/.

Witschge, L. 2018. Should we be worried about 'killer robots'? April 2018, https://www.aljazeera.com/indepth/features/worried-killer-robots-180409061422106.html.

Understanding the Threat Posed by COTS Small UAVs Armed with CBR Payloads

N. R. Jenzen-Jones

1 Overview

The potential threats posed by commercial off-the-shelf (COTS) unmanned aerial vehicles (UAVs) are increasingly well-understood. COTS UAVs have been acknowledged as a powerful force-multiplier in asymmetric conflicts, and the capabilities offered by these systems have been readily adopted by a range of state and non-state armed actors. Until recently, the advantages offered by COTS UAVs fell primarily into the intelligence, surveillance, target acquisition, and reconnaissance (ISTAR) realm. Unmanned aerial systems have been successfully employed in this role by non-state actors to support a range of violent actions, from adjusting indirect-fires to vectoring vehicle-borne improvised explosive devices (VBIEDs) towards their targets (Friese et al. 2016).

Increasingly, both state and non-state actors have used COTS small UAVs to execute direct strike missions using improvised air-delivered munitions (IADMs) and modified conventional munitions. Analysis of the use of these improvised and craft-produced aerial munitions in recent years has shown that, in conjunction with COTS or 'hobbyist grade' UAVs, they can provide a modest precision strike capability to belligerents (ARES 2016, 2017a, b; Jenzen-Jones and Wright 2018a). In addition to the use of UAVs to deliver conventional (e.g. explosive, incendiary) payloads, there are logical concerns that these platforms may be used as a means of delivery for

N. R. Jenzen-Jones (✉)
Armament Research Services (ARES), Newcastle upon Tyne, UK

Armament Research Services (ARES), Perth, Australia
e-mail: nic@armamentresearch.com

M. Martellini, R. Trapp (eds.), *21st Century Prometheus*,
https://doi.org/10.1007/978-3-030-28285-1_9

chemical, biological, or radiological (CBR)[1] payloads. There exists a potential for attacks of this nature to be conducted by different types of malicious actors, including individuals, small groups, and transnational terrorist organisations. At present, the users of COTS small UAVs for violent purposes have mostly been non-state actors typically classed as militias, several of whom are also considered terrorist organisations by many western nations. The so-called Islamic State has been foremost amongst these, conducting numerous strike missions with armed UAVs (Jenzen-Jones and Wright 2018a). However, COTS small UAVs have also been used in a concerning manner by individuals, including Yasuo Yamamoto, who landed a small UAV carrying slightly radioactive sand[2] on the roof of the residence of Japanese Prime Minister Shinzo Abe (BBC 2015; Sekiguchi 2015).

Despite the absence of any recorded UAV attacks delivering CBR agents, it is prudent to consider this a very real threat.[3] Non-state actors have shown that they can indeed acquire, produce, weaponise, and deliver CBR agents (see, for example, Ballard et al. 2001; Strack 2017; Bunker 2019). Mitigating against this threat requires an analysis of the different factors which could enable a UAV-borne CBR attack. Whilst the potential for UAVs to deliver CBR payloads has been acknowledged for some time,[4] it is only in recent years that their ease of operation, technical characteristics, and widespread commercial availability have combined to elevate their use in this manner from the realm of the possible to the probable.

2 COTS Small UAVs

The proliferation of armed attacks by small UAV platforms—and the potential for such attacks to incorporate CBR payloads in future—is directly tied to the rise in the availability and popularity of COTS UAVs. These systems can be broadly defined as

[1] The term 'CBR' is derived from the more widespread term 'CBRN', which encompasses chemical, biological, radiological, and nuclear weapons, and is based on military terminology. Military terms previously in vogue have included 'ABC' (atomic, biological, chemical) and 'NBC' (nuclear, biological, chemical). Another term, 'WMD' (weapon of mass destruction), is now also in common use, but can also encompass high explosives and other effects that can result in high casualty rates (Smith 2018a). For the purpose of this chapter, the term 'CBR' is preferred as it refers discretely to the agents under consideration.

[2] The sand, which the perpetrator had collected from Fukushima Prefecture, was contaminated with trace amounts of the radionuclide caesium-137. Whilst Yamamoto's stated goal was to protest Japanese nuclear energy policy, he indicated in blog entries that he understood his actions to constitute 'a terror act' (Asahi Shimbun 2015).

[3] A related threat is that of attacks using COTS small UAVs to delivering high explosive munitions targeting sites which contain CBR materials, occasioning their release. In 2018, for example, a UAV flew within the security perimeter of the Bugey nuclear plant in Lyon, France (De Clercq 2018). Such a threat is beyond the scope of this chapter.

[4] See, for example, Ballard et al. (2001).

powered aerial vehicles weighing less than 25 kg[5] that do not carry a human operator, use aerodynamic forces to provide vehicle lift, can fly autonomously or be piloted remotely, can be expendable or recoverable, are commercially available, and can be purchased by members of the general public (Bone and Bolkcom 2003; FAA 2015; Friese et al. 2016).[6] Both fixed-wing and rotary-wing UAVs are now commonplace in the consumer market.

Reports from 2018 indicate that the total global UAV market may be worth more than 50 billion USD by 2025, up from an estimated 11.5 billion in 2016. This rapid growth in the sector as a whole has also been reflected in the commercial portion of the market, with increased competition resulting in "better services and enhanced products" being delivered to users (The Insight Partners 2018). Whereas military users have historically employed highly sophisticated UAVs that are orders of magnitude more expensive than their commercial counterparts, COTS UAVs initially gained traction as hobbyist projects and novelties. Nonetheless, their rapid advancement has led to some blending of the two markets. Current market trends have borne UAVs which can be easily adapted to the requirements of asymmetric forces, and violent actors benefit from increasing access to UAVs targeted at the commercial and 'prosumer' markets, which are, in many cases, similar or identical to systems considered 'tactical UAVs' by many militaries.[7] COTS UAVs feature increasingly advanced cameras, speed, and control capabilities. High-end UAVs with ranges of around 8 km, flight durations in excess of 30 minutes, maximum altitudes of 5000 m or more, and payload limits of 4 kg or greater are now available to private individuals[8] (Smith 2018a). Tracking capabilities and radar developments allowing for their use in all environments can contribute to enhanced strike capabilities. Commercial UAV technology companies often receive military funding, and many producers focus on both the civilian and defence markets (ARES 2018b; Zwijnenburg and Postma 2018). As military and law enforcement markets develop, technologies developed for these applications often filter down to commercial and hobbyist users (ARES 2018b; Friese et al. 2016).

These markets also enable greater access to the parts needed to produce mission-specific UAVs. UAVs assembled from component parts may allow for access to the most effective designs with limited exposure, using equipment procured via different sources in different areas within a given terror network. Instructional websites

[5]This gross take-off weight stipulation means that most examples would fall within the US Department of Defense's 'Group 1' and 'Group 2' classifications (0–20 and 21–55 lbs, respectively) (DoD 2011).

[6]For a more thorough discussion of the component parts of this definition, see Friese et al. (2016).

[7]The initial military uses for 'prosumer'-type tactical UAVs were primarily related to reconnaissance, providing small-unit formations with a better understanding of the ground in front of them (Smith 2018a). These roles remain important today, and the technology has been similarly adopted by non-state actors (Friese et al. 2016).

[8]In the case of larger UAVs that do not fall within the 20 kg limit of a 'small' UAV, various agricultural sprayer models exist that could be modified to disperse chemical or biological agents. Some of these can carry a liquid payload of up to 50 kg, however they remain restricted in terms of range and flight duration (Smith 2018a).

and word-of-mouth allow for a cottage industry of "from scratch" UAV development through idea sharing, dissemination of designs, and publication of lessons learned. Experienced builders can covertly source components to enhance UAV capabilities, achieving end-states that may otherwise be regulated. Accordingly, UAVs can be manufactured with characteristics to fit within specific operational requirements. The natural lag time in responding to threats which violent non-state actors enjoy when acquiring a new threat mechanism or compromising the established defences of the global community enhances this threat. Nonetheless, with the recognition that their products have been used by terrorist networks, manufacturers such as DJI have implemented controls and restrictions designed to limit the value of UAVs to violent actors, including enhanced geofencing and buyer restrictions (Rees 2018). In some cases, manufacturers have even integrated software and firmware tracing mechanisms (Rassler 2018). In many jurisdictions, acquiring larger and more capable types of UAV is difficult, and often regulated by mandatory licencing and registration schemes (Smith 2018a).

3 Conventional Munitions and Riot Control Agents Deployed from COTS Small UAVs

In late January 2017, the Islamic State (IS) media outlet in Ninawa province, Iraq, released footage entitled "The Knights of the Dawawin," which documents the use of UAVs in an attack against Iraqi Army forces (Rassler et al. 2017). This video, and others released subsequently, depict the aerial delivery of IADMs and modified conventional high explosive (HE) munitions from COTS small UAVs. Whilst these attacks were by no means the first such delivery of HE IADMs from these platforms, they heralded a sharp increase in the battlefield use of such weapons; in February 2017 alone, IS is believed to have conducted some 200 such attacks. The rate had dropped by the spring, but attacks continued sporadically through the rest of the year (ARES 2017a). Most of the attacks made use of modified conventional munitions, with the majority of these based around 40 × 53SR mm grenade projectiles, especially the M383 HE model.[9] These were primarily sourced via battlefield capture from Iraqi forces, who make use of the 40 × 53SR mm MK 19 and Iranian *Nasir* automatic grenade launchers (Jenzen-Jones and Wright 2018a). The Islamic State also operated a sophisticated production programme, developing and manufacturing their own scratch-built IADMs which shared parts commonality with other craft-produced munitions in their arsenal (Jenzen-Jones and Wright 2018a). In Iraq and elsewhere, other conventional munitions are also adapted, including 40 × 46SR mm and 30 × 29B mm grenade launcher cartridges or projectiles, and projectiles taken from 20, 23, and 30 mm cannon rounds (ARES n.d., 2017b).[10]

[9] The M383 weighs 340 g, contains 55 g of RDX, and produces a 15 m fragmentation radius (Jenzen-Jones and Wright 2018a).

[10] In at least one case, propaganda leaflets were dropped by an IS UAV (ARES n.d.).

State security forces have also employed COTS UAVs in the strike role. In Iraq, for example, government forces quickly followed the Islamic State's lead, fielding their own improvised munitions by early 2017, when the Iraqi Federal Police employed hobbyist-type UAVs to deliver modified 40 mm HE grenades during fighting in Mosul (ARES n.d.). This was quickly followed by standardised production of modified grenade launcher and cannon munitions by Iraq's Technical Directorate for Military Production (TDMP), known to supply the Interior Ministry's Popular Mobilization Units (Jenzen-Jones and Wright 2018a; ARES n.d.). Research institutes in modern western states have sought to test the capabilities of COTS small UAVs in the strike role, and the threat has been considered by national preparedness organisations.[11]

In Ukraine, pro-Russian separatists used UAVs to deliver munitions as early as 2014 (Ferguson and Jenzen-Jones 2018). From 2015, a series of attacks using ZMG-1 type incendiary grenades in Ukraine have sometimes had catastrophic consequences, with attacks on two depots in 2017 being particularly destructive and showcasing the significant military and psychological value of these delivery platforms (Mizokami 2017a, b). Other designs have incorporated rocket launchers and recoilless weapons. The Ukrainian government has also tested various designs developed by private companies in Ukraine, including rotary-wing UAVs which can carry anti-tank rockets and even guided missiles (ARES n.d; Jenzen-Jones and Wright 2018a; Wendle 2018). In April 2018, Ukrainian forces recovered what appeared to be a rudimentary hand-launched sacrificial UAV, which had been used by separatists near Donbass. Separatist forces have also experimented with deploying individual PTAB-2.5M submunitions from UAVs (ARES n.d.). In Libya, a DJI Matrice 100 suspected to have been operated by the Libyan National Army was downed in Derna in April 2018. Footage allegedly present on the UAV's camera shows it delivering craft-produced ordnance, possibly of an HEDP type (ARES n.d.; Jenzen-Jones and Wright 2018a).

In Venezuela, President Nicolás Maduro was the target of an assassination attempt which was executed using COTS small UAVs. In the failed attack, two UAVs flew toward Maduro as he was delivering a speech in Caracas. Although many details of the incident remain unconfirmed, it was reported that neither UAV made it to their intended target; both functioned some distance from Mr. Maduro, injuring seven people in the crowd (The Economist 2018). This incident, coupled with the rapid integration of COTS small UAVs into the battlefield, is perhaps the point of debarkation for the use of these type of devices as a method of assassination.

These brief examples highlight the two primary approaches to using COTS small UAVs in the strike role, which hinge on the expendability of the UAVs themselves. In most of the Islamic State's attacks on Iraqi forces, UAVs acted as miniature 'bombers,' releasing small air-dropped munitions. This of course allows for the use of the UAV on the battlefield beyond its strike role (e.g. for ISTAR purposes), and in future attacks. In the attempt on Maduro's life, the UAVs were treated as expendable, flying into a non-permissive environment with no intention of being recovered.

[11] See, for example, NDPC (2016).

The 'sacrificial' method,[12] with the UAV acting as both the vehicle and the payload carrier and initiator at the target, shows a different conceptualisation of how an attack might be delivered; in other words, by a single-use, manually guided weapon system. For some time, the VBIED had been conceived of as a 'poor man's guided missile'—the COTS small UAV may well supplement or supplant them in that role.

COTS small UAVs are also increasingly used to deliver riot control agents (RCA). These may be deployed via air-dropped RCA munitions or directly from airframes, in the form of chemical aerosol sprays or powders. Israeli forces operating along the border with Palestine have employed both delivery mechanisms, and at least four different configurations of UAVs and delivery system (ARES 2019; Hilton 2018; Rogoway 2018). These range from simple, remotely-operated 'cages' that hold armed RCA grenades, to more elaborate aerosol sprayers and self-contained munition-dispensing units, such as the Ispra Ltd. 'Cyclone' (Ispra 2015; ARES 2019). Companies in several other countries, including China, France, India, South Africa, Turkey, and the United States have also experimented with or deployed C-UAVs delivering RCAs (Crowley 2015; Friese et al. 2016; ARES 2019).

4 Chemical, Biological, and Radiological Payloads

CBR payloads may vary considerably according to the type and nature of the agent in question. In many ways, considering CBR agents collectively can be misleading, as the nature, duration, pathways, and delays associated with the various harmful effects of these agents can vary substantially between categories. Early chemical weapons agents, such as chlorine or phosgene, were borrowed directly from industry for their known hazardous properties. Later generations, such as mustard and nerve agents, were developed specifically for military purposes and with limited uses within the civilian chemical industry. Chemical agents may be acquired from existing stocks, both military and industrial, or manufactured from precursors. Biological agents, as their name implies, are taken from nature. There are three principal forms: viruses, bacteria, and toxins.[13] Most agents were initially identified as a result of research into diseases and other public health concerns. 'Weaponised' variations[14] on

[12] UAVs of this type are often improperly referred to as 'suicide drones' or similar. Some of these types of systems may be considered to overlap with so-called 'loitering munitions' (ARES 2018b).

[13] Whilst most naturally occurring toxins are classified as biological weapons, a subset (such as chemically toxic metals) or all are sometimes considered chemical weapons. The term 'biochemical weapons' is sometimes encountered, further blurring the categories. There exists no bright line between the two, but toxins are covered by both the Biological Weapons Convention and the Chemical Weapons Convention, and some are even listed in the CWC schedules (e.g. ricin, saxitoxin) (Ganesan et al. 2010; Pitschmann 2014; OPCW 1993).

[14] Variations may include 'formulated agents', a term implying a higher standard of production and development. Formulated agents may incorporate vitamins, opportunistic infection inhibitors, surfactants, cushioning materials, and other additives, and may be processed in such a way to ensure optimal compatibility with desired delivery methods and pathways of attack (Weber 2012).

these naturally-occurring agents were then developed to allow for their manufacture, storage, and deployment outside of endemic environments. Biological agents may be acquired from nature, or from research or medical facilities. In the future, it may be possible to synthesise biological agents using advanced gene manipulation techniques. Radiological material has little military value other than in the development of nuclear weapons. However, there are many medical and industrial processes that utilise the properties of the radioactive decay of these materials (Smith 2018a). Acquisition of radiological material could be from industrial, medical, or academic facilities. In all cases, there are significant international and national safeguards in place to restrict the acquisition, movement, and use of CBR material. These have both safety and security goals.

There remain yet other threats that are still emergent, some of which may not be seen to fall within the remit of counter-WMD agencies. In a March 2019 interview, Vayl S. Oxford, director of the US Defence Threat Reduction Agency (DTRA) specifically raised the threat of synthetic biology and legal narcotics when asked about his concerns regarding the rapid pace of technological change in today's world. He wrote:

> The use of new technology like additive manufacturing and synthetic biology is worrisome, both legitimately and illegitimately. The questions I pose—what do we want to tackle in synthetic biology and how do we find the nefarious actions versus the legitimate actions? We do have a chemical/biological office that is investing in synthetic biology research so we can monitor where that could become more of a threat vector than just the chemical/ biological legitimate use. Another threat that we are facing is fentanyl. This is a purely legal substance but used in the wrong way, they can be very dangerous. We have seen on the news that based on improper they can be extremely dangerous. However, fentanyl is not classified as WMD because it is not considered a weapon. It is important for us to focus more analytics effort on fentanyl, but until it is declared as a weapon or an improvised threat, we will be waiting on the Department for guidance. Congress is adamant about what we define as an improvised threat, so it is important we spend our resources and time on specific threats that are within our mission space (Hummel 2019).

As UAVs become increasingly commonplace in warfare—and in everyday life—new and novel uses for these platforms will no doubt evolve. For non-state actors, this will likely include the delivery of non-explosive payloads, including CBR agents. In late 2018, Islamic State affiliates uploaded two digital 'posters' to various distribution channels which show a common model of COTS small UAV carrying unidentified payloads, digitally superimposed against skylines of New York and Paris. Text on the poster reads: "Sender: The Islamic State" (ARES n.d.). On several channels, text that accompanied the postings suggested that not only were such attack vectors now available in Western countries, but also that there may be the capability to deliver lethal payloads other than explosive devices (Smith 2018a). Just a few months prior, another Islamic State propaganda image was released which shows a hooded figure holding a device spewing green smoke against the San Francisco skyline. Text on the image reads "We will make you fear the air you breath [sic]" (ARES n.d.). This, in turn, followed a July 2018 Islamic State video calling for biological terrorism attacks on western targets, and rudimentary instructions for contaminating water supplies and food (MEMRI 2018).

Despite these threats, the use of CBR agents in terror attacks over the last 30 years has been limited on a global scale. The sarin nerve agent attack on the Tokyo subway, executed by Japanese doomsday cult *Aum Shinrikyo* in 1995, is perhaps the most well-known of these. Aum also experimented with dispersing sarin from a radio-controlled helicopter prior to their subway attack (Ballard et al. 2001). In many cases, the psychological effect of a chemical or biological attack on civilians has proved much more impactful than the effects of the agent itself (Cruickshank 2018). The 2001 anthrax and 1984 Rajneeshee *Salmonella* attacks in the United States created substantial panic and generated political effects well beyond their direct human impacts. A US government report estimates that, in the event of a CBRN attack, the public health apparatus could expect five psychological casualties for every physical casualty (Warwick 2001).

In contrast to the relative absence of CBR agents in terror attacks, the use of such agents in warfare has intensified. In active conflict zones, CBR use is now more commonplace than at any time in recent history. This is almost entirely due to the increased use of both original-purpose chemical weapons (e.g. sulfur mustard and sarin) and repurposed toxic industrial chemicals (TIC; e.g. chlorine)[15] in Syria (Smith 2018a; ARES 2018a). Both Syrian government and Islamic State forces standardised production of TIC munitions over previous years. Syrian government designs included mass-produced improvised rocket-assisted munitions (IRAMs) and IADMs. Both IRAMs and IADMs have been used extensively by government forces in Syria, primarily delivering payloads of industrial chlorine (Jenzen-Jones and Wright 2018b). Nonetheless, as a percentage of the total number of attacks made by Syrian government and Islamic State forces operating in Syria, the use of chemical weapons remains very low.

5 CBR Agents: Physical Forms, Exposure Pathways, and Effects

The effectiveness of a given attack depends largely on matching the available agent to a method of delivery (as determined, in part, by its physical form) and a type of target (and its vulnerable exposure pathways). Attacking a water reservoir with an agent neutralised or diluted beyond effectiveness by water, for example, will not result in optimal outcomes for the attacker. Similarly, targeting a crowd of people with a powdered biological agent which is only effective when ingested is unlikely to be an efficient use of resources. The target's defences against a particular agent would also need to be considered. Unbroken skin is an effective barrier against almost all biological agents,[16] for example (Weber 2012). CBR agents may be

[15] Note that when TICs are weaponised, they are considered to be chemical weapons under the Chemical Weapons Convention (ARES 2018a).

[16] Some notable exceptions are known. These are withheld on security grounds.

acquired by non-state actors in a variety of physical forms. These will be more or less suitable for weaponisation and employment from a COTS UAV according to a number of factors. Chemical agents are most likely to be encountered in the form of a gas, liquid, aerosol (microfine liquid drops suspended in air), smoke, or vapour (liquid in its gas phase). Biological agents are generally in the form of liquids, aerosols (microfine liquid or solids),[17] or solids. Biological agents may also be transported and transmitted by biological vectors, such as fleas, which could in turn be transported by UAVs. Radiological material is typically encountered in solid form, commonly as a powder or as small metal pellets or rods. They may be held in a gas or liquid suspension.

Generally speaking, the method of employment for a given agent will depend on its physical form. In some cases, non-state actors may be able to alter the properties of an agent, and may do so in order to optimise it for use in a particular delivery system or against a particular type of target. When released from a UAV, liquid agents will tend to fall to the ground. As well as presenting a liquid hazard, off-gassing will create some vapour. This vapour will be of a similar temperature to the local air and will typically remain close to the source. It may re-condense and fall back to the ground in a nearby location, presenting a secondary liquid hazard. Liquid absorbed or adsorbed may have a reduced rate of off-gassing and present a liquid hazard for a considerable period of time. When an agent is disseminated as a vapour, it will typically generate an agent cloud which will expand, cool, and grow denser. In many cases, this tends to retain its form. If its vapour density is greater than the surrounding air, it will sink and flow over the surface. Low winds can allow the cloud to retain its volume and form for long periods of time. If its vapour density is less than the air, it will rise and dilute. Increased turbulence will increase dilution; thereby reducing concentration. Aerosols—finely divided solid particles and/or liquid droplets suspended in air or another gas—behave similarly to vapours, but they are typically less dense and the higher heat of their initial release may cause the cloud which is formed to initially rise before descending. Solid material, commonly weaponised as fine powders, will fall to the ground relatively quickly. In some cases, turbulence may delay this. These can present an enduring hazard (Smith 2018a).

Radiological dispersion devices (RDD), including so-called 'dirty bombs', use different methods to scatter radiological material. A dirty bomb combines a conventional explosive with a radiological agent. Whilst the effects of the conventional explosive are likely to be significantly more damaging when the device is employed, dirty bombs are acknowledged to pose additional threats, including psychological impacts, contamination of the affected area, and associated second- and subsequent-order effects. Other RDDs could feature particles of radioactive material suspended in a liquid, or dispersed in the smoke plume of an incendiary device. The effects of an RDD not accompanied by an explosive detonation may go unnoticed for some

[17] Aerosolisation (or 'atomisation') may be achieved through a variety of methods, including explosion, expulsion, hydraulic, airblast, or mechanical (Weber 2012).

time. Radiological materials cannot be neutralised or decontaminated; they can only be removed and stored until they naturally decay (Smith 2018a). Certain radionuclides, such as caesium-137 and strontium-90, are of particular concern, as they can penetrate deeply into construction materials such as concrete at affected sites—even chemically bonding with such materials—significantly complicating and prolonging the radiological hazard (Farfán et al. 2011).

Violent non-state actors including terrorists have shown an ongoing interest in acquiring radioactive materials (USNRC 2018a). The IAEA's Incident and Trafficking Database recorded an average of 35 radiological source thefts every year between 2008 and 2015. The report noted that 'A small number of these incidents involved seizures of kilogram quantities of potentially weapons-usable nuclear material, but the majority involved gram quantities' (IAEA 2017).

Most CBR attacks conducted by a COTS small UAV are likely to see the agent in question dispersed in the air over a target, or by a device dropped on or near the target. A number of factors will limit the effectiveness of attacks of this type, but wind will be amongst the most important in almost all cases. Strong winds may render COTS small UAVs inoperable or difficult to pilot, and the winds at the moment of attack will directly affect how an agent is dispersed or how a munition falls. In particular, where CBR agents are dispersed over a target, wind—and other environmental factors such as terrain, temperature, and humidity—will have a profound impact on the attack's effectiveness. These environmental factors affect the rate of fall of an agent, the direction in which an agent is carried, evaporation it may be subject to, and so on (Weber 2012; Smith 2018a).

Assuming the CBR material reaches the designated target, it must then enter the human body. There are four ways substances may enter the human body. The two most immediate in the course of a CBR attack delivered by are UAV are likely to be inhalation and absorption (typically via dermal contact). Attacks utilising agents in the form of a gas or vapour are the most likely to be successful at accessing this pathway; aerosolised agents must be of the correct particle size to be effective.[18] Absorption may take place where the agent interacts with the exterior surfaces of the body. Agents of all physical forms may affect the body via dermal contact or other bodily contact. Depending on the situation at the time of an attack, much of potential victims' bodies will be covered, typically by clothing. Depending on the agent, clothing may provide some protection. Gas is unlikely to be adsorbed and most clothing will act as a barrier against solids. Vapour and liquid aerosol may be absorbed, however. Most clothing fabric will not provide substantial protection against liquids, and may even enhance the dermal absorption risk by holding the liquid against the skin (Smith 2018a). Absorption may also be the primary pathway for exposure after the agent has been dispersed; for example, a powdered agent may be dispersed to coat tools, vehicles, or work surfaces which are later touched (NRC 2000, ch. 4). Where food or water supplies, in particular, are attacked, ingestion

[18] Too large, and particles will be trapped by the body's natural defences (e.g. nasal hair or mucus in the airway); too small and they will be immediately breathed out (Smith 2018a).

may occur. Certain solids may dissolve in liquid sources such as water; other liquids or fine particulate may mix into the same. Gases may in some cases be absorbed. Whilst injection is generally conceived of as delivery by a hypodermic needle, in the case of CBR attacks delivered by UAVs it is more likely to occur when fragmentation or sharpened projectiles (e.g. darts or flechettes) pierce the skin. In many cases, physical trauma may pose a more immediate or more substantial risk to the victim.

An agent may also be characterised in terms of its transformation and transport between environmental media. For example, substances may accumulate in the area of an attack, or be transported (e.g. by rainwater runoff) to secondary environments where they may cause further harm. The physical characteristics of the agent are the primary factors influencing how it is transformed or transported upon reaching a target area. A gas is likely to be dispersed relatively quickly by natural air currents, for example, but if denser than the surrounding air it may flow into subterranean structures such as basements. Mobile liquid agents, with the characteristic of thin oil or water, can flow into micro-fine holes and gaps in soil, vegetation, or man-made structures, such as tarmac or concrete. However, they are more likely to evaporate and will typically be more susceptible to heat. More viscous liquids may be less likely to flow to secondary environments, but may off-gas for longer periods of time and present an additional hazard (Smith 2018a). In some cases, this may result in the agent entering the body through unforeseen or unintended exposure pathways. An aerosolised agent may be intended to enter the body through inhalation, for example, but may instead settle on foodstuffs and later enter the body through ingestion. Alternatively, these 'second-order' effects may be intentional; see the discussion of 'indirect off-target attacks', below. When conducting an exposure assessment, specialists will first examine the immediate environment of a victim, before looking for other pathways and secondary environments (NRC 2000, ch. 4).

The toxicity, latency, and persistency of a given agent are the key factors in quantifying its effects and potential impacts. Toxicity is an indicator of how poisonous a substance is to a biological entity. Any chemical can be toxic if absorbed or consumed in a large enough quantity. The toxicity of CBR material is often expressed in terms of the quantity that would be expected to be lethal to 50% of affected persons (the 'median lethal dose', or LD_{50}). Table 1 shows the median lethal dose for selected substances, and uses a classification system to place substances into toxicity categories (Trautmann 2001; USNRC 2018b). Biological pathogens are difficult to classify in terms of toxicity. Their expected mortality rates vary, but much is dependent upon the victim and other situational factors. Exposure pathways are often variable and unreliable as well. A few anthrax spores would theoretically be lethal, for example, but frequently would not enter the body of a target in the intended way. Some biological toxins, such as *Botulinium*, are many times more toxic than most chemical warfare agents (NRC 2000, ch. 3). Radiological effects are highly dependent on the dose received and the rate at which it is received (Smith 2018a). In general, the more toxic the material, the less of it is required to cause harm. However, caution should be exercised in assessing theoretical risks. Hypothetical scenarios may make unreliable assumptions, such as even agent distribution, accuracy of animal testing data, and environmental factors such as terrain or meteorological conditions. It should also

Table 1 Median lethal dose and toxicity category for selected substances

Substance	LD_{50} (mg/kg; p.o.)	Toxicity category
Sucrose	30,000	Practically non-toxic (*not classified*)
Ethanol	7000	Slightly toxic (*not classified*)
Acetaminophen	2000	Moderately toxic (*Category 5*)
Acetylsalicylic acid	1000	Moderately toxic (*Category 4*)
Caffeine	200	Very toxic (*Category 3*)
Nicotine	50	Extremely toxic (*Category 2*)
Cyanide	10	Extremely toxic (*Category 2*)
Ricin	1	Super toxic (*Category 1*)
Sarin	0.12	Super toxic (*Category 1*)
VX	0.015	Super toxic (*Category 1*)
Aflatoxin	0.003	Super toxic (*Category 1*)
Botulin	0.001	Super toxic (*Category 1*)
Ionising radiation	4–5 sv[a]	N/A (*N/A*)

Notes: GHS category classification for acute toxicity (p.o.) provided in italics and parentheses
Sources: Trautmann (2001); USNRC (2018b); NRC (2000), ch.3; Smith (2018a); and United Nations (2011), p. 109.
[a]Approximately 400–450 rem, received over a very short period (USNRC 2018b). Note that the toxicity of radiological materials is difficult to compare to chemical and biological agents. It will vary with both the level of radioactivity and the size of the source, as well as other factors such as distance and duration of exposure. For example, a large quantity of a low-activity radionuclide could have a similar dose rate to a smaller quantity of a more active source

be noted that the more toxic agents will also pose increased risk to the adversary in their acquisition, processing, and preparation phases.

The term 'latency' is used to describe the time it takes to recognise the effects of an agent on a victim. This may manifest in symptoms that a victim is able to describe, or in physical signs observable by others (including those present on a deceased victim). Some chemical agents, such as cyanide, can act on a victim in seconds. Others, such as mustard agents, may be visible after a couple of hours. Biological agents tend to act over days or weeks; in many cases, early signs and symptoms are

Table 2 Summary of typical agent properties by type

Chemical		Biological		Radiological	
Sources:	Mostly synthetic (some natural)	Sources:	Mostly natural (some synthetic)	Sources:	Mostly synthetic (some natural)
Toxicity:		Toxicity:		Toxicity:	
Latency:	Minutes to hours	Latency:	Days	Latency:	Days to weeks
Persistency:	Minutes or weeks-months	Persistency:	Hours to years	Persistency:	Days to centuries
Transmissibility:	Contact	Transmissibility:	Contact & proximity	Transmissibility:	Contact
Visual signs of release:	Possible	Visual signs of release:	Unlikely	Visual signs of release:	Possible
Characterisation of release:	Probable	Characterisation of release:	Possible	Characterisation of release:	Probable

Source: adapted from Smith (2018a). (Images via FEMA)

non-specific and may not be readily identified as the result of a biological attack. Radiological effects are highly variable, dependent primarily on the dose received and the duration over which it was received. The effects of high doses delivered rapidly are generally observable within days, whereas recipients of frequent low doses may not present symptoms for years. Some specialists have suggested that radiological weapons have a profound psychological impact due to people's 'inability to perceive the presence of radiation with the ordinary human senses and to concerns about perceived long-lasting radiation effects' (Smith 2018a; Salter 2001).

Agents can also vary substantially in terms of their persistency; that is how long they remain viable and capable of acting as intended. Most chemical agents can be broadly divided into two groups: Those that remain stable for only a few minutes, and those that can remain hazardous for weeks or months. Some biological agents may survive outside a living host for many years, whereas others may die within minutes or hours of exposure. Humidity, temperature, and UV radiation are the primary factors affecting the persistency of biological agents (Weber, 2012). Radiological material decays exponentially according to its sub-atomic characteristics. The time it takes for half of a radionuclide's atoms to decay is expressed as its 'half-life'. These range from days to millennia, although many common medical radionuclides last between a few days and 30 years (Smith 2018a; Li 2008, pp. 257–259; Perez et al. 2008, pp. 366–367). Table 2 summarises some of the key typical agent properties by type.

6 Acquisition of CBR Agents by Non-state Actors

There are several reasons for the very low rate of incidence of CBR attacks. At a fundamental level, the acquisition, production, and/or weaponisation of CBR materials by non-state actors is challenging. Some non-state groups will choose to eschew the damaging optics and limited battlefield effectiveness of CBR weapons. Groups that do not wish to attract the wrong kind of national or international attention may also avoid these weapons; large-scale procurement of CBR materials is likely to enter the sight picture of various global intelligence and law enforcement agencies. Nations which are party to international instruments such as the Chemical Weapons Convention (CWC)[19] or the Biological Weapons Convention (BWC)[20] are subject to reporting requirements, and typically go to great lengths to secure stockpiles of controlled agents. Even in edge cases where CBR precursors, components, or weapons were stolen from states not party to, or not abiding by, such instruments, there is a good chance a country would self-report. This would likely be motivated, at least in part, by the fear of being implicated in supporting terrorist acts.

If a CBR attack is to be planned and executed using high-grade materials, the act of acquiring the component materials would almost invariably rely on more covert means; accessing these via established criminal networks or a willing state- or quasi-state sponsor (Meulenbelt and Nieuwenhuizen 2016; Smith 2018a). Even in these scenarios, the acquisition of CBR material constitutes a high risk, moderate reward end-state. Criminal networks, for the most part, are (or become) porous and susceptible to law enforcement and intelligence penetration and countermeasures. Acquisition of CBR materials or weapons via state sponsorship poses extreme risk to the supplier nation; in many cases, this is likely to directly affect its future economic wellbeing, or even its sovereignty. Individuals may also be directly targeted in response to supporting the use or proliferation of CBR weapons. A variety of individuals, organisations, and states have been sanctioned by national and multinational bodies in recent years. Both Russian and Syrian entities have been subject to sanctions for the use of chemical weapons in the United Kingdom and Syria, respectively, for example.[21] Sources of CBR material are increasingly secured and monitored by specialised components within national governments. In the United States, for example, certain radiological materials are controlled by licensing at the state and federal levels, monitored by various agencies, and tracked in a National Source Tracking System, administered by the US Nuclear Regulatory Commission (USNRC 2018a).

[19] Properly the 'Convention on the Prohibition of the Development, Production, Stockpiling and Use of Chemical Weapons and on Their Destruction'. See OPCW (1993).

[20] Properly the 'Convention on the Prohibition of the Development, Production and Stockpiling of Bacteriological (Biological) and Toxin Weapons and on Their Destruction'. See UNODA (1975).

[21] See, for example, EU Council (2018, 2019), Wintour (2018), and Wroughton and Zengerle (2018).

Another path is for non-state actors to produce their own chemical or biological materials or, in the case of some biological pathogens, 'harvest' these from endemic areas. This scenario is often particularly challenging to the would-be CBR terrorist in terms of the coordination and consolidation of required equipment and specially-trained individuals. These factors, and the frequent reluctance of such highly-trained and educated individuals to risk their life and liberty for a terrorist agenda, significantly decrease the feasibility of such scenarios (Gellman 2003; Strack 2017). As a result, several actors have turned towards adapting industrially- and commercially-available agents, including toxic industrial chemicals (TIC), to nefarious ends. In Syria, this has been regularly observed in the forms of attacks using chlorine, a readily-available TIC.[22] Chlorine gas has proved a chemical agent of choice for primarily logistical reasons. It is easy to produce industrially or acquire commercially, and it has a variety of plausible legitimate uses (Jenzen-Jones and Wright 2018b). Perhaps most importantly, chlorine is critical to water purification process and regularly produced and acquired for that purpose; accordingly, it is difficult for international sanctions to restrict chlorine. It is also relatively non-persistent, and can be difficult to identify as a weapon even a few hours after an attack. This property, as well as it commonplace use in civilian areas, has resulted in it being a somewhat 'deniable' weapon. The relatively slow onset of incapacitation and the dispersion over wide areas explains why most incidents in Syria have not produced large numbers of fatalities (Jenzen-Jones and Wright 2018b; Smith 2018b).[23]

Perhaps the least-likely scenario is one in which a non-state actor stumbles upon a chance opportunity to acquire CBR material. This occurred after the fall of Mosul in 2016, when Islamic State forces discovered a stock pile of cobalt-60 stored in an unsecured location (Hummel 2016). Similarly, in December 2013, a cargo truck transporting medical equipment which contained cobalt-60 was hijacked by armed assailants from a petrol station in Mexico (Archibold and Gladstone 2013). Consequently, there have been justifiable fears from the global community that the radionuclide might be employed in an improvised radiological dispersion device; to date, this has not happened. The reasoning behind the Islamic State's decision to refrain from incorporating cobalt-60 into their arsenal has not yet been deciphered. Most likely, it was driven by a combination of factors: a lack of technical expertise, limited means to weaponise the material, and safety inhibitions of technicians. Such an RDD, or 'dirty bomb', would also have been of limited use in the type of open, relatively conventional warfare that IS was

[22] As a weapon, chlorine gas functions as a choking agent, attacking a victim's respiratory system. Concentrations of 1–3 parts per million (ppm) will begin to irritate they eyes, nose, and throat. At 30 ppm, chlorine gas will typically induce shortness of breath, coughing, and chest pain, whilst at 60 ppm, victims may suffer from pulmonary edema, as the body responds to chlorine's corrosive effects on the alveoli. Once concentrations reach 400 ppm, exposure is likely to prove fatal after 30 minutes; at 800 ppm, victims may survive only a few minutes (Smith 2018b).

[23] One exception to this general rule is the 7 April 2018 Douma incident involving TIC IADMs. The combination of atypical symptoms (like extreme pupil dilation) and the high fatality rate (49) may indicate that another agent with greater toxicity was involved (Jenzen-Jones and Wright 2018b).

engaged in at the time. It is logical to assume they would prioritise the use of simpler, more reliable, and more readily-available weapon systems, such as SVBIEDs; and prioritise research and development into conventional munitions, such as IADMs to be deployed from COTS small UAVs.

7 Potential Targets for CBR Attacks[24]

The use of CBR materials in attacks conducted by non-state actors has had a significant impact on not only direct and indirect victims, but also on global security postures. The use of TIC (chlorine), mustard, and nerve agents in Syria has generated significant military and political responses globally, for example.

Were a major CBR attack facilitated by UAVs to take place in a highly-developed, industrialised nation, it would demonstrate a level of capability and intent hitherto unseen (Smith 2018a).

In traditional military practice, the selection of a target precedes the selection of the methods and means of warfare used to engage that target. This process of weapon-target matching, sometimes known as 'weaponeering', is a fundamental element of the targeting process for armed forces (Cross et al. 2016). Non-state actors, however, are almost invariably limited by the methods and means available to them. The choice of target will be primarily determined by factors of access and vulnerability. A CBR attack may also be valued for its psychological impacts, and so CBR agents may be selected over conventional munitions which would be considered more effective by traditional military logic. Nonetheless, many of the same principles apply, and perfunctory or detailed assessments may be made of methods of engagement (dispersion), target composition, population density, soil/terrain type, time of attack, weather, and so on.[25] The characteristics of the delivery method and delivery platform may also be adjusted to achieve the desired effects on target. For example, an attacker planning to strike a precision target with a chemical agent delivered from a fixed-wing UAV may select an explosively atomised air-delivered munition rather than an aerosol spray. This may, in turn, necessitate a change in platform selection; in practical terms, the accuracy and precision of small bomb-type munitions delivered from fixed wing UAVs is lower than when those munitions are delivered from stable, multi-rotor type designs. Of course, the attacker would then be limited in terms of maximum range—trade-offs are inherent to this decision-making process.

CBR agents as delivered by COTS small UAVs may be used for either on-target or off-target attacks. On-target attacks are those in which the munition is employed

[24] Note: given the particularly sensitive nature of discussing the potential targets of CBR attacks and the relative effectiveness of attacks against different target types, some information has been withheld on security grounds. Legitimate parties are invited to contact ARES for further information at: contact@armamentresearch.com.

[25] For a more detailed discussion of the targeting process, see Cross et al. (2016), pp. 40–46.

Table 3 Exposure pathways and methods of attack

Attack method	Common exposure pathways	Uncommon exposure pathways
On-target (direct)	Inhalation, absorption	Ingestion, injection
On-target (indirect)[a]	Ingestion	Inhalation, absorption, injection
Off-target (direct)	Inhalation, absorption	Ingestion, injection
Off-target (indirect)	Ingestion, absorption	Inhalation, injection

[a]As noted, indirect on-target attacks are likely to be very rare

with the intent to directly and immediately affect the target, whereas off-target attacks deliberately intend to affect the target from a stand-off distance (effects usually being delayed by minutes or hours), or by utilising second- and subsequent-order effects, typically characterised by a substantial interval of time between delivery and effect (hours, days, or months). The latter type of attack is sometimes referred to as an 'indirect off-target attack'.[26] For example, in attacking an outdoor music or sporting event, an on-target attack may rely on dispersing a liquid agent from low altitude directly on to the target, whereas an off-target attack might use multiple UAVs to release an aerosolised agent upwind of the target site. An indirect off-target attack could be executed by an earlier contamination of the food or water supply for the site. Off-target attacks are frequently dependent on favourable meteorological and environmental conditions, and typically require more substantially more agent than on-target releases[27] (Weber, 2012). Given the limited payload capacity of small UAVs, effective off-target attacks would be much harder to realise. Table 3 shows the exposure pathways common to the different types of attack.

If an actor's primary aim is to inflict maximum casualties on a population, then biological agents would likely be the preferred choice. Despite the reputation of chemical and radiological weapons, large quantities of these agents are typically required to produce large-scale casualties. Perhaps the only notable exception are nerve agents, which pose significant challenges to acquire and employ. The toxicity, transmissibility, and latency of biological agents make them the most suitable type of CBR for causing mass-casualty events. Whilst impossible to quantify, the medium- to long-term psychological effects of such an attack would no doubt also be significant. However, the comparative 'invisibility' of such an attack may render chemical or radiological agents more appealing to certain actors. A low-latency chemical agent would have particularly dramatic and immediate effect.[28]

The sheer quantity and variety of extant radiological agent sources and the low technical barrier to crude weaponisation (as in a dirty bomb) may suggest this

[26] 'Indirect on-target' attacks are theoretically possible (e.g. a UAV flying overhead in full view of the target and dropping contaminated consumables that people subsequently ingest), but in practice likely to be a rarity.

[27] One estimate for an attack with a nominal biological agent put the difference at 'a few grams' versus 'hundreds of grams' (Weber 2012).

[28] Biological agents that cause particularly horrible illnesses (for example, haemorrhagic fever viruses, such as Ebola or Marburg) may be selected for similar reasons (Riedel 2004).

category is the easiest type of CBR agent for a non-state actor to acquire and weaponise. Whilst the preferable materials of this type would almost always have to be stolen by a non-state actor, a range of viable radionuclides are commonly used in hospital radiation therapy and industrial radiography. Nearly 3,000 cases of theft, loss, and unauthorised possession of radioactive materials were reported to the IAEA between 1995 and 2015 (Green 2016).

CBR agents may also be utilised to render an area contaminated, resulting from either an on-target or off-target attack. This area denial effect is particularly suited to radiological material, as dispersed by an RDD, in terms of persistency, although chemical and biological agents may also generate substantial clean-up costs. Given the abundance of caution modern governments exhibit when it comes to CBR events, even a minor attack can have massive financial impacts. The British government announced in 2018 that the costs of decontamination in Salisbury following an alleged Russian attack using a Novichok chemical agent were expected to run into the tens of millions of pounds (Kerbaj 2018). This figure does not include the immediate costs of the police response (estimated at more than 7.5 million pounds) nor the subsequent investigations (BBC 2018). The 2001 anthrax letter attacks in the United States are estimated to have cost around 320 million dollars to clean up (Schmitt and Zacchia 2012). Clearly, the financial impacts of any CBR event could be significant. Accordingly, financial damage may be a primary motivating factor for hostile actors; in these cases, specific target selection and conventional damage metrics (i.e. casualties) may be of reduced importance.

8 Barriers to a Successful CBR Attack by COTS Small UAVs

There are several reasons a violent non-state actor or individual may elect to employ a COTS small UAV as the delivery platform of choice for a CBR attack. Many of the specific advantages of such a platform are withheld from this article on security grounds.[29] Broadly speaking, however, some of the key advantages could include precision, access to targets, difficulty of interception, optimised dispersion, reduced risk of detention, anonymity, and a demonstration of 'modern' capabilities. There are also substantial barriers to a successful CRB attack by COTS small UAVs. As noted previously, there are a number of challenges that must be overcome to weaponise most CBR agents. It is not the intention of this discussion to address those technical challenges in detail, however it should be reemphasised that existing UAVs which have been adapted for dispersing other types of agents—such as agricultural or riot control agents—could potentially be adapted to disperse CBR payloads (Friese et al. 2016).

Even should CBR material be successfully acquired and weaponised by a non-state actor, there are further obstacles standing in the way of a successful attack

[29] Legitimate researchers or goverment representatives are invited to contact ARES for further information at: contact@armamentresearch.com.

with such agents. For a potential terrorist user, key constraints of most UAVs, particularly multi-rotor types, are the limited payload, range, and flight time. A $1,000 USD hobbyist UAV can typically lift no more than 1 kg (2.2 lbs) and has a maximum flight duration of 20–30 minutes. A $5,000 USD prosumer UAV can perhaps increase that lift to 8 kg with a slightly longer duration. To get meaningfully greater payloads of up to 50 kg (110 lbs), commercial crop-spraying designs are available; but their flying time can be significantly reduced—sometimes to less than 15 minutes. In many countries these systems are not available to civilians or are strictly regulated (Friese et al. 2016; Smith 2018a).

Along with regulations regarding the possession or operation of UAVs, many countries are deploying control measures to ensure that UAVs do not threaten the security of sensitive sites; be those security facilities, critical infrastructure, or areas where large numbers of people may congregate (e.g. sports stadia). At the global level, such critical infrastructure measures align with UN Security Council Resolution 2341 (2017), that directs member nations to conduct a threat analysis of critical infrastructure. Counter-UAV systems are increasingly being deployed, including a range of kinetic, electromagnetic, directed energy, and novel designs (Friese et al. 2016; Jenzen-Jones 2017a, b). The UK MoD, for example, has reportedly ordered a directed energy (laser) counter-UAS system which offers the capability to detect targets at ranges of up to 5 km (Smith 2018a). Portable jamming devices have been employed in combat in Iraq and Syria, countering the COTS small UAVs used by the Islamic State. Batelle DroneDefender, Radio Hill Dronebuster Block 3, and Raysun MD1 Multicopter Defender rifle-configuration jammers have been observed, as have larger units fitted to light tactical vehicles and deployed around bases, including the Blighter AUDS system and DIY examples produced by Iraqi popular mobilisation units (PMUs) (Fulmer and Jenzen-Jones 2017; Jenzen-Jones 2017a).[30]

Geofencing systems are another important development in preventing UAVs from accessing restricted airspace. In most cases, the term 'geofencing' describes a software- or firmware-level virtual GPS perimeter that is programmed to prevent a given UAV from being piloted into restricted airspace. In current practice, these areas include airports, critical infrastructure, and military facilities (Dixon, 2017). In most cases, such systems are integrated by the UAV manufacturer. In some cases, commercial users are exempt from some or all geofencing. Private or commercial users may also be required to 'self-authorise', disabling geofencing protocols by providing identifiable data, such as credit card or driver's license details (GPS World 2018; Waddell 2017). Non-state actors have already developed several methods to defeat geofencing restrictions, however, and in the case of a few specific models of UAV, a 'cat and mouse' game between manufacturers/geofencing developers and malicious actors has developed (Jenzen-Jones 2017a).

[30] It is important to add that, despite the relatively profusion of modern 'drone jammer' rifles, most Islamic State UAVs which were successfully engaged during the battle of Mosul were brought down with machine gun fire (Fulmer and Jenzen-Jones 2017).

'Identify, friend or foe' (IFF) systems are used to identify aircraft for both civilian and military applications.[31] They consist, at the most basic level, of an interrogator and a transponder, the latter transmitting a unique identifying signal when interrogated. So-called 'mini' format IFF transponders—some weighing less than 1.5 kg— are already in service around the world (Rees 2013). Smaller 'micro' transponders—some weighing as little as 100 g—have also been developed and these have been examined for use with small UAVs by military forces (Armistead 2018; US Navy 2014; Sagetech 2018). Such developments could normalise the use of geofencing and IFF on commercial, government, and other small UAVs, providing more robust and reactive gatekeeping of airspace. Depending on the IFF mode that is used, however, there remain concerns about the security and integrity of the signal (Gellender 2014).

C-UAV technology and associated employment tactics are being employed on a greater scale than ever before. However, several non-state actors have already taken innovative steps to circumvent existing countermeasures and tactics (Jenzen-Jones and Wright 2018a). These range from sophisticated modification of components, firmware, and software to simple 'kludges', such as switching off or removing GPS modules, or covering broadcast antennae in makeshift Faraday cages (Peitz 2018; Jenzen-Jones 2017a). Non-state actors may also seek to hack UAVs, compromising or acquiring aerial systems by exploiting vulnerabilities. Jamming control frequencies, accessing video feeds or other telemetry, overwhelming onboard processors and precipitating malfunctions, spoofing GPS coordinates to encourage landing in an area controlled by the malicious actor, and even 'hijacking' complete control, are all techniques which have been exploited in current UAVs (Moskvitch 2014; Glaser 2017).[32] Even US military UAVs have not been immune from these attacks; UAV video feeds were hacked by Iraqi insurgents (MacAskill 2009).[33] One security professional even suggested that, in the future, UAVs with automated access to restricted areas (for example, with appropriate geofencing or IFF credentials) might be stealthily hijacked along their normal route, outfitted with CBR devices, and used to carry such devices into a secure area without knowledge of their controlling party.[34] Despite rapid evolutions, the protective framework against hostile UAV attacks remains loosely organised, inconsistently applied, and poorly tested (Jenzen-Jones 2017a).

Even if a violent group can acquire and weaponise CBR material; purchase or build a UAV capable of carrying and deploying the payload; and avoid any counter-

[31] The name is somewhat misleading, as the system can only truly identify 'friendly' or other responsive aircraft from unknown (unresponsive) aircraft.

[32] Similar techniques have exploited vulnerabilities in commercial airliners. See, for example, Biesecker (2017).

[33] Non-state actors may also exploit cyber vulnerabilities to access classified or sensitive documents regarding UAVs or CBR agents online, or purchase such documents off other disreputable parties. In 2018, for example, a vendor on the dark web was offering sensitive documents related to the US military's Reaper UAV (Brewser 2018).

[34] Author interview with confidential security source, January 2019.

UAV measures in place, this still may not necessarily allow them to conduct an attack in the manner they desire. CBR materials have a vary significantly in their physical characteristics, effects, intoxication/infection pathways, and persistency, as well as encompassing a broad range of differing storage and dispersal requirements. Chemical agents tend to intoxicate relatively quickly, typically within minutes to hours, and signs can be relatively obvious. Most chemical warfare agents are liquids, so effective distribution would require a sprayer system to disperse the agent onto exposed skin. Biological weapons are notoriously difficult to weaponise, requiring finely-tuned dispersion systems that will increase the likelihood of a suitably quantity of a given agent reaching a targeted area. Several days may pass before infection with biological agents becomes apparent, posing serious concern to health professionals responsible for the identification and treatment of asymptomatic diseases. Radiological effects, unless resulting from a massive and recognisable source such as a nuclear core meltdown, are also likely to take time to manifest in a victim. Conceivably this could take years and be masked by other unrelated and chronic causes. An important additional consideration is the variable potency of CBR material. Whilst there are lots of 'book values' that might state that a few micrograms or microscopic spores are sufficient for a fatal dose, experience has shown that there are a large number of real-world variables that can significantly alter those values. Given the relatively small payloads of the majority of UAVs, small quantities of CBR material may have a much more limited effect than anticipated under 'field' conditions.

9 Conclusions

Whilst there is currently a low probability of a UAV attack with CBR agents, it is essential that we remember that the technology to outfit a rotary- or fixed-wing UAV with a dispersal system is commercially available, groups can readily acquire a number of less-toxic noxious agents or skin irritants, and a range of 'soft targets' are vulnerable to attack. The impacts of a 'hoax' CBR attack may be nearly as serious as an authentic one, particularly if such an attack were combined with real-time propagation of false narratives designed to instil fear, confusion, and panic in the target population. Alternatively, attacks using readily available agents may target links in the critical infrastructure chain that are not immediate priorities for hardening, such as power transmission and distribution, large open-air water reservoirs, and farms.

Even where CBR agents cannot be successfully acquired or weaponised, violent actors may use UAVs armed with conventional payloads to strike facilities which house such materials. In 2018, members of the French branch of the environmental advocacy organisation Greenpeace flew two UAVs into the airspace of the Bugey Nuclear Power Plant near Lyon, France (De Clercq 2018). Whilst no substantial damage was caused—one UAV harmlessly crashing into the side of a building located within the secure site—it demonstrated the ease with which the airspace immediately above a critical infrastructure facility could be penetrated. Additionally, despite claims from the activists that the intent was to raise awareness of this danger, non-lethal, seemingly 'harmless' probing attacks may be observed or replicated by violent actors to refine emerging tactics and improve UAV design (Smith 2018a). Although the threat of a CBR terror event using UAVs is lower than other more established means of attacks, ongoing assessments of the feasibility of such attacks and the appropriate defensive posturing to engage such threats must remain part of the counter-terror dialogue. The possibility of a group opportunistically acquiring high-potency CBR materials should likewise not be discounted, nor should the threat posed by hoax attacks or attacks using unrestricted or otherwise readily available agents, including TICs. Due to the availability of certain precursors and TICS, the organisational knowledge held by some non-state actors, and the availability of existing chemical-dispersion technology for UAVs, chemical payloads are the most likely to be weaponised and employed (Binder et al. 2018; Smith 2018a). From the vantage point of a terrorist organisation, a cost-benefit analysis will lean towards the continued use of conventional munitions at present. Although CBR attacks constitute a minute percentage of all terror attacks—and even smaller percentage of battlefield actions—and whilst most attacks launched from COTS small UAVs have employed conventional arms and munitions, the CBR threat should not be discounted. CBR incidents may not be common, but they still pose a significant physical, psychological, and economic threat, and COTS small UAVs may well be used to facilitate such attacks in future.

Acknowledgements The author would like to acknowledge the generous contributions of his colleagues at Armament Research Services (ARES), including Jerry Smith, Larry Friese, Sean Flachs, and Daniel Hughes. Mr. Smith, in particular, provided detailed analysis of the CBR threats outlined herein, drawing on his considerable expertise. Thanks are also due to those who shared their input and feedback on the condition of anonymity. All errors remain those of the author alone.

References

Archibold, Randal C., and Rick Gladstone. 2013. Truck with radioactive load is recovered in Mexico. *New York Times*. 4 December. https://www.nytimes.com/2013/12/05/world/americas/Radioactive-Cargo-Mexico.html.

ARES (Armament Research Services). 2016. *Emergence of UAV-delivered munitions in Iraq*. Confidential briefing paper. Perth: ARES.

———. 2017a. *An assessment of the use of UAV-delivered munitions by the Islamic State*. Confidential. Perth: ARES.

———. 2017b. *Unmanned threats update: The use of COTS UAVs by non-state actors*. Confidential. Perth: ARES.

———. 2018a. *Suspected CW attacks on East Ghouta, January & April 2018*. Confidential report. Perth: ARES.

———. 2018b. *Unmanned ambitions: Mapping the rise of military UAV industries worldwide*. Confidential report. Perth: ARES.

———. 2019. *Israeli use of COTS UAVs to deliver RCA: March-April 2019 analysis*. Confidential report. Perth: ARES.

———. n.d. *Conflict materiel (CONMAT) database*. Confidential. Perth: ARES.

Armistead, Scott E. 2018. *DSIAC Technical Inquiry (TI) response report: Small-micro IFF transponder survey*. Report number DSIAC-2019-1032. Draft.

Asahi Shimbun. 2015. UPDATE: Man who operated drone found in Tokyo was protesting Abe's nuclear policy. 25 April. Archived at: https://web.archive.org/web/20150427224620/http://ajw.asahi.com/article/behind_news/social_affairs/AJ201504250021.

Ballard, T., J. Pate, G. Ackerman, D. McCauley, and S. Lawson. 2001. *Chronology of Aum Shinrikyo's CBW activities*. Monterey Institute of International Studies. Available at: https://www.nonproliferation.org/wp-content/uploads/2016/06/aum_chrn.pdf.

BBC (British Broadcasting Corporation). 2015. Japan radioactive drone: Tokyo police arrest man. *BBC News*. 25 April. https://www.bbc.co.uk/news/world-asia-32465624.

———. 2018. Salisbury nerve agent attack 'cost police force £7.5m'. *BBC News*. 4 June. https://www.bbc.com/news/uk-england-wiltshire-44353580.

Biesecker, Calvin. 2017. Boeing 757 Testing shows airplanes vulnerable to hacking, DHS says. *Avionics International*. 8 November. https://www.aviationtoday.com/2017/11/08/boeing-757-testing-shows-airplanes-vulnerable-hacking-dhs-says/.

Binder, Markus K., Jilliam M. Quigley, and Herbert F. Tinsley. 2018. *Islamic state chemical weapons: A case contained by its context?. CTC Sentinel*. March 2018.

Bone, Elizabeth, and Christopher Bolkcom. 2003. *Unmanned aerial vehicles: Background and issues for congress*. Report for Congress: RL31872. 25 April. Available from http://fas.org/irp/crs/RL31872.pdf.

Brewster, Thomas. 2018 A hacker sold U.S. military drone documents on the Dark Web for Just $200. *Forbes*. 11 July. https://www.forbes.com/sites/thomasbrewster/2018/07/11/a-hacker-sold-u-s-military-drone-documents-on-the-dark-web-for-just-200/#788829213dc8.

Bunker, Robert J. 2019. *Contemporary chemical weapons use in Syria and Iraq by the Assad regime and the Islamic state*. Carlisle: Strategic Studies Institute (SSI), US Army War College. https://ssi.armywarcollege.edu/pubs/display.cfm?pubID=1400.

Cross, Kenneth, Dullum Ove, N.R. Jenzen-Jones, and Marc Garlasco. 2016. *Explosive weapons in populated areas: Technical considerations relevant to their use and effects*. Perth: Armament Research Services. http://armamentresearch.com/wp-content/uploads/2016/06/ARES-Special-Report-Explosive-Weapons-in-Populated-Areas-May-2016_web.pdf.

Crowley, Michael. 2015. *Tear gassing by remote*. London: Remote Control Project.

Cruickshank, Paul. 2018. A view from the CT Foxhole: An interview with Hamish de Bretton-Gordon, Former Commander of U.K. CBRN Regiment. *CTC Sentinel*. August 2018. pp. 5–9.

De Clercq, Geert. 2018. Greenpeace crashes Superman-shaped drone into French nuclear plant. *Reuters*. 3 July. https://www.reuters.com/article/us-france-nuclear-greenpeace/greenpeace-crashes-superman-shaped-drone-into-french-nuclear-plant-idUSKBN1JT1JM.

Dixon, Drew. 2017. Geofencing stops drones in their tracks. *Government Technology*. 1 August. http://www.govtech.com/public-safety/Geofencing-Stops-Drones-in-Their-Tracks.html.

DoD (United States Department of Defense). 2011. *Unmanned aircraft system airspace integration plan*. Version 2.0. Washington, D.C.: United States Department of Defense.

EU Council (Council of the European Union). 2018. Chemical weapons: The council adopts a new sanctions regime. Press release: 15 October. http://www.consilium.europa.eu/en/press/press-releases/2018/10/15/chemical-weapons-the-council-adopts-a-new-sanctions-regime/.

———. 2019. Chemical weapons: The EU places nine persons and one entity under new sanctions regime. Press release: 21 January 2019. http://www.consilium.europa.eu/en/press/press-releases/2019/01/21/chemical-weapons-the-eu-places-nine-persons-and-one-entity-under-new-sanctions-regime/.

FAA (United States Federal Aviation Administration). 2015. Clarification of the applicability of aircraft registration requirements for Unmanned Aircraft Systems (UAS) and request for information regarding electronic registration for UAS. https://www.federalregister.gov/articles/2015/10/22/2015-26874/clarification-of-the-applicabilityof-aircraft-registration-requirements-for-unmanned-aircraft.

Farfán, E.B., et al. 2011. Assessment of (90)sr and (137)cs penetration into reinforced concrete (extent of "deepening") under natural atmospheric conditions. *Health Physics* 101 (3): 311–320.

Ferguson, Jonathan, and N. R. Jenzen-Jones. 2014. *Raising red flags: An examination of arms & munitions in the ongoing conflict in Ukraine*. Research Report No. 3. Perth: ARES. http://armamentresearch.com/Uploads/Research%20Report%20No.%203%20-%20Raising%20Red%20Flags.pdf.

Friese, Larry, N. R. Jenzen-Jones, and Michael Smallwood. 2016. *Emerging unmanned threats: The use of commercially-available UAVs by armed non-state actors*. Special Report No. 2. Perth: Armament Research Services (ARES). http://armamentresearch.com/wp-content/uploads/2016/02/ARES-Special-Report-No.-2-Emerging-Unmanned-Threats.pdf.

Fulmer, Kenton, and N. R. Jenzen-Jones. 2017. Improvised air-delivered munitions in Syria & Iraq: A brief overview. *Counter-IED Report*. Spring/Summer 2017. pp. 1–10.

Ganesan, K., S.K. Raza, and R. Vijayaraghavan. 2010. Chemical warfare agents. *Journal of Pharmacy and Bioallied Sciences* 2 (3): 166–178.

Gellender, ed. 2014. Past, present and future of secondary radar. Presentation. 2 July. https://ieee.li/technical/past-present-and-future-of-secondary-radar/.

Gellman, Barton. 2003. Al Qaeda near biological, chemical arms production. *Washington Post*. 23 March. http://www.washingtonpost.com/wp-dyn/content/article/2006/06/09/AR2006060900918.html?noredirect=on.

Glaser, April. 2017. The U.S. government showed just how easy it is to hack drones made by Parrot, DBPower and Cheerson. *Recode*. 4 January. https://www.recode.net/2017/1/4/14062654/drones-hacking-security-ftc-parrot-dbpower-cheerson.

GPS World. 2018. *DJI will unlock geofencing for enterprise drone users*. 17 July. https://www.gpsworld.com/dji-will-unlock-geofencing-for-enterprise-drone-users/.

Green, Andrew. 2016. Radioactive Source Security Working Group to help improve security programmes for Member States. *International Atomic Energy Agency*. Press release. 5 May. https://www.iaea.org/newscenter/news/radioactive-source-security-working-group-to-help-improve-security-programmes-for-member-states.

Hilton, Daniel. 2018. Drones over Gaza: How Israel tested its latest technology on protesters. *Middle East Eye*. 16 May 2018. https://www.middleeasteye.net/news/drones-over-gaza-how-israel-tested-its-latest-technology-protesters.

Hummel, Stephen. The Islamic State and WMD: Assessing the future threat. *CTC Sentinel*. January 2016. pp. 18–21.

Hummel, Kristina. A view from the CT Foxhole: Vayl S. Oxford, Director, Defense Threat Reduction Agency. *CTC Sentinel*. March 2019. pp. 10–14.

IAEA (International Atomic Energy Agency). 2017. IAEA Incident AND Trafficking Database (ITDB): Incidents of nuclear and other radioactive material out of regulatory control. 2017 Fact Sheet. https://www.iaea.org/sites/default/files/17/12/itdb-factsheet-2017.pdf.

Ispra Ltd. 2015. Cyclone riot control drone system. Company brochure.

Jenzen-Jones, N.R. 2017a. *Countering the non-state UAV threat*. Confidential report. Perth: ARES.

———. 2017b. *Emerging unmanned threats: The continued use of COTS UAVs by non-state actors*. Presentation at Counter UAS Conference, London, May 2017.

Jenzen-Jones, N.R., and Galen Wright. 2018a. Improvised, craft-produced and repurposed munitions deployed from UAVs in recent years. *Counter-IED Report*. Spring/Summer 2018. pp. 53–64.

———. 2018b. Improvised chemical munitions in Syria, January 2017–August 2018. *Counter-IED Report*. Autumn 2018. pp. 87–98.

Kerbaj, Richard. 2018. Novichok clean-up to cost tens of millions. *The Sunday Times*. 20 May. https://www.thetimes.co.uk/article/novichok-clean-up-to-cost-tens-of-millions-3j8pn8fk2.

Li, Zuofeng. 2008. Physics and clinical aspects of brachytherapy. In *Technical basis of radiation therapy: Practical clinical applications*, ed. Levitt et al., 4th ed. Berlin: Springer.

MacAskill, Ewen. 2009. US drones hacked by Iraqi insurgents. *The Guardian*. 17 December. https://www.theguardian.com/world/2009/dec/17/skygrabber-american-drones-hacked.

MEMRI (Middle East Media Research Institute). 2018. Pro-ISIS Media outlet circulates video calling for biological attacks in the West. 20 July. https://www.memri.org/tv/pro-isis-video-calls-for-biological-attacks-in-the-west/transcript.

Meulenbelt, Stephanie E., and Maarten S. Nieuwenhuizen. 2016. *Non-state actors' pursuit of CBRN weapons: From motivation to potential humanitarian consequences*. Geneva: ICRC. https://www.icrc.org/en/download/file/24548/irc97_17.pdf.

Mizokami, Kyle. 2017a. Kaboom! Russian UAV with Thermite Grenade blows up a billion dollars of Ukrainian Ammo. *Popular Mechanics*. 27 July 2017. https://www.popularmechanics.com/military/weapons/news/a27511/russia-UAV-thermite-grenade-ukraine-ammo/.

———. 2017b. Another Ukrainian Ammo Dump goes up in massive explosion. *Popular Mechanics*. 27 September 2017. https://www.popularmechanics.com/military/weapons/news/a28412/ukrainian-ammo-dump-explosion/.

Moskvitch, Katia. 2014. Are drones the next target for hackers. *BBC Futures*. 6 February. http://www.bbc.com/future/story/20140206-can-drones-be-hacked.

NDPC (National Disaster Preparedness Consortium). 2016. Exponential growth in drone technology leads to weaponization of small unmanned aerial systems. *The NDPC News*. Fall 2016. https://www.ndpc.us/pdf/NDPCNews_Fall2016.pdf.

NRC (National Research Council). 2000. *Strategies to protect the health of deployed U.S. forces: Detecting, characterizing, and documenting exposures*. Washington, D.C.: The National Academies Press.

———. 1993. *Convention on the prohibition of the development, production, stockpiling and use of chemical weapons and on their destruction* [CWC]. https://www.opcw.org/fileadmin/OPCW/CWC/CWC_en.pdf.

Peitz, Dirk. 2018. Die billigste Luftwaffe der Welt. *Die Zeit*. 8 August. https://www.zeit.de/digital/2018-08/drohnen-venezuela-anschlag-waffe-kauf/komplettansicht.

Perez, Carlos A., Robert D. Zwicker, and Zuofeng Li. 2008. Clinical applications of low dose rate and medium dose rate brachytherapy. In *Technical basis of radiation therapy: Practical clinical applications,* ed. Levitt et al., 4th ed. Berlin: Springer.

Pitschmann, Vladimír. 2014. Overall view of chemical and biochemical weapons. *Toxins* 6 (6): 1761–1784.

Rassler, Don. 2018. *The Islamic State and Drones: Supply, scale, and future threats*. West Point: Combatting Terrorism Center. https://ctc.usma.edu/app/uploads/2018/07/Islamic-State-and-Drones-Release-Version.pdf.

Rassler, Don, Muhammad al-Ubaydi, and Vera Mironova. 2017. The Islamic State's Drone documents: Management, acquisitions, and DIY Tradecraft. *CTC Blog*. 31 January. https://ctc.usma.edu/ctc-perspectives-the-islamic-states-drone-documents-management-acquisitions-and-diy-tradecraft/.

Rees, Mike. 2013. Raytheon's Mini IFF transponders to be used on Korean Air UAVs. *Unmanned Systems News*. 19 June. https://www.unmannedsystemstechnology.com/2013/06/raytheons-mini-iff-transponders-to-be-used-on-korean-air-uavs/.

———. 2018. Intel demonstrates remote drone identification solution. *Unmanned Systems News*. 20 August. https://www.unmannedsystemstechnology.com/2018/08/intel-announces-new-open-standard-for-remote-drone-identification/.

Riedel, Stefan. 2004. Biological warfare and bioterrorism: A historical review. *Baylor University Medical Center Proceedings*. 17 (4): 400–406.

Rogoway, Tyler. 2018. Israel uses drone racers to down incendiary kites and drones to dispense tear gas over Gaza. *The Warzone*. 14 May. https://www.thedrive.com/the-war-zone/20853/israel-uses-drone-racers-to-down-incendiary-kites-and-drones-to-dispense-tear-gas-over-gaza.

Sagetech. 2018. XP family of transponders. Accessed: 25 Jan 2019. https://sagetech.com/xp-transponders/.

Salter, C.A. 2001. Psychological effects of nuclear and radiological warfare. *Military Medicine* 166 (12 Suppl): 17–18.

Schmitt, Ketra, and Nicholas A. Zacchia. 2012. Total decontamination cost of the anthrax letter attacks. *Biosecurity and Bioterrorism: Biodefense Strategy, Practice, and Science* 10(1).

Sekiguchi, Toko. 2015. Drone found at Japan Prime Minister Shinzo Abe's Office. *Wall Street Journal*. 22 April. https://www.wsj.com/articles/drone-found-at-japan-prime-ministers-office-1429694098.

Smith, Jerry. 2018a. *UAV + CBR: Hype or horror?* Confidential report. Perth: ARES.

———. 2018b. *Chlorine as a Weapon*. Confidential Report. Perth: ARES.

Strack, Columb. 2017. The evolution of the Islamic State's chemical weapons efforts. *CTC Sentinel*. October 2017. pp. 19–23.

The Economist. 2018. A failed UAV attack shows that Nicolás Maduro is vulnerable. 8 August 2018. https://www.economist.com/the-americas/2018/08/09/a-failed-UAV-attack-shows-that-nicolas-maduro-is-vulnerable.

The Insight Partners. 2018. *Unmanned Aerial Vehicle (UAV) market to 2025 – Global analysis and forecasts by component by type and application*. Summary. Available at: https://www.researchandmarkets.com/research/vx2jd5/global_unmanned?w=5.

Trautmann, Nancy M. 2001. *Assessing toxic risk*. Arlington: National Science Teachers Association.

United Nations. 2011. *Globally harmonized system of classification and labelling of chemicals (GHS)*. 4th ed. New York & Geneva: United Nations. https://www.unece.org/fileadmin/DAM/trans/danger/publi/ghs/ghs_rev04/English/ST-SG-AC10-30-Rev4e.pdf.

UNODA (United Nations Office for Disarmament Affairs). 1975. *Convention on the prohibition of the development, production and stockpiling of bacteriological (Biological) and toxin weapons and on their destruction* [BWC]. http://disarmament.un.org/treaties/t/bwc/text.

US Navy (United States Navy). 2014. Micro identification friend or Foe (IFF). Navy SBIR FY2014.2. https://www.navysbir.com/14_2/2.htm.

USNRC (United States Nuclear Regulatory Commission). 2018a. Backgrounder on Dirty Bombs. 25 May. https://www.nrc.gov/reading-rm/doc-collections/fact-sheets/fs-dirty-bombs.html.

———. 2018b. Lethal Dose (LD). 6 July. https://www.nrc.gov/reading-rm/basic-ref/glossary/lethal-dose-ld.html.

Waddell, Kaveh. 2017. The invisible fence that keeps drones away from the president. *The Atlantic*. 2 March. https://www.theatlantic.com/technology/archive/2017/03/drones-invisible-fence-president/518361/.

Warwick, M. C. 2001. Psychological effects of weapons of mass destruction. *The Beacon* (National Domestic Preparedness Office newsletter) (3):1–8.

Weber, Michael E. 2012. *Biological weapons and employment of biological agents*. Vol. 1. Self-published.

Wendle, John. 2018. The fighting UAVs of Ukraine. *Air & Space Magazine* (Smithsonian). February 2018 issue. https://www.airspacemag.com/flight-today/ukraines-UAVs-180967708/.

Wintour, Patrick. 2018. UK sanctions against Russia: What impact will they have?. *The Guardian*. 14 March. https://www.theguardian.com/politics/2018/mar/14/the-uk-sanctions-imposed-on-russia-by-theresa-may.

Wroughton, Lesley, and Patricia Zengerle. 2018. U.S. says to issue chemical weapons-related sanctions against Russia. *Reuters*. 6 November. http://www.reuters.com/article/us-usa-russia-sanctions/u-s-says-to-issue-chemical-weapons-related-sanctions-against-russia-idUSKCN1N-B2M7.

Zwijnenburg, Wim, and Foeke Postma. 2018. *Unmanned ambitions security implications of growing proliferation in emerging military drone markets*. Utrecht: PAX.

Education and Training as a Disruptive Dual Use Technology

J. I. Katz

> "Knowledge is the first and most important ingredient for a covert weapons program"[1]

1 Introduction

When we think of a disruptive technology, we usually think of a specific invention and its application. Obvious examples in the field of nuclear weapons include the atomic and hydrogen bombs, but also their cruise and ballistic missile delivery systems, heat-resistant shells that permit weapons launched by ballistic missiles to survive re-entry into the atmosphere, targeting systems, airplanes, transporter-erector-launchers (TELs) and submarines that carry, and conceal, these weapons in the air, on the ground and under the sea, and many more. Analogous specific inventions exist in the fields of chemical, biological and cyber weapons: specific toxic chemicals, like nerve agents, biotechnological means of manipulating the genomes of infectious agents, and cybernetic inventions that enable malware to damage, destroy, or redirect the computer systems on which most modern technology depends.

[1] Obeidi and Pitzer (2004, epigraph pp. 228–229). This book also contains information relevant to the 2003 Iraq war: "the long term intention was to wait out the inspection process and try to build a bomb later, when the world was no longer watching" (Ibid. p. 152). Iraq had no WMD in 2003, as was learned after the war, but it intended to build them. The war was premature, but not unnecessary.

J. I. Katz (✉)
Department of Physics and McDonnell Center for the Space Sciences, Washington University, St. Louis, MO, USA
e-mail: katz@wuphys.wustl.edu

M. Martellini, R. Trapp (eds.), *21st Century Prometheus*,
https://doi.org/10.1007/978-3-030-28285-1_10

A great deal of attention has, properly, been paid to the importance of controlling the distribution of the technologies specific to WMD production and delivery, and of dual use technologies with actual or potential peaceful applications that also can be utilized in weapons or weapons systems. In this essay I look further up the supply chain, to the training of people capable of utilizing those technologies and equipment and of designing and developing them themselves. Even if a technology were available as a turn-key facility, its effective use would require significant expertise.

2 Education and Training

Education is usually considered benign, and made available to anyone interested and academically qualified. Its benignity is a misconception and its availability can be dangerous. In many technical fields education and training are dual use technologies whose distribution needs to be controlled. Some education, particularly at higher levels, takes place in universities and technical institutes. Equally, perhaps more, important is the technical training that is necessary to make the technology actually do its job. The engineer and physicist who design a nuclear weapon are useless without the technicians who can fabricate its parts in the shop. The computer scientist who invents sophisticated malware is ineffective without a hacker who can get it into the target system. Technical training takes place on the literal or figurative shop floor, either in a formal apprenticeship or informally, as a new recruit is "shown the ropes" by more experienced people.

As an example, a would-be nuclear proliferator may acquire a nuclear reactor to make plutonium or centrifuges to enrich uranium, or chemical processing gear with which to fabricate reactor fuel elements for irradiation, or to separate plutonium from those irradiated fuel elements, or to reduce enriched uranium hexafluoride to metallic uranium. But further up the knowledge supply chain are the people who could develop those technologies themselves. Even to use purchased or stolen technology effectively requires a significant level of expertise, though more at the technician than the engineer level. While many Western companies have had no compunctions about supplying these nuclear-critical and dual use technologies to likely proliferators, even were export controls on them effective there would still be the threat of indigenous development of the technology by the proliferator's scientists, engineers and technicians. The former Iraqi and present (suspended, at the time of writing) Iranian nuclear weapons programs were largely based on purchased or stolen Western technology, but the North Korean program, the only one of these that has succeeded in making a nuclear weapon and building missiles of intercontinental range, has been largely indigenous.

Education in the technologies of WMD is a dual use strategic good, even though it is not a tangible object. In order to prevent proliferation of WMD it is as important to prevent the education of personnel from likely proliferators in relevant technical fields as it is to prevent their acquisition of dual-use or weapons-related tangible objects. For nuclear weapons these range from electronic components to entire nuclear reactors. For chemical and biological weapons they include bioreactors and DNA-manipulating reagents. For cyber weapons they include code, independent of its physical realization provided it is machine-readable.

The export of dual use education is usually carried out in the open, and is plainly visible. That is because it usually takes place at universities and technical schools in technologically advanced and free countries in Europe or the Americas. This is not always the case—explicit Soviet training of Chinese and North Korean nuclear scientists was covert, and in a dictatorship even basic facts (for example, how many foreign students, and from which countries, are being trained in the various branches of engineering) that would be freely accessible in an open country are hard to ascertain. But in free countries it is obvious, simply by observing the nationalities of students enrolled in relevant departments, either from departmental sources or when they occasionally appear in the student newspaper for unrelated reasons. These students, given a choice, generally prefer to pursue their educations in free countries because life is so much better there—who would have chosen to travel to the Soviet Union for education, if the United States or Western Europe had been an option? Not all such students may even realize that they are being trained for a WMD program—offered the opportunity to pursue advanced engineering training abroad, what Iraqi, Iranian or Pakistani would not jump at it? Finance is rarely an obstacle—Western universities pay their graduate students stipends that are enough for a modest but adequate life, much more pleasant than life back home.

The students need not be witting. Were they told "We are sending you abroad for technical training so you will return and build our nuclear weapons program", some of them would surely inform Western counter-intelligence services, and the host countries would likely stop issuing student visas. Instead, they are likely told "We are sending you abroad to acquire skills that will contribute to the growth of our industry and economy, and you will occupy respected, valued and well-rewarded positions when you return." That would appeal not only to a sense of patriotism, but also to natural self-interest. That's something they could put on their visa and student applications in response to questions: "Why do you want to study X? Why do you want to enroll at institution Y?" Why do you want a student visa? It even promises that the applicant does not expect to turn a student visa into permanent residency.

3 Case Histories

These points are best illustrated with specific examples:

Iran Before its 1979 revolution, Iran had a program to develop nuclear power and an underlying nuclear technology base, including technical training. At the time, Iran was not considered a threat; its government was a good international citizen without ambitions outside its borders. There is no evident connection between the training and technology transferred then (many of the people trained emigrated following the revolution) and Iran's post-revolution nuclear weapons program.

The weapons program began following the end of the Iran-Iraq war in 1989, under the cover of a civilian power program. The power program includes the power reactor at Bushehr, a project that was begun in 1974, interrupted and damaged in that war, that was completed by Russia and finally went on-line in 2012. Its fuel is under IAEA safeguards, to be removed to Russia after irradiation. If those safeguards are maintained, Iran will not have an opportunity to extract the plutonium it contains, and there is no indication that Iran is developing the ability to do so.

The significance of this power reactor for the Iranian weapons program is that the reactor justifies Iranian acquisition of dual use nuclear training and technology. Many of the skills and technologies required to handle and fabricate fissile metals and isotopes that are required in nuclear power are also necessary in a weapons program. The physics of neutrons in a reactor bears some resemblance to that of neutrons in a bomb—neutrons are still neutrons. Power reactors make plutonium but are not ideally suited for that purpose, for removing irradiated fuel rods for extraction of plutonium conflicts with routine operation and refueling for power generation. However, the understanding of reactor engineering and operations developed in a nuclear power program is also applicable to plutonium-production reactors.

In addition to the power reactor at Bushehr, Iran has built, probably with some foreign help (NTI 2017), a lower power (40 MW) heavy water reactor, ostensibly for research and medical isotope production, at Arak, along with a plant to make heavy water. Such a reactor is well suited to make plutonium. Under the Joint Comprehensive Plan Of Action, the original Arak reactor was disabled, and it will be redesigned and rebuilt in a manner that makes less plutonium. The expertise and equipment required to build other natural uranium heavy water reactors remain, as does the heavy water plant, although once a reactor is filled with heavy water it does not need significant additional quantities.

Iraq The Iraqi nuclear weapons program is believed to have begun in the early 1970s (NTI 2015). By the late 1970s large numbers of Iraqi students were studying various branches of engineering at major western universities. Iraq had (and has) little industry other than petroleum and petrochemicals. Yet in the late 1970s there were many Iraqi students at UCLA studying chemical, electrical and mechanical engineering (UCLA doesn't even have a department of petroleum engineering). There were few evident job opportunities for them in Iraq, yet the Iraqi government was sending them to study abroad.

The reason could readily be intuited: Saddam Hussein was building a nuclear weapons program, from the ground up. This was an intelligence indicator evident to anyone on the UCLA campus (where I was on the faculty). Even though Saddam Hussein bought the Osirak reactor from France for the purpose of breeding plutonium, he would need an extensive technical base to extract plutonium from the irradiated fuel, as well as to design and fabricate a nuclear weapon with it. France would sell him the reactor, under a fig leaf of peaceful use, but he would have to use indigenous resources to complete his weapon, because there could be no fig leaf for that. After the destruction of Osirak in 1981 Iraq turned to uranium enrichment, which would require expertise in several branches of engineering, depending on which method (centrifuge, electromagnetic, even gaseous diffusion) were used. The most important indigenous resource would be trained people.

Mahdi Obeidi, who led the Iraqi centrifuge program prior to the First Gulf War, has given a detailed account showing how foreign technical training was turned to a proliferator's use (Obeidi and Pitzer 2004). Growing up in Iraq, he was sent to the Colorado School of Mines from 1962 to 1967 on an Iraqi government scholarship for a M.S. program in petroleum refining engineering. This was likely long before Iraq developed nuclear ambitions or considered them feasible; he was sent for the peaceful purpose of contributing to the Iraqi oil industry. But, once an engineer, always an engineer, and the scientific foundation and quantitative habits of mind of one branch of engineering are transferable to other branches.

By the early 1970s, Iraq was actively interested in nuclear weapons, using research, medical isotope production and power generation as an excuse for its interest in nuclear technology. Petroleum refining doesn't have much to do with gas centrifuges for enriching uranium, but they demand ultra-strong materials, and a country contemplating a gas centrifuge program needs materials engineers. If you are looking for candidates for advanced training in one branch of engineering, you will find them among other engineers. After working in the petroleum industry for several years, in 1972 Obeidi was sent to the University of Swansea for a Ph.D. in materials engineering. After receiving it in 1975, he was assigned to work on materials for nuclear reactors. In 1976 he was sent to Italy for further training, and there developed a personal contact that later provided him a copy of the hard-to-obtain (but not officially classified) Zippe report on gas centrifuges. Obeidi worked on materials for nuclear reactors, including the Osirak reactor whose purpose was the production of plutonium.

Following the destruction of Osirak in 1981 Iraq concluded that its nuclear weapons program had to have a low profile; a reactor cannot be hidden, but uranium enrichment can be. Obeidi was first assigned to work on gaseous diffusion, an energy-hogging technology generally considered obsolete. In 1987 he was appointed head of the gas centrifuge program. At least three of his five department heads were educated in the West. Until the effort was shut down in the First Gulf War in 1991, it received much technical help from European companies (and the University of Virginia), some of it naive and more of it mercenary (Obeidi and Pitzer 2004). Formal academic training is not the only means by which information can be transferred.

Pakistan The most damaging WMD proliferator was A. Q. Khan (Langewiesche 2005), who stole URENCO centrifuge technology, and proliferated not only to his home of Pakistan, but to Libya, Iran and North Korea. After undergraduate education in Pakistan, in 1961 he went to Berlin and then Delft to get a Masters degree in metallurgy, married a Dutchwoman, and in 1972 got a Ph.D. in metallurgical engineering at the Catholic University of Leuven. It appears to be purely accidental that he was then hired by a URENCO subcontractor designing gas centrifuges. In contrast to the Iraqi engineering students a few years later, he does not appear to have been sent for foreign training for nefarious reasons; rather, it was a natural career path for an ambitious young man interested in engineering. The opportunities at home, both for advanced training and employment, were limited.

All would have been well had he not been turned into a Pakistani nationalist by Pakistan's defeat, surrender and dismemberment in its 1971 war with India. In 1974 the war was followed by India's first nuclear test. Khan essentially became a "walk-in" to Pakistan's then embryonic nuclear weapons program, writing directly to Prime Minister Bhutto who placed him in a leadership role. Khan began to systematically collect centrifuge design information and even sample parts that were sent to Pakistan, to which he went in 1975, never to leave.

In the case of Pakistan only a single person, A. Q. Khan, has been identified as the channel by which technology (gas centrifuges) enabling nuclear weapons was transferred. This transfer was not directly enabled by his technical training in metallurgy. That training was only the path to his access to essential information. But without a degree in metallurgy and research in maraging copper alloys (not useful in centrifuges) he would never have gained access to centrifuge technology by employment at a URENCO subcontractor. And without other Pakistani engineers trained abroad, Khan's stolen designs could never have been turned into operating centrifuges.

North Korea Much less is known about the contribution of technology transfer to North Korea's nuclear weapons program because of the closed nature of its society. The program received some assistance from the Soviet Union in its early stages, but this assistance appears to have ended in the mid-1960's, and its subsequent development was indigenous (NTI 2018).

4 Conclusions

It is necessary to treat education and training in technical fields as dual use commodities. An application for a student visa to a Western country from a country that is a proliferation threat should be considered an application to export the science and technology the student plans to study, and denied if that science and technology would further the development of WMD.

References

Langewiesche, William. 2005. *The wrath of Khan – How A. Q. Khan made Pakistan a nuclear power—And showed that the spread of atomic weapons can't be stopped.* The Atlantic (November 2005), https://www.theatlantic.com/magazine/archive/2005/11/the-wrath-of-khan/304333/. Accessed 24 Mar 2019.

NTI. 2015. *Iraq – Nuclear.* https://www.nti.otrg/learn/countries/iraq/nuclear/. Accessed 22 Mar 2019.

———. 2017. *Arak nuclear complex*. https://www.nti.org/learn/facilities/177/. Accessed 26 Mar 2019.

———. 2018. *North Korea – Nuclear*. https://www.nti.org/learn/countries/north-korea/nuclear/. Accessed 27 Mar 2019.

Obeidi, Mahdi, and Kurt Pitzer. 2004. *The bomb in my garden: The secrets of Saddam's nuclear mastermind.* Hoboken: Wiley.

Part II
Evolving Risk Mitigation Strategies and Technologies

Detection and Identification Technologies for CBRN Agents

Olivier Mattmann

1 Introduction

Today, the world faces threats that range from urban bombings and natural disasters to intentional or accidental releases of chemical, biological, radiological, or nuclear (CBRN) materials, with or without the aid of explosives (CBRNe). These threats may be attributed to states, terrorist organizations, non-state political groups, organised crime organisations, or individuals; or they may result from accidental releases from nuclear facilities, biological research, diagnostic laboratories, or chemical industry facilities; or natural exposure to biological agents. The complexity and unpredictability of these threats is high. Many potential CBRN events are of low probability but very high impact. Policymakers are frequently reluctant to take serious counter measures due to this lack of certainty.

Although CBRNE threats are synonymous to weapons of mass destruction it should be noted that the term destruction is consistent with the nuclear element and biological events with pandemic potential. Chemical weapons, certain biothreat agents and radiological dispersal devices (dirty bombs) are expected to provoke disruption but not destruction when released. It should also be noted that chemical, radiological/nuclear and explosives threats represent acute emergencies while a bioterrorism attack or a pandemic are gradually developing emergencies which peak depending on the incubation period, virulence, toxicity and other factors.

The CBRNe detection and identification capability provides the ability to detect CBRNe materials at points of manufacture, transportation, and use. This capability includes the detection of CBRNe materials through instrumental monitoring rather than monitoring their effects (i.e. signs or symptoms) in humans and animals, which is addressed by the public and animal health systems. The importance of training,

O. Mattmann (✉)
Hotzone Solutions Group, The Hague, The Netherlands
e-mail: olivier.mattmann@hotzonesolutions.org

M. Martellini, R. Trapp (eds.), *21st Century Prometheus*,
https://doi.org/10.1007/978-3-030-28285-1_11

communication, and close coordination with key partners including intelligence, law enforcement, public safety, public health, and international partners, is recognized as a critical enabler for this capability.

Since many toxic chemicals act quickly, rapid detection is needed to prevent lethal or incapacitating results following release. Sample collection, sample processing and information processing are vital to provide identification and warning of chemical exposure. In biological incidents, it is important to locate "Patient/Animal Zero" [the first sick person or sick/dead animal] and identify contacts that might have been infected before the initial symptoms were observed. It is equally important to identify the source released by a dirty bomb including type of radiation emitted and duration of half-life, which might range from days to mega-years. Collected data should always be compared with natural background values to avoid false positive alarms and mobilization of response mechanisms. It would be ideal to establish early warning/monitoring systems but this is usually done during specific events (e.g., the Olympic Games), for specific targets (e.g., monitoring near nuclear power plants), or in the aftermath of a natural or man-made event.

Detection can be accomplished at a designated location (point detection) or from a distance (standoff detection), although the latter is particularly difficult for low volatility agents (such as VX). The sensitivity of a detector is crucial for detecting lethal concentrations. Equipment must be reliable, able to provide identification quickly with a low false alarm rate and high accuracy, be low in weight but ruggedized, and be integrated into an alarm system so that warnings can be distributed and proper action can be taken as soon as possible. A spectrum of variables (including location, persistence, properties and concentration of the agent, relative humidity, environmental temperatures) can influence the performance of the detector. In addition, accurate detection is required to know when the environment is safe for normal operations. Identification can be performed on site but usually requires more elaborate laboratory techniques and bench-top equipment.

Although protection against chemical agents is available, continuous wearing of personal protective equipment degrades performance, making personnel reluctant to use them unless it is mandatory. Similarly, many prophylactic measures are most effective if implemented before exposure (e.g. pyridostigmine pretreatment for nerve agents), while many counter measures must be initiated within minutes after exposure (e.g. atropine and pralidoxime).

Chemical agent detectors must function in demanding, real-world environments where cost, portability [weight], and time are important factors (Sferopoulos 2008). They must also be operational 24/7, be widely deployable, ruggedized and able to be networked. At present, the most challenging aspect for the detection and identification of chemical agents is the differentiation of the agent of interest from other chemicals present in the environment. Furthermore, detection sensitivity is a necessary factor as detectors are required to provide advanced warning and must detect concentrations of chemical agents well below the IDLH[1] levels. Detection capability

[1] Immediately dangerous to life or health air concentration values.

involves factors such selectivity, sensitivity, response time and false alarm rates whilst detector performance includes factors such as warm-up time, calibration requirements, portability, power requirements and ongoing costs associated with training and maintenance. A desirable detector would be one that can detect both chemical warfare agents and toxic industrial chemicals selectively within an acceptable time, thus enabling an effective medical response. As yet, no “ideal” detector that meets all the above-mentioned requirements is commercially available.

Therefore, operationally, responders should deploy at least two detectors using at least two different technologies, to cover for possible breakdown or malfunction and to ensure that at least one technology is capable of detecting the threat on time. It is not about saving money; it is about the prevalence of logic!

Understanding chemical/radiological detection and detectors In order to evaluate both the equipment capabilities and their suitability to given operational requirements certain common terms need to be defined and clarified:

- **Detection** is the ability to distinguish a person, an object or a substance from the background. Detection can be point/on-site or remote. Detection requires sampling (close contact with soil or liquids; suction of air) that can be performed either manually or automatic (remote monitoring systems). Detectors can be handheld, portable, mounted on a vehicle, unmanned aerial/ground vehicle combined with a global positioning system or a sea buoy (i.e. for radiation warning).
- **Recognition** is the ability to classify the target-object’s class (animal, human, vehicle, organic, or inorganic).
- **Identification** is the ability to describe the target-object in detail (a man with a gun, a dog, a nerve agent or ammonia).
- **Monitoring** is the regular systematic purposeful observation and recording/giving feedback of activities taking place in a given geographic area or referring to a given group of people or animals before, during or after an incident. According to the European Nuclear Society, a monitoring area is a radiation protection area for which fixed dose limit values are valid and which is subject to monitoring according to stipulated regulations. In adaptation to the Euratom basic standards, a monitoring area is a radiation protection area that does not belong to the controlled area and in which persons in a calendar year may receive an effective dose of more than 1 millisievert or organ doses higher than 15 millisievert for the eye lens or 50 millisievert for the skin, hands, forearms, feet and ankles (ENS n.d.).
- **Reporting** is the process that enables the gathered information to be used in decision-making.
- **Sensitivity** is determined by the lowest concentration of an agent that can be detected with confidence. For chemical agents, the more toxic a chemical is the more sensitive the detector needs to be (Sun and Ong 2005). The sensitivity of a detector may be dependent upon a number of factors including the agent and environmental and operational conditions (Guide 100-04, Saver 2005).

- **Selectivity** is the ability of a detector to respond only to the targeted agent in a sample. A selective detector must be able to separate targeted compounds, over a broad range of concentrations, from any other substance which may be present in a sample (Sferopoulos 2008). The major disadvantage associated with selective detectors is that they are limited in the number of compounds that they can detect. Presently there is no single detector that is one-hundred-percent selective. For example, a chemical warfare agent detector based on flame photometric detection will only respond to phosphorus and sulfur compounds, but not to a chemical agent that does not contain phosphorus or sulfur. For field applications, non-selective detectors may be more suitable if a broad-spectrum early warning system is desired or if the environment is clean (no previous exposure to any agent). Non-selective detectors may be utilized to provide an initial survey of an area in a civilian scenario, given that these detectors can respond to various chemicals simultaneously and that chemicals used by terrorists are generally unpredictable. However, if a nonselective detector produces a response, it would be necessary to survey the suspect area with a more specific detector to identify or discriminate potential chemical agents from other compounds present.
- **False alarms** occur if a detector responds when a chemical agent is not present – false positive, or if it fails to respond to a chemical agent that is present – false negative. False positive alarms are usually observed when an interferent is present, which may be a chemical molecularly similar to the target agent, or a substance which may contain elements that are also present in the target agent. False negative alarms are more problematic than false positive alarms because the failure to produce an alarm may lead to dangerous situations. The failure of a detector to alarm to a chemical agent that is present may be due to any number of reasons including operator error, changing environmental conditions, humidity effects, detector malfunction such as software quirks, and the presence of chemical interferents which may mask normal detection capabilities. In a similar way, it is important to differentiate between natural background radiation and artificial radiation. Natural Background Rejection (NBR) technology is used to eliminate fluctuating natural background levels while measuring radiation. Naturally occurring radiation has a distinct "signature" that differs radiation from artificial or man-made sources. NBR leverages this signature to quickly differentiate between natural and artificial radiation by stripping away any natural background radiation that is registering during a scan, delivering a more accurate result of artificial radiation levels. Unfortunately, operators using instruments without NBR often set their alarm thresholds higher to eliminate the nuisances caused by natural radiation, causing false alarms, or they ignore alarms due to their frequency, potentially missing out on hidden or shielded sources. Using an instrument with NBR allows to keep an alarm threshold extremely low without the need to worry about false alarms from non-threatening sources – so when the alarm goes off, it is time to take action.

Understanding biological detection Detection of hazardous pathogens (categories "A" and "B") is totally different compared to the detection of chemical and radiological agents. First of all, it is important to discriminate if the material under examination or suspicion is organic or inorganic. This is the case for the infamous anthrax letters that continue to ignite panic and require a state of mobilization. Then there is a difference in the study of purified biological agents compared to environmental and clinical samples. A third issue is that remote monitoring especially of open spaces is not as effective as it should be and it is biased by a number of influencing factors such as velocity of winds, environmental temperatures, relative humidity, rain and snow, and atmospheric pollution, just to name a few. When used in confined spaces, the quantity and content of the air examined might provide more accurate results. Laboratory tests available in many forms provide accurate detection and identification of the offending agent but it should be noted that it is of outmost importance to start treatment protocols based solely on suspicion instead of waiting for laboratory confirmation – the latter might cost the life of the patient despite the speed that certain technologies provide. Every hour counts when biological warfare agents have been released or suspected. This is why fast detection methods have been developed and some of them are already available in the market, such as SpinDx (SANDIA 2013), RapiDx (SANDIA 2011), or disposable matrix devices.

2 Chemical Detection/Identification Technologies

Various technologies enable detection of chemicals of special interest and chromatography has a prominent position in the long list of field and bench detectors.

2.1 Chromatography

Gas chromatography (GC) Gas chromatography is an instrumental technique for separating components of a mixture based on their boiling point, polarity, and affinity to a gas chromatography column. Since samples must be vaporized prior to GC separation, this technique is commonly used for analysis of volatile and semi-volatile organic compounds but it is not generally applicable for materials with boiling points greater than 300 °C or for materials that decompose at high temperatures.

A range of detectors can be used in GC, such as mass selective detection (Vekey 2001), flame photometric detection (Chasteen 2009a), photoionization detection (Stauffer et al. 2008), electron capture detection (El Sohly et al. 2008) or flame ionization detection (JoVE 2019). GC instruments are typically bench top models, but field instruments have become more common in recent years, ranging from those designed to be mounted in vehicles to hand-held devices. Gas Chromatography is used extensively in forensic science, and for solid drug dose (pre-consumption form) identification and quantification, arson investigation, paint chip analysis, toxicology, to identify and quantify various biological specimens and crime-scene evidence, and more.

Ion Chromatography (IC) Ion chromatography (or ion-exchange chromatography) separates ions and polar molecules based on their affinity to an ion exchanger (Lamont-Doherty Earth Observatory n.d.). It works on almost any kind of charged molecule—including large proteins, small nucleotides, and amino acids. Ion chromatography is typically used to separate compounds generally considered non-volatiles, therefore making their analysis non-conducive to techniques such as GC/MS. A number of different detection systems are compatible with ion chromatography. These commonly include conductivity, UV/visible spectroscopy and mass spectrometric detectors. When coupled with mass spectrometry, IC can yield additional molecular elucidation by providing mass-to-charge (*m/z*) data, thus increasing confidence of ionic identification. Typically, IC systems are bench top instruments. One advantage of ion chromatography is that only one interaction is involved during the separation as opposed to other separation techniques; therefore, ion chromatography may have higher matrix tolerance. Another advantage is the predictably of elution patterns, based on the presence of the ionizable group (Miles 1997). Disadvantages include the constant evolution of the technique which leads to the inconsistency from column to column (Neubauer 2009). A major limitation is that it is limited to ionizable groups.

Liquid Chromatography (LC) Liquid chromatography (Long 2019) is often used for non-volatile analytes including in the analysis of environmental, biological, and pharmaceutical samples. It is capable of separating analytes ranging in size from ions to large polymers and biological molecules including proteins. Other techniques such as ion chromatography (IC), size exclusion chromatography (SEC), gel permeation chromatography (GPC), and gel filtration chromatography are specialty forms of LC.

LC may be combined with a variety of detectors. Most common are ultraviolet-visible, fluorescence, and mass spectrometric detectors. More specialized detectors, including electrochemical, evaporative light scattering, and refractive index detectors are available for specialty analysis.

LC instrumentation is most commonly found in benchtop models, and is used widely in industry and environmental applications.

2.2 Colorimetric Detection

Colorimetric detectors indicate the presence of a target chemical through a chemical reaction that results in a color change (Sun and Ong 2005). Colorimetric test kits are available for a wide variety of CBRNE targets ranging from the M8/M9 paper for CWA detection to the SPX 300 for trace explosives detection. The tests are usually analyte and/or class specific and are well suited for presumptive testing. Results are typically examined by visual comparison of the resulting color to a key in the test kit, but some vendors offer electronic readers that provide more objective results. Detection limits are typically in the part-per-million to part-per-thousand (ppm-ppt) range, depending on the analyte.

Typically, colorimetric detection has a small form factor, can be easily transported by hand, and generally require little or no power supply. Kit cost varies depending on the analyte and the number of reagent packets in the kit. Some require one-time use consumables such as cartridges, reagents, papers or swabs for each target chemical.

2.3 Elemental Analysis

Elemental analysis determines the elemental and sometimes isotopic composition of a material. It allows to determine the ratio of elements of a sample, and to work out a chemical formula that fits this ratio. This method helps determine if a sample is a desired compound, and confirms the purity of a compound. Elemental analysis can be qualitative and/or quantitative.

Quantitative analysis determines the mass of each element or compound present. Methods include gravimetry, optical atomic spectroscopy, and neutron activation analysis.

Qualitative analysis helps determining which elements exist in a sample – a variety of technologies is available, including atomic absorption (Sevostianova n.d.), atomic fluorescence (Sanchez-Rodas et al. 2010), inductively coupled plasma mass spectrometry (ThermoFisher Scientific n.d.-a), inductively coupled plasma optical emission spectroscopy (ThermoFisher Scientific n.d.-b), microwave plasma atomic emission spectroscopy (Agilent Technologies 2011), X-ray diffraction (Dutrow and Clark n.d.), and X-ray fluorescence (Wirth and Barth n.d.).

2.4 Molecular Spectroscopy

Amplifying Fluorescence Polymers (AFP) Chemical sensors based on amplified fluorescence quenching of solid-state conjugated polymer films (Thomas et al. 2007) consist of a glass capillary tube whose interior is coated with a polymer film comprised of a conjugated backbone with pentiptycene groups. In operation, the AFP polymer is excited to a fluorescent state using an external light source. An air pump is then used to draw the sample through the capillary; compounds containing nitroaromatic groups (TNT, RDX, etc.) bind to the polymer, causing fluorescence to be quenched. The change in fluorescence is read by a photometer positioned axially to the capillary. Whilst in conventional fluorescent-based sensors, the signal arises due to 1:1 analyte to chromophore binding, in AFP, the polymer chains are electronically conjugated and binding at any point along the polymer chain results in complete quenching of all sites on the polymer. Hence, a single analyte molecule activates multiple chromophores, resulting in an amplification of 100 to 1000 times. AFP is primarily used for vapor sensing such as to detect landmines and other explosive devices. AFP technology is relatively new, and employed in a limited number of sensor devices nearly all of which are handheld.

Atomic Emission Spectroscopy (AES) Atomic emission spectroscopy (Thakur 2007) – more commonly referred to as emission spectroscopy – examines the wavelengths of photons emitted by atoms or molecules during their transition from an excited state to a lower energy state. Each element emits a characteristic set of discrete wavelengths according to its electronic structure; by observing these wavelengths the elemental composition of the sample can be determined. AES is most useful for samples that are dissolved or suspended in aqueous or organic solutions. AES is typically employed as a bench top instrument. It is a mature technology available from a number of manufacturers.

Far-Infrared/Terahertz (THz) Spectroscopy Far-infrared or Terahertz spectroscopy (Ghann and Uddin 2017) identifies chemicals based on the interaction of molecules with electromagnetic radiation in the far-infrared/terahertz region (20–400 cm^{-1}). Molecular absorbances of far-infrared light cause molecular rotations that can be classified by chemical functional groups. The resulting THz spectrum is characteristic for a given molecule. THz spectra of pure materials can be searched against reference databases, whilst data from unknown mixtures require examination by an experienced THz spectroscopist. THz spectroscopy can quickly differentiate polymorphic differences when compared to other optical spectroscopy methods, lending itself to analysis of pharmaceuticals, consumer products or other material that has multiple polymorphs. The technique is most often used in the analysis of inorganic and organometallic species, although condensed phase biological samples have also been investigated. Bench top systems are most common. Instruments are available from a number of manufacturers.

Hyperspectral Imaging (HSI) Hyperspectral imaging collects electromagnetic spectral information over a two- or three-dimensional space (Schelkanova et al. 2015). In general, the electromagnetic range that can be exploited ranges from visible to long wave infrared (LWIR). The advantages of HSI include the ability to analyze samples at stand-off distances and over potentially large areas such as battlefields and chemical release plumes. HSI has been applied for environmental monitoring and in geology, mining, agriculture, surveillance, and chemical detection. A variety of instruments are available including small handheld devices, larger stationary and vehicle-mounted configurations. The technology is still maturing, but a number of instruments are on the market.

Mid-Infrared Spectroscopy (MIR) Mid-infrared spectroscopy (Chalmers 2013) identifies chemicals based on the interaction of molecules with electromagnetic radiation in the mid-infrared region (400–4000 cm^{-1}). Infrared spectroscopy identifies chemicals based on the absorption of specific wavelengths of mid-infrared light which causes molecular rotations and vibrations. These can be classified by chemical functional groups. The resulting MIR spectrum is characteristic for a given molecule. Attenuated total reflection (ATR-MIR) is the most popular technique today – it allows for the fast analysis of solid and liquid materials. MIR spectra can be searched against large reference spectral databases, making this technique a powerful tool for identification of chemical unknowns. The most common detectors are benchtop instruments, but a growing number of manufacturers have released portable and handheld models. The technique is mature with a large number of companies having MIR instruments on the market.

Near Infrared (NIR) Spectroscopy Near-infrared spectroscopy (Davies n.d.) identifies chemicals based on their interaction with electromagnetic radiation in the near-infrared region (4000–10000 cm^{-1}), resulting in a spectrum characteristic for that molecule. It differs from mid-infrared spectroscopy in that the molecular absorbances are of higher-energy, creating vibrations of higher orders (i.e., combinations and overtones) than those of the mid-infrared region. NIR instruments are employed to analyse food, pharmaceuticals, and consumer products. They are fast and can be operated with minimal training. The spectra can be searched against reference spectral databases for pure materials, but mixture analysis requires a chemometric approach configured by an expert spectroscopist. Bench top systems are most often used for analyzing condensed phase materials. In the last two decades, portable and handheld instruments have come into vogue, incorporating cloud-based reference libraries and enabling devices the size of a USB drive.

Nuclear Magnetic Resonance (NMR) Spectroscopy NMR spectroscopy (Radboud University and Faculty of Science n.d.) uses radio-frequency pulses (1–1000 MHz) to excite molecules placed within a strong magnetic field (0.5–20 Tesla). Molecules containing NMR-active nuclei produce detectable signals as they relax back to the ground spin state. The most common nucleus analyzed is ^{1}H because of its high natural abundance, high sensitivity, and occurrence in organic molecules. Other common nuclei amenable to NMR spectroscopy include ^{13}C, ^{31}P, ^{19}F, and ^{15}N.

NMR can analyze neat liquids or dissolved organic solids (preferably in deuterated solvents) with concentrations down to 100 mM. Chemical identification is based upon processing raw data and fitting the resultant spectrum against theoretically simulated data. Spectral libraries are not required. Identification is most confident when the sample contains a single chemical. Chemicals best analyzed with NMR include drugs, explosives, chemical weapons, solvents, and other organic molecules (with molecular weights below 500 amu). A number of commercial NMR instruments are available as the technique is very mature.

Photoacoustic Detection (PA) Photoacoustic spectroscopy (Quan and Davis 2017) utilizes the principles of radiation absorption and the photo-acoustics for molecular detection. Samples – solids, semi-solids, liquids, or gases – are placed into a closed cell containing a non-absorbing gas. The sample is then irradiated and energy absorbed by the solid is released when the surrounding gas molecules, which are heated by the solid, subsequently relax back to the initial vibrational state. The excess heat energy causes a change in the pressure of the gas cell, which results in an acoustic wave, which can be detected with a sound-measuring device such as a microphone or a piezoelectric sensor. Visible and ultraviolet light may be used as radiation sources, but the most modern instruments rely on infrared radiation. Advantages of PA detectors include high sensitivity and stability, wide dynamic range, and quick response. Their main disadvantage is poor specificity and susceptibility to interference.

As a result, PA detectors are not recommended for screening of unknown samples. PA has been used in studies of biological systems, pharmaceutical analysis, nanoparticulate materials, and materials science. PA instruments are available in both benchtop and portable models. The technique is mature, and increasingly used with the inclusion of infrared radiation sources. Commercial instruments are readily available.

Raman Spectroscopy (Raman) Raman spectroscopy (Nanophoton n.d.) measures the energy of photons generated by inelastic scattering of monochromatic excitation photons. During excitation, the photons interact with the electron cloud of an analyte and a small percentage of these photons undergoes a change in energy as a result of the interaction. This energy change corresponds to molecular vibrations, rotations, and other low-frequency modes forming a spectrum. Raman spectroscopy is often described as a complimentary technique to Fourier-transform infrared spectroscopy – FTIR (Jasco website n.d.) because both techniques are used to probe vibrational energy levels although different selection rules allow FTIR and Raman to interrogate different sets of molecular vibrations. Raman scattering is an inherently low probability phenomenon and as such, the Raman scattering cross-section of a particular analyte will generally be several orders of magnitude less than the corresponding FTIR absorption cross-section. Practically, this means that the sensitivity of Raman spectroscopy is lower than FTIR spectroscopy. Raman instruments are available in a variety of configurations. The most common instruments are bench top models but a wide variety of portable and handheld models are also available commercially.

Surface Enhanced Raman Spectroscopy (SERS) SERS (Semrock website n.d.) utilizes the unique optical properties of nanostructured metallic substrates to enhance the intensity of the native Raman signal of an adsorbed analyte by several orders of magnitude. Nanostructured substrates composed of noble metals (primarily silver and gold) are capable of supporting localized surface plasmon resonance (LSPR). LSPR is induced by the resonant excitation of surface-bound substrate electrons, which generates a highly enhanced localized electromagnetic field. Raman scattering intensity is governed by the polarizability of the analyte being interrogated as well as the localized electromagnetic field experienced by the analyte as a result of laser excitation. LSPR results in a significant increase in magnitude of the local electromagnetic field near the surface of a nanostructured substrate. Adsorption of an analyte onto an LSPR-supporting nanostructured substrate followed by excitation with a laser of appropriate wavelength leads to an increase in Raman signal intensity by several orders of magnitude and a significant increase in sensitivity compared to traditional Raman spectroscopy. A number of commercial substrates are available and cost is dependent upon substrate configuration.

Ultraviolet-Visible (UV-Vis) Spectroscopy Ultraviolet-visible spectroscopy (Reusch 2013) is a technique that identifies chemicals based on the interaction of molecules with electromagnetic radiation in the ultraviolet-visible region (10,000–33,333 cm^{-1} or 300–1000 nm). Molecular absorbances of UV-Vis light cause electronic transitions within molecules. These transitions typically occur with the presence of a transition metal ion, or a conjugated organic molecule. There are UV-Vis reference databases, but there is much less specificity in a UV-Vis spectrum compared to MIR data, making it a secondary analytical technique for unknown chemicals. Point analysis ("bench top") systems are used for analyzing liquid phase materials, although a few systems are also capable of analyzing gases and solids. UV-Vis instrumentation has been scaled down to hand-held size for specific colorimetric applications, but most research-grade instruments are of larger size, and are often combined with NIR instrumentation. Instrument are available that have either single light paths or multiple beams for simultaneous analysis of reference samples. Additionally, some instruments are capable of making time-resolved measurements for kinetics experiments. UV-Vis is an extremely mature analytical technique. Many instruments are available commercially.

2.5 Sensors

Catalytic Bead Sensor (CAT) Catalytic bead sensors (Adamovica 2017) sense the presence of gas at around 20% Lower Explosive Limit (LEL). Two platinum coils are embedded in a bed of alumina and the voltage difference across them is measured using a Wheatstone bridge circuit configuration. One wire contains a catalyst that makes it an oxidative environment while the other wire inhibits oxidation. When an oxidizable gas is present in the sensor, oxidation occurs at one wire and not at the other. This induces a voltage difference across the wire and an alarm is triggered to indicate the presence of a gas at around 20% LEL. These sensors are useful for the detection of flammable and combustible toxic industrial chemicals (i.e. ammonia; hydrogen; acetylene). The sensitivity of catalytic detectors is typically affected by two things —contamination or poisoning of the active bead, or blockage of the flame arrestor which gas must pass through to reach the beads.

Electrochemical Detection (EC) Electrochemical detectors (Antec Scientific n.d.) detect chemicals in the gas phase by measuring changes in current across a target-specific electrode. When a target gas contacts the electrode an electrochemical reaction takes place, producing a change in current. A variety of EC sensors and detectors are available for identifying gases such as HS_2, CO_2, TIC/TIMs, explosives, and other chemicals of interest.

Most EC-based detectors can be configured with sensors for specific targets. Most units can detect one to five gases simultaneously. The most promising electrochemical systems use broad-response sensors capable of detecting multiple gases simultaneously.

Flame Ionization Detection (FID) A flame ionization detector (Chasteen 2009b) operates on the principle that combustible chemicals produce ions during combustion, which can be detected as a current. In practice, a target gas stream is directed into a flame (typically a hydrogen-air flame), which results in the combustion of target chemicals and the production of positively charged ions and electrons. The resulting current is sensed by electrodes placed close to the flame. FID is a non-specific detector since any molecule with carbon and hydrogen atoms will produce a response.

Flame Photometric Detection (FPD) Flame photometric detectors (UMASS n.d.) are a type of gas chromatograph that observes the emission characteristics of samples when burned in a flame. Excited elements emit light at particular wavelengths. This light is detected by a photomultiplier tube. From the spectral response, specific elements can be detected within the sample.

Micro Electro Mechanical Systems (MEMS) A Micro Electro Mechanical System (Prime Faraday Technology Watch 2002) incorporates all of the essential components of a chemical detection system into a single micro-scale device. The actual MEMS sensing mechanism may range from 1 to 100 micrometers in size, allowing the size of the overall chemical detector to be quite small compared to most conventional detection systems.

Products utilize a sensor platform in which polymer-filled micro-machined capacitors (also referred to as chemi-capacitors) measure the dielectric constant of an array of selectively-absorbing materials. The interaction between target analyte and polymer modifies the dielectric properties of the polymer, resulting in a change in capacitance. Analyte selectivity is achieved by populating the array with different polymers that exhibit a variety of capacitive properties, which in theory results in a signature output pattern from the array. In reality, there is often not enough difference in capacitive property among the polymers to distinguish individual chemicals. Polymer films can also swell in high humidity and may suffer very long recovery times following an exposure event. Polymer film MEMS sensors tend to respond and recover faster to high vapor pressure chemicals such as acetone and ethanol, while recovery times following exposure to low vapor pressures chemicals can be several minutes to hours.

Photo Ionization Detection (PID) Photoionization detectors (MSA n.d.) are commonly used for detection of volatile hydrocarbons and other organic vapors. Gas molecules are ionized by ultraviolet light and the resulting ion current is directly proportional to the gas concentration. PID systems provide a rudimentary degree of chemical specificity because only chemicals with ionization potentials below the energy output of the PID lamp can be detected.

Pulsed Flame Photometric Detection (PFPD) Pulsed flame photometric detection (The Sam Houston State University n.d.) is almost identical to flame photometric detection except the flame is pulsed on and off in order to increase the signal to noise ratio and the selectivity while decreasing gas requirements.

Surface Acoustic Wave Detection (SAW) SAW detectors (Senseor n.d.; Sferopoulos 2008) rely on chemically selective polymers that absorb target gases, producing a measurable change in a property of the polymer (resistance, capacitance, etc.). SAW sensors are piezoelectric crystals that detect the mass of chemical vapors absorbed into chemically selective coatings on the sensor surface. This absorption causes a change in the resonant frequency of the sensor. Many SAW sensor coatings have unique physical properties that allow reversible adsorption of chemical vapors.

2.6 Ion Mobility Spectrometry Methods

Ion Mobility Spectrometry (IMS) Ion mobility spectrometry (Sferopoulos 2008, Warwick and Dunn 2011 pp. 15–35) identifies compounds based on the time required for ionized molecules to drift through an electric field. Sample molecules in the gas phase are ionized using either a corona discharge source or a radiation source such as ^{63}Ni or ^{241}Am. The ions then enter a field-free drift region with an electric gradient, where they drift toward a collector. The time required for an ion to reach the collector depends on several factors, including shape, mass, and charge. In theory, unknowns may be identified via comparison of their drift profiles to an on-board library. IMS is widely used in field portable chemical detectors due to its small size, portability, low power requirement, sensitivity, and quick analysis time (less than a few seconds in most cases). Disadvantages of IMS include relatively high cost, high false positive rate (i.e., low target specificity), and limited number of target chemicals. Most of the systems were designed specifically to detect CWAs, but several IMS manufacturers have broadened their product line to include detection capability for TIC/TIMs, narcotics, and explosives.

Differential Ion Mobility Spectrometry (DMS) and Field Asymmetric Ion Mobility Spectrometry (FAIMS) FAIMS (FAIMS website n.d.) and DMS (Maziejuk et al. 2015, pp. 283–29) use the same principle: sample molecules in the gas phase are ionized and the resulting ions pass through a drift region to which a high-frequency waveform is applied. Identification of compounds is achieved by measuring the difference in ion mobility at high electric field relative to low field. This technique is similar to ion mobility spectrometry (IMS), with the exception that a high-frequency waveform is applied. Advantages of FAIMS/DMS over IMS include higher sensitivity, better selectivity, and reduced size.

Ion Trap Mobility Spectrometry (ITMS) Ion trap mobility spectrometry (Rapiscan Systems n.d.) detects chemicals based on migration of ions through an electric field. ITMS is very similar to IMS in that both techniques use ion mobility for identification, but differs from IMS in the source-drift tube interface design. Morpho Detection, the sole vendor of ITMS systems, claims that the ITMS interface, which does not require a membrane or shutter grid to isolate the source from the drift tube, provides greater sensitivity.

2.7 Mass Spectrometry (MS)

Mass spectrometry (Mellon 2003, pp. 3739–3749) identifies a chemical by measuring characteristic molecular ions or fragments produced as the chemical is ionized. Sample molecules are ionized with an electron beam, and the resulting molecular and fragment ions then pass through a mass analyzer where their masses are measured. A separation technique such as gas chromatography is typically added to the front end of the MS to separate mixtures into individual components. Mass spectrometry is generally considered the benchmark for identification of unknown organic chemicals because it is highly sensitive and selective, and mass spectra are easily searchable against vast reference databases.

Sector Field Mass Spectrometry Sector Field Mass Spectrometry is a method that employs a static electric and/or a magnetic sector in order to elucidate the masses of the constituents of a mixture (Clarke 2017, pp. 1–15).

Quadrupole Mass Spectrometry Quadrupole mass spectrometers (LCGC n.d., p. 640) are perhaps the simplest of mass spec instruments. Functioning effectively like a radio, the quadrupole MS – so-named because of its four charged rods, which run parallel to the flight paths of the ions it measures – filter ions by mass-to-charge ratio (m/z) by altering the voltages in the rods. Quadrupole mass spectrometers are available in a variety of formats, including a single quadrupole mass spectrometer, a triple quadrupole MS (three quadrupoles arranged in tandem), and a Q-TOF (a triple quad variant in which the third quadrupole is replaced with a time-of-flight mass analyzer).

Tandem Mass Spectrometry (MS/MS) Tandem mass spectrometers (Nationalmaglab n.d.-a) are capable of multiple rounds of mass spectrometry, usually three, with one mass spectrometer used for some form of molecular fragmentation. MS/MS analytical methods typically call for monitoring ion transitions rather than single ions. This allows for extremely specific and selective identification but cannot be used for the identification of unknowns. In atomic mass spectrometry, the principle is the same except the collisional-induced dissociation in the second chamber is used to remove molecular interferences from the analyte mass of interest. Tandem mass spectrometry can also be performed over time in a single mass spectrometer, such as a quadrupole ion trap, but this approach is currently limited to molecular mass spectrometry applications.

Isotope Ratio Mass Spectrometry (IRMS) IRMS is a specialized version of MS that employs mass spectrometric methods to determine the relative abundance of isotopes in a single sample.

Ion Trap Mass Spectrometry (IONTRAPMS An ion trap mass spectrometer (Nationalmaglab n.d.-a, n.d.-b; Sferopoulos 2008) identifies molecules by ionizing and fragmenting them, then measuring the mass-to-charge ratio (m/z) of the observed ions to create a full-scan mass spectrum. Ions are created by electron ionization and are passed into the ion trap portion of the mass spectrometer where they can be manipulated further (for MS/MS measurements) or released to the detector. Because ions and molecules have time to interact with each other in the ion trap, ions and/or molecules can react to form new ions that are not normally seen in quadrupole mass spectra for the same compound. Ion trap mass spectrometers can be smaller than other types of mass spectrometers and do not require as high a vacuum, so they can typically start up from an off position much faster than other types of mass spectrometers (~15 min). Air and water in the system do not affect the analysis as much as they do other types of mass spectrometers. The scan range of an ion trap mass spectrometer is approximately one order of magnitude (e.g. m/z 40–400).

Time-of-Flight Mass Spectrometry (TOF-MS) A TOF determines mass by measuring the time it takes for an ionized molecule to travel the length of a vacuum chamber to the detector (Sferopoulos 2008; Shimadzu n.d.). Mass detection by TOF is more accurate than that of other mass detectors and it is capable of detection over a large range, from 50 to over 150,000 Da. TOF is commonly used with matrix-assisted laser desorption/*ionization* (MALDI) ionization (Nationalmaglab) but can be used with most ionization techniques with incorporation of an ion trap. TOF detection can be used for primary detection or as part of a MS/MS system for the detection of compound fragments.

Orbitrap The newest addition to the family of high-resolution mass spectrometry analyzers, Orbitrap (Zubarev and Makarov 2013) is an ion trap mass analyzer consisting of an outer barrel-like electrode and a coaxial inner spindle-like electrode that traps ions in an orbital motion around the spindle. The image current from the trapped ions is detected and converted to a mass spectrum using Fourier transformation of the frequency signal.

2.8 Thermal analysis

This is a branch of materials science where the properties of materials are studied as they change with temperature (Wunderlich 2001, pp. 9134–9141). Several methods are commonly used, distinguished from one another by the property which is measured. Examples are:

- *Differential Scanning Calorimetry (DSC)* – displays heat absorbed or released by a sample, as compared with a reference material
- *Thermo Gravimetric Analysis (TGA)* – measure the change in mass of a sample as the sample is heated at a controlled rate in a controlled atmosphere
- *Thermo Mechanical Analysis (TMA)* – displays mechanical characteristics of materials in response to fluctuations in temperature such as heating or cooling

Devices that use these methods, either alone or in combination (i.e. simultaneous thermal analysis (STA) that combines DSC and TGA), have applications in emergency response (i.e. confined space monitors, LEL sensors), gas analysis, or laboratory analysis.

3 Biological Detection/Identification Technologies

It is important to understand *why* certain pathogens – CDC categories "A" and "B", see (CDC 2012) – possess bioterrorism potential and why it is important to invest in detection technologies that will rapidly clarify the nature of a threat and guide the response needed to spare lives and mitigate consequences. Bioterrorism agents must have certain characteristics to be effective, including infectivity, pathogenicity, virulence, transmissibility, stability, toxicity, incubation period, and lethality (Walper et al. 2018).

3.1 Key Concepts

Infectivity is the ease with which a microorganism establishes itself in the host; it is not necessarily related to **pathogenicity**, which is the pathogen's ability to cause disease in the host, nor to **virulence**, which refers to the severity of that disease.

Lethality, referring to the ease with which the pathogen causes death, is different still. Consider the incapacitating agent Brucella, causing brucellosis, which requires only 10–100 bacteria to establish itself in a host, making it highly infective, but is lethal in fewer than 5% of cases, even if untreated (Dembek 2011).

Toxicity reflects the severity of the illness precipitated by a toxin.

Incubation period is the time that elapses between exposure to a bioagent and the appearance of symptoms.

Transmissibility of the pathogen directly from one person to another either through casual contact, e.g., Ebola, or intimate contact, e.g., HIV, or indirectly through vectors such as mosquitoes or fleas, e.g., malaria or plague, is required to seed an epidemic.

Stability refers to a pathogen's ability to survive environmental factors and plays a critical role in determining the nature and effectiveness of the dispersal method.

Morbidity refers to a person or the number of people in a population who have a given disease.

Mortality generally refers to death on a large scale.

3.2 Biological Detection Methods and Equipment

Most biological detection equipment involves significant (hours-days) incubation periods and a number of processing steps (centrifugation, filtration, dielectrophoresis, immunogenic separation, nucleic acid extraction, and concentration) for the identification of threat materials. Since time is life, the use of magnetic nanoparticles, for example, as a means of scavenging a bioagent from an analytical sample and concentrating it with a magnetic field can vastly simplify the subsequent detection process (Fornara et al. 2008). On the other hand, the human body, with all the symptoms of disease that it presents, is an excellent indicator of the presence of a bioagent, and the identification of a diseased patient may well be the impetus to collect the clinical samples necessary to identify a growing public health threat. Nevertheless, it is very encouraging that there is a very wide spectrum of techniques (not all are mentioned here), both field and bench, that support the detection and identification of biothreats.

Cell cultures Culture methods are still generally considered the gold standard for diagnosis of infection with many bacterial pathogens (Meenakshi 2018). There is a variety of culturing media:

Differential Media – bacteriological growth media (i.e. blood agar) that contain specific ingredients to allow one to distinguish selected species or categories of bacteria by visual observation.

General Media – containing substantial nutrients (e.g. nutrient broth or agar and trypticase soy broth or trypticase soy broth agar) and will support the growth of a wide variety of microorganisms.

Selective Media – formulated to contain ingredients (e.g. antibiotics) that inhibit the growth of selected species or categories of bacteria while allowing other organisms to grow (e.g. mannitol salt agar).

Specialty Media – media that are optimized, chemically defined formulated for specific uses or cell lines for very specialized applications (e.g. stable isotope labeling with amino acids in cell culture).

Viral Culture – cell culture systems (in vials or tubes) are live cell monolayers that support the growth of viruses.

Bacterial culture takes a long time relative to disease progression even for bioagents that are easily cultured such as *B. anthracis* (1–2 days), *Y. pestis* (>24 h for barely visible colonies), *B. mallei* (2 days, under enhanced circumstances), and *B. pseudomallei* (3–5 days). However, some bacterial species grow poorly on culture medium: *F. tularemia* is typically identified with chocolate agar, but grows poorly even under optimized conditions, *M. tuberculosis* requires 9 days or more, and Brucella diagnosis takes up to 21 days via culture (with a highly variable 10–90% accuracy), whereas the agent responsible for Q Fever cannot be reliably cultured.

Electrophoresis Electrophoresis is a way of separating biological molecules in an electric field based on the fact that different molecules have different natural electric charges associated with them (Beck 2018). This causes the different components of a sample to move at different rates under the influence of an electric field. Examples of various types of electrophoresis are:

- *Denaturing Gradient Gel Electrophoresis (DGGE)* – used to separate and detect DNA molecules that differ by as little as a single nucleotide.
- *Polyacrylamide Gel Electrophoresis (PAGE)* – used to separate proteins according to their electrophoretic mobility which is determined by molecular weight, conformation, and charge.

- *Pulsed-Field Gel Electrophoresis (PFGE)* – addresses the limitations of conventional gel electrophoresis to separate large fragments of DNA that are not adequately separated using standard gel electrophoresis.

Flow Cytometry Flow cytometry is used to analyze the physical and chemical characteristics of particles in a fluid as it passes through at least one laser (Robertson 2018). Cell components are fluorescently labelled and then excited by the laser to emit light at varying wavelengths. The greater the amount of fluorescence emitted the larger the source (i.e. more of the target antigen). In addition, multiple fluorescently tagged antibodies of varied colors can be used to determine if multiple antigens are present in a matrix.

Immunology Certain immunoassays use antigens to test for the presence of matching antibodies in solution. Examples include:

- *Enzyme-linked Immunosorbent Assay (ELISA)* – the gold standard method for detecting and quantifying antigens and antibodies. While simplicity of the assay is a key advantage, the inability to detect multiple parameters simultaneously (multiplex) is a significant disadvantage.
- *Hand Held Immunoassay (HHA)* (Abingdonhealth n.d.; Med.Navy.Mil n.d.) – Immunosensor devices are usually manufactured in the form of a test strip and use the specific interaction of target protein and antibodies to elicit a color deposition when the two interact. The devices operate on the principal of lateral flow immunochromatography in which the various background components are separated across an absorbent membrane into discrete regions. These devices allow for rapid presumptive identification of both biological organisms and toxins in 15 min or less and require minimal training. No sample preparation is required. Results may be read visually, although strip reading devices can be used to improve consistency of analysis. The devices are disposable, contain no moving parts, have a long shelf life, and operate in a wide range of temperatures. There are well-validated devices for numerous BW targets. Unlike nucleic acid-based detection devices, the immunoassays have the ability to directly detect toxins. A disadvantage is that the devices are typically not as sensitive as molecular based detection devices, and they only provide qualitative, not quantitative data.
- *Western Blot (WB)* – identifies proteins using antibodies specific to proteins that have been first separated from one another according to their size by gel electrophoresis. The Western blot test is performed after the gel-electrophoresis.

Microscopy Microscopy is a standard method for the characterization of biological agents. There are three well-known branches of microscopy: optical, electron, and scanning probe microscopy, along with the emerging field of X-ray microscopy. Various types of microscopy include:

- *Atomic Force Microscopy (AFM)* – a type of high-resolution scanning probe microscopy.
- *Fluorescence Microscopy (FM)* – combination of state-of-the-art optical systems with computerized control and the capability to obtain digital images.

- *Phase Contrast Microscopy (PCM)* – employs a contrast-enhancing optical mechanism that allows microorganisms to be viewed in their natural state without the use of deleterious killing, fixing and staining procedures required for standard optical microscopy.
- *Scanning Electron Microscopy (SEM)* – a powerful technique that can achieve 500,000 fold magnification. Biological samples must be coated in gold prior to analysis by SEM to preserve the integrity of the sample and prevent it from changing or decaying during the SEM process.
- *Transmission Electron Microscopy (TEM)* – an electron beam passes through a thinly-sliced resin-embedded sample to create a two-dimensional image of the sample thus achieving magnifications of 1,000,000 times or more.

Polymerase Chain Reaction (PCR) Real-time Polymerase Chain Reaction (PCR) amplification is generally regarded as the most sensitive and specific means of detecting potential biological warfare pathogens (Terelii and Tüzüni 2014). Results can usually be obtained in about an hour, including a separate sample preparation step to manually extract DNA signatures prior to PCR analysis. Well-characterized DNA extraction kits are available for rapid extraction (20–30 min) of target DNA from a variety of complex liquid and solid matrices including food, water, and tissue. Samples are prepared by the end user and the instrument performs PCR amplification of the target sequences. Thermal cycling produces additional copies of target sequences, and the fluorescent levels in the reaction change as additional copies are produced. During thermal cycling, the instrument detects copies of the target sequences by measuring changes in fluorescence. A major disadvantage of nucleic acid-based detection devices is their inability to directly detect non-nucleic acid targets such as toxins (Favrot 2015). However, toxins may be indirectly detected via the presence of residual DNA fragments of the organism used to produce the toxin. Residual DNA fragments from the organism used to produce a toxin are typically present at low levels even in the purest of toxin preparations, and fragments are most likely present in higher concentrations in crude preparations.

Other molecular technologies include: Capillary Electrophoresis Sequencing (CES); Digital PCR (DPCR); Evanescent Wave Fiber-Optic Biosensor (EWFO); Lawrence Livermore Microbial Detection Array (LLMDA); Microarray (MA); Next Generation Sequencing (NGS); Northern Blot (NB) RNA; Single Molecule Real-Time Sequencing (SMRT); Southern Blot (SB) DNA; Zero-Mode Waveguide (ZMW).

Whole Genome Sequencing Along with PCR, genomic sequencing is one of the major scientific developments of the past century (Priest 2017). Although it generally remains too time-consuming and expensive to run for large sample sets or constant environmental monitoring, sequencing technology continues to develop at an incredible pace with decreasing cost, increasing read lengths, and sample throughput growing at a nearly exponential rate (Vincent et al. 2017).

Pyrosequencing replaced Sanger as the current method for DNA sequencing, as it represents both a 100-fold increase in throughput and a 10-fold decrease in cost when compared to the electrophoresis-based Sanger method. Pyrosequencing is done by the real-time monitoring of phosphate release following base incorporation into a growing DNA strand. Certain commercial ventures already employ this technology (Roche 454 system; Illumina Solexa system; SOLiD; IonTorrent). Next-generation sequencing (NGS) and whole-genome sequencing (WGS), as their names imply, utilize a form of PCR to generate millions of short nucleic acid fragments that can then be assembled into a full genome sequence. NGS and WGS have significant potential for monitoring the molecular epidemiology of known viral pathogens and for the identification of unknown or poorly characterized viral species. For well characterized viral agents, NGS/WGS can be used to monitor genetic drift and variability, to track transmission through populations, and to identify genetic variations that contribute to resistance to antibiotic or antiviral therapeutics (Visser et al. 2016).

Direct Mass Spectrometry One of the best examples of the instrumentation-based approach, MS has long been used to determine the chemical species present in a sample. Unlike the other approaches in this category, MS detects (bio) molecules on the basis of their intrinsic mass-to-charge (m/z) ratio, which provides a unique biochemical signature for each bio-analyte. The method works well for small molecules, but not for larger biomolecules such as proteins necessitating new softer ionization techniques such as matrix-assisted laser desorption/ionization (MALDI) and electrospray ionization (ESI) (Ho and Reddy 2010). These systems are often coupled with time-of-flight (TOF) mass analyzers.

Peptide mass fingerprinting (PMF) Also known as protein fingerprinting, PMF is an analytical technique for protein identification in which the unknown protein of interest such as a toxin (Bergström 2016), is first cleaved into smaller peptides, whose absolute masses can be accurately measured with a mass spectrometer such as MALDI-TOF or ESI-TOF.

Chromatography with MS Detection Although direct MS analysis is highly useful in the detection of biothreat agents, samples often require additional preparation, purification, and especially target enrichment steps. This is particularly important in whole cell bacterial detection where the target pathogen may only be present in trace amounts within the complex background of benign bacterial species and other nucleic acid materials found in biological or environmental samples. Gas chromatography (GC), liquid chromatography (LC), and affinity chromatography are the most prominent methods to fractionate and concentrate samples before detection by MS.

Waveguide Based Sensors These sensors typically rely on the capture of a target biomolecule to the sensor surface, usually through an attached biorecognition element such as antibodies or nucleic acid hybridization. This capture alters the physical parameters of the system, which results in a change of the interrogating waveform. Six generalized classes of waveguide biosensors are available: surface plasmon resonance (SPR), fiber optic biosensors (FOBS), quartz crystal microbalances (QCM), surface acoustic waves (SAW), microcantilevers (MCL), and magnetoelastic/magnetorestrictive (ME/MR) resonators.

Surface Plasmon Resonance (SPR) SPR can sometimes offer a quick and portable (handheld) alternative for differential analyte detection while functioning as an effective screening tool, especially for toxins such as mycotoxins, saxitoxin, TTX, ricin [200 ng/mL in 10 min] (Feltis et al. 2008). A typical SPR system consists of a thin gold surface on top of a highly refractive glass surface. Light is reflected off the metal film, which excites the surface plasmons of the metal, forming an evanescent wave that can be displayed and recorded (Mariani and Minunni 2014). SPR has been used to detect a variety of pathogenic bacteria and related proteins, including *B. anthracis*, *M. tuberculosis, S. typhimurium,* and *E. coli O157.*

Fiber Optic Biosensors (FOBS) FOBS are another class of waveguide-based biosensors, using the optical propagation of an evanescent wave traveling through a fiber optic cable via total internal reflection to detect target pathogens. One of the most mature demonstrations of FOBS has come from the Naval Research Laboratory in Washington DC, which has led the development of several fully integrated systems over the past 20 years, including the RAPTOR (Anderson and Rowe-Taitt 2000) and the NRL Array Biosensor (Taitt et al. 2008). In one demonstration of the RAPTOR's multiplexing capabilities, six biohazard agents were simultaneously detected, including the bacterial biothreats *B. anthracis* (7.1 × 104 CFU/mL), *B. abortus* (1.5 × 105 CFU/mL), and *F. tularensis* (7.3 × 106 CFU/mL), while demonstrating a false-negative rate of 2.4% and a false-positive rate of 5.2%.

Quartz Crystal Microbalances (QCM) QCM are a class of piezoelectric-based devices, often in the form of targeted immunosensors, in which mass changes are detected by changes in frequency of the quartz resonator. QCM can be used for biological or chemical sensing techniques in air or liquid environments, making it a potentially robust and versatile sensor. These devices can be highly sensitive, capable of detecting mass change as low as 1 $\mu g/cm^2$. QCM devices have been demonstrated for a wide range of biothreat agents including for tularemia detection, *B. anthracis*, Salmonella, *E. coli* O157, and other E. coli antigens (Hao et al. 2011; Li et al. 2011).

E-nose systems These systems are colorimetric arrays which can detect volatile organic compounds (VOCs) released by bacterial growth in contained environmental setups such as culture. The presence and concentration of the VOC profile can provide a species-specific signature to identify the target bacteria (Carey et al. 2011). They have been used to identify *M. tuberculosis* and foodborne pathogens, along with discriminating between other biothreats. In one study, they were used to differentiate between strains of *B. anthracis*, *M. pseudotuberculosis*, *E. coli* and *Y. pestis* (Lonsdale et al. 2013).

Magnetic Sensors Magnetic-based separation techniques, particularly using magnetic micro- and nanoparticles, have been around for a long time and their benefits are well documented (Jamshaid et al. 2016). Magnetic beads have been shown to be effective when combined with apoferritin nanoprobes, in both detection and separation of biothreat agents (Seo et al. 2016). There are also two classes of wave-based magnetic sensors, magnetoelastic (ME) and magnetoresistive (MR), each of which has been used in various biothreat detection applications.

Disposable matrix devices Disposable matrix (NCRJS n.d.), also known as tickets or kits, usually involve dry reagents, which are reconstituted when a sample is added. There are one-step assay formats, as well as more complex formats involving multiple steps that are performed using one or more reagents. Ticket assays can be automated using instrumentation to perform the manual assay steps and provide a semiquantitative test readout. Rapid handheld assays with greater sensitivity, specificity, and reproducibility are under development for a wide range of bacterial agents and toxins. These assays have excellent stability characteristics, and test results are easy to obtain.

- *Hand-Held Immunochromatographic Assays (HHAs)* are simple, one-time-use devices that are very similar to the urine test strips used in home pregnancy tests. There are currently 10 live agent assays in production, four simulants, and five trainers (only saline solution is needed to get positive results). These tests provide a yes/no response, and are used in virtually all fielded military biological detection systems, in developmental systems, and by a number of consequence management units. Power is not required to use HHAs manually.
- *BTA™ Test Strips* are detection strips that are manufactured by Tetracore LLC and distributed by Alexeter Technologies, LLC. The technique (lateral flow immunochromatography) uses monoclonal antibodies that are specifically attracted to the target substance. Above a certain concentration, the antibodies and target substance combine to form a reddish band that appears in a window. The test is positive if two colored lines appear, one colored line indicates a negative. This technique provides fewer false positives in environmentally collected samples. Anthrax and ricin assays are available, with other assays in development.
- *Sensitive Membrane Antigen Rapid Test (SMART)* is a ticket-based system for detecting and identifying multiple analytes. It detects antigens in the sample by immunofocusing colloidal gold-labeled reagents (leveled antibodies) and their corresponding antigens onto small membranes. Positive results (formation of a red dot) are detected by an instrument that measures the membrane reflectance. An automated ticket-based system can be used to perform the SMART immunoassays. A variety of rapid tests are available, including SMART-II; BADD; ABICAP (Zasada et al. 2015).

Virulence and resistance markers Since the sequencing of the first bacterial virus by Sanger *et al.* in 1977, more than 60,000 sequencing projects have been published (Genomes OnLine Database). These include more than 49,000 bacterial genome projects, the sequences of 61.2% of which are available in public databases. Thanks to the various next-generation sequencing (NGS) technologies and platforms that are commercially available, genome sequencing had a significant impact on clinical microbiology by enabling the development of various sequence-based tools, notably molecular detection, serological and genotyping assays. In addition, by providing access to the complete gene repertoire of a strain, NGS also provides a unique way to decipher the virulence potential and predict the antibiotic resistance pattern of clinical isolates. Identification and characterization of virulence factors, notably toxins, and antibiotic resistance markers of pathogens are crucial in understanding bacterial pathogenesis and their interactions with the host, and in the development of novel drugs, vaccines and molecular diagnostic tools. In addition, detecting such virulence or resistance markers may help improving outbreak monitoring and therapeutic management. Prior to genome sequencing, identification of virulence factors successively relied on biochemical approaches, or systematic molecular screening of a panel of genes demonstrated to play a role in pathogenesis using molecular cloning and/or mutagenesis. Over the past two decades, thanks to genomics combined to functional analyses (transcriptomics and proteomics), the rate of virulence factor discovery has increased dramatically. Bacterial virulence factors in genomes may be identified by homology search with known virulence genes. For more detail on virulence and resistance markers see (Bakour et al. 2016).

Over the past years, the emergence of multidrug-resistant (MDR) bacteria has become a major public health concern worldwide. Recent studies demonstrate that antibiotic resistance encoding genes pre-existed the introduction of antibiotics. Several mechanisms have been described through which bacteria become resistant to antibiotics: production of natural or acquired enzymes that metabolize the antibiotic; producing antibiotic-modifying enzymes; modifications of the antibiotic target that prevent its binding; membrane impermeability and overexpression of efflux systems. Several databases dedicated to antibiotic resistance markers are currently available, with many bioinformatics tools created and available online.

Zoos' role in a national detection/alert system In 2018 the World Health Organization listed "Disease X" among the diseases most in need of research and development. Disease X is not a specific illness, but rather a hypothetical epidemic that could be caused by a pathogen (contagious strain of a virus or bacterium) that we don't yet realize affects humans. Experts have warned that we're not prepared to find such a disease at its likely animal source, or spot it quickly when it starts making people sick. In the summer of 1999, veterinary pathologist Tracey McNamara – head pathologist at the Bronx Zoo – noticed that crows were dropping dead on zoo grounds (Gannon 2018). Her investigation showed she was dealing with a novel disease. She wondered if the deaths could be connected to a spate of human deaths from an unusual form of encephalitis in New York City. But she had a hard time getting a diagnosis for the crow disease because government agencies didn't deal with disease samples from zoo animals. The disease turned out to be the West Nile virus, which hadn't been seen in the Western Hemisphere before. Two decades later, the lessons from the West Nile outbreak are only slowly being learned, and developing countries could be especially vulnerable to an outbreak of an emerging infectious disease that originates in wildlife.

Next Generation Analysis Beyond modifications to the assay itself, improvements in analysis that move beyond a simple “naked eye” evaluation can have a dramatic effect on the achievable limits of detection, sensitivity, and general utility of lateral flow assays (LFAs). Feng et al. utilized Google Glass as their rapid diagnostic test reader to achieve a deployable system for high-throughput analysis (Feng et al. 2014).

A smart application was written for Google Glass that could qualitatively and quantitatively assess a number of different rapid diagnostic tests. As long as these tests contained so-called “Quick Response” codes (i.e., a digital barcode), more than one test at a time could be processed by Google Glass’s camera. The collected images were digitally transmitted to a server for processing before the results were returned to the user. The information was also geo-tagged and stored on a spatiotemporal map to provide real-time statistics. Such devices could enable healthcare professionals to monitor and track outbreaks, spurring rapid preventative measures while also providing testing capabilities in resource limited environments.

4 Mixed CB Agents

One of the results of the FP7-SEC MIRACLE EU project was the ability to analyze samples containing a mixture of CBRN agents or toxic industrial materials (Voborova et al. 2015). A simple, rapid and efficient ultrafiltration procedure for separating and safely handling samples that contain a mix of chemical and biological agents (CB mixed sample) in liquid and solid matrices has been developed and thoroughly validated by the Centre for Applied Molecular Technologies/Defense Laboratory Department – Biological Threats (CTMA/DLD-Bio) at the Catholic University of Louvain in Belgium. The efficacy of ultrafiltration for separating a CB mixed sample into its C and B counterparts has been demonstrated by using quantitative performance indicators for both types of compounds. While allowing a separate analysis of each compound in optimal safety conditions, this method also improves the processing of unknown CBRN samples – an important contribution towards enhancing preparedness and consequence management in the fight against terrorism. After unknown samples have been processed so they only contain separated CBRN agents, the detection and reliable identification of biological warfare agents (BWA) must be performed rapidly, using specific and highly sensitive tests (usually PCR or quantitative PCR (qPCR)). The CTMA/DLD-Bio has been developing ultra-rapid amplification of nucleic acids technologies as the so-called ‘isothermal’ method that enables the amplification of targeted nucleic acids at a constant optimal temperature ranging between 37° and 65 °C, using a compact portable Twista fluorimeter instrument. Weighing one kilo, this basic equipment can replace the larger, heavier and more expensive conventional thermal cyclers currently used. These weight and cost advantages mean that isothermal amplification methods are attracting increased interest for the rapid detection of BWAs, and for diagnosing infectious diseases and their causative agents in general. There are several isothermal amplification methods, including Loop-mediated isothermal amplification (LAMP), Transcription-Mediated Amplification (TMA) and Recombinase Polymerase Amplification (RPA). In addition, CTMA/DLD-Bio is currently developing a novel multiplex microarray assay to detect the main BWA pathogens in a single test. Such simultaneous testing methods usually require heavier equipment and are time consuming.

5 Radiological/Nuclear Detection and Identification Technologies

Radiation detection is of crucial importance not only following the detonation of a dirty bomb (improvised radiological dispersal device) but also during the preparation phase when radioactive materials are transported or handled. Special technologies support detection and identification of the type of materials emitting radiation:

- *Geiger-Müller Detectors (GM)* identify gamma and beta radiation (Short 2015). They count the number of gamma rays or beta particles entering the detector per second. They do not provide identification information unless coupled with alternate detectors. Some GM tubes also detect alpha radiation sources.
- *Helium-3 detectors* operate on a principle of absorption (Siebach 2010). Helium-3 absorbs neutrons and emits high-energy particles that are measured as a current and converted to neutron counts. Helium-3 detectors do not identify specific neutron sources.
- A *scintillation detector* is often portable. The scintillation detector is not as versatile as the GM meter, although it can be used to look for contamination from some radioactive materials (MedPhys n.d.). The scintillation detector's active portion for detecting radioactivity is a solid crystal (the scintillator) with which the radioactive emission must interact. This limits the use of the detector to gamma rays and high-energy beta particles since medium- and low-energy beta particles cannot penetrate the crystal and, therefore, cannot interact.
- *Scintillators* (Krammer n.d.) can be made in different sizes, and the thickness of the scintillator determines its ability to absorb and detect certain radiation emissions. A thin scintillator is an excellent choice for low-energy gamma rays and high-energy beta particles. The ray or particle will be absorbed within the thin scintillator and the light produced by this interaction will pass through the remaining thickness to allow the gamma ray to be detected. A high-energy gamma ray is likely to pass right through the thin scintillator without interacting. A thick scintillator is the choice for radionuclides emitting high-energy gamma rays, as it is thick enough to absorb the gamma ray but not too thick to prevent the light that is produced from being detected.
- *Semiconductor detectors* typically identify gamma radiation sources. These detectors are based on Cadmium Zinc Telluride (CZT) semiconductors (KROMEK website n.d.) that directly convert x-ray or gamma-ray photons into electrons; the radiation deposits energy at some point in the crystal lattice where it generates pairs of charge carriers. By application of an electric field, the charge carriers get swept to the cathode and anode of the device where they induce a current pulse that can be detected.
- *Solid-state detectors* such as CsI (Cesium Iodide), or NaI (Sodium Iodide) emit light based on the excitation energy transferred to them by the radiation source (CERN n.d.). The emitted light is captured by photodiodes which produces a current that is measured by the detector.

The above technologies can be found in a variety of commercial products (Ridha n.d.; Miglierini 2004) such as in *personal dosimeters* (devices estimating the radiation dose deposited in an individual wearing the device), *identifiers* (devices that measure differences in radiation emission energies to determine the specific type of radioisotope present), *personal radiation detectors* designed to detect, and alert a minimally trained wearer when radiation exceeds natural background levels, *fixed port screening equipment* (Kwak et al. 2010) for high-throughput screening for all types of radiation, and *survey meters* (Vivekanand n.d.) that detect the presence of radioactive contamination in the form of alpha, beta, gamma, or neutron particles or x-rays. First responders use the device to determine if an emergency resulted in radioactive contamination. The devices allow the operator to determine the type of radiation and the amount so as to properly handle the situation.

6 Explosives Detection/Identification Technologies

Improvised explosive devices (IED) are still among the most prolific and devastating weapons used by terrorists. IED types are numerous and limited mainly by the ingenuity and intent of the bomber or organization concerned,. The effectiveness of an IED depends on the environment, access to finances, technology and key materials, know-how, and opportunity. Lethality of the main charge is often linked to technological expertise of the bomb maker.

The Middle East is the epicenter of IED use, and the so-called Islamic State, Al Qaeda in the Arabian Peninsula (AQAP) operating principally out of Yemen, Al Qaeda in the Islamic Maghreb (AQIM) operating in North West Africa, Boko Haram mainly operating in Nigeria, and Al Shabaab in East Africa all manufacture and deploy relatively crude IED. Several sophisticated detection devices ranging from refined versions of traditional metal detectors to vehicle-mounted radar sensors have been developed and used with considerable success.

However, technology tends to be effective against technology. Unsophisticated IED are less easy to detect with high tech solutions. For example, contraptions using simple wooden soleplates to trigger an IED can be amongst the hardest to detect and defeat. This has forced the development of the "AtN (attack the network) approach" which focuses on all levels to defeat IED systems by attacking the terrorist human networks, defeating the device, and preparing the force (AJP- 3.15 2018).

Various technologies are used to identify the constitution of explosives, including gas chromatography, ion or liquid chromatography, colorimetric, micro-electro-mechanical systems (MEMS), x-ray diffraction and fluorescence and various molecular spectroscopy methods – especially ion mobility spectrometry used to screen packages, vehicles, clothing, and other items for trace residues of *explosives*. Combined chemical-explosives portable/handheld detectors are available commercially.

Dogs continue to be the gold standard against which other explosives detection methods are judged. Trained canines provide a reliable and time-proven method for detecting concealed explosives. There is in principle no explosive compound that dogs cannot be trained to sniff out (Tomšič 2003).

The canine olfactory mechanism is an extremely sensitive chemical sensor. Experiments have shown the dog's olfactory sense to be from hundred thousand to one million times more sensitive than that of a human. A dog interprets the world predominantly by smell, and its brain is dominated by an olfactory cortex. While a dog's brain is only one-tenth the size of a human brain, the part that controls smell is, proportionally speaking, approximately 40 times larger than in humans. Dogs possess anywhere from 125 million to 300 million olfactory receptors in their noses, compared to about 5 million in humans. These receptors occur in special sniffing cells deep in a dog's snout (Rebmann et al. 2000). Canine detection, however, suffers from drawbacks such as the high cost associated with continuous training, their narrow attention span and inability to work round the clock, and behavioral and mood variations. When a dog is overheated and actively panting, its sense of smell is reduced by as much as 40% as it uses the air to cool itself rather than for smelling (MacDonald et al. 2003).

In a study appearing in *Analytical Chemistry* (Ong et al. 2017), scientists reported a new, more rigorous approach to training dogs and their handlers based on real-time analysis, with the aid of mass spectrometry, of what canines actually smell when they are exposed to explosive materials.

Researchers from the University of Rhode Island are working on a sensor that measures the energy that is released as an explosive molecule breaks down. Nanowires in the sensor act as a catalyst to cause that decomposition so the explosive can be detected (Mcdermott 2016). The system also measures the electrical properties of the catalyst as it interacts with the explosive vapor, as a second check. It detects both nitrogen-based and peroxide-based explosives.

Duke researchers have taken the first steps toward building an artificial "robot nose" made from living mouse cells (Hitoshi et al. 2018). A prototype was based on odor receptors grown from the genes of mice that respond to target odors, including the smells of cocaine and explosives. Existing "E-noses" use various chemical compounds instead of receptor stem cells to detect smells. Human, dog and mouse genomes contain around 20,000 genes, which contain instructions to create proteins that smell, taste, feel, move and do everything that our bodies do. About 5% of mouse genes have been identified as instructions to make odor receptors. In contrast, humans only use about 2% of their genes to make odor receptors.

7 Conclusions

The huge variety of technologies that in principle can be employed for CBRN detection makes technology selection difficult. As an example, a list of criteria for the evaluation of different types of chemical agent detectors would usually include (for details see Annex 1):

- Detectable agent(s)
- Detection limits
- Response time
- Simultaneous detection of other materials/agents
- Portability and weight
- Power requirements
- Operational life time
- Operational temperature range

The only solution is an in-depth study of the pros and cons of each technology in relation to the mission and the threats that a specific organization faces. It is wise, too, to remember that a single piece of equipment is no piece of equipment, and that no current detector fulfils all the desired criteria for a successful and on-time detection, irrespective of price, size or looks.

Annex I – Comparison of Various Detection Technologies

Selected from: Rodi Sferopoulos (2008): A review of chemical warfare agent (CWA) detector technologies and commercial-off-the-shelf items. Human Protection and Performance Division, Defense Science and Technology Organization, Australian Government. DSTQ-GD-0570 (2008)

Assessment criteria for available IMS detectors

Criteria	CAM	APD 2000	Multi-IMS	Raid-M-100	IMS 2000	GID-3	Sabre 4000	LCD-3
Detected agents	Blood, Blister, Choking and Nerve agents plus selected TICs[39]	GA, GB, GD, VX, I ID, L, Pepper spray & Mace[64]	Nerve, Blister, Blood and choking agents[46]	GA, GB, GD, GF, **VX**, HD, HN, L, AC TICs: Chloride, Cyanide, SO_2, Toluene diisocyanate, Arsine[53]	GA, GB, GD, Blister agents including L[56]	GA. GB. GD, GF, VX, L. Programmable, TIC detection optional[59]	GA, GB, GD, GF, VX, Vesicants, TICs. Drugs & Explosives[62]	Nerve, blister & blood agents, TICs[64]
Limits of detection	LODs in line with or exceed the NATO requirements[6,39,43]	V agents – 4ppb; G agents – 15ppb; H – 30ppb; L – 200ppb[44]	Nerve 0.01 0 1 mg/m^3; Blister-0.5–2.0 mg/m^3; Blood/choking- 20–50 mg/m^3[3,46]	Low ppb up lo several ppm[64]	Nerve – 20 µg/m^3; Blister 200 µg/m^3[3,64]			LOD in line with or exceed the NATO requirements[64]
Simultaneous detection	No[59]	Yes for nerve and blister agents. To detect irritants the mode must be manually changed[68]	Yes[46]	Yes[53]	Yes[56]	Yes for nerve, blister, blood and choking agents and TICs[59]	No[63]	Yes[64]
Portability and weight	Hand-held, weight <2 kg with battery[6,39,40]	Hand-held, weighs <3 kg with batteries[44,45]	Hand-held, weigh <800 g with battery[46]	Hand-held, weighs <3 kg	Hand-held, weighs <3 kg with battery[56,58]	Vehicle mounted and use by dismounted troops, weighs <7 kg with battery[59]	Hand-held, weighs <3.5 kg with battery[63]	Small. lightweight, weighs <500 g[64]

Criteria	CAM	APD 2000	Multi-IMS	Raid-M-100	IMS 2000	GID-3	Sabre 4000	LCD-3
Power requirements	Single 6V rechargeable-lithium-sulfur dioxide battery or 12V power supply[6,39,40]	6 standard 'C' alkaline batteries[44]	Rechargeable lithium-ion battery[46]	Rechargeable lithium- ion battery pack[53]	Lithium manganese dioxide batteries or power suppl[56]	Lithium-sulfur dioxide battery, rechargeable battery or mains power supply[59]	Rechargeable lithium- ion battery or mains power supply	4 × AA Lithium iron Disulphide or 4 × AA Alkaline Manganese Dioxide batteries or power supply[64]
Operational life	14 h continuous at 20°C[6,39,40]	6–8 h[44,45]	10 h[46]	Minimum battery life of 6 h intermittent use in a 24 h period at 10°C to 49°C[32]	16 h, however has an auto shutdown feature after 15 mins[56]	14 h continuously[60]	4 h[62]	Alkaline batteries –30 h above 10°C, Li batteries –40 h above 10°C[65]
Operational temperature range (°C)	−25 to +55[6,39,40]	−30 to +52[44]	−30 to +50[44]	−30 to +50[51]	from −25[56]	−30 to +50[59]	0 to +45[63]	−31 to +55[64]

Assessment criteria for available FPD detectors

Criteria	AP2C	AP4C	MINICAMS
Detected agents	G, V and H agents[71]	CWAs and 49 of the 58 chemicals on NATO's TIC list[78]	Detects and alarms to all chemical warfare agents, precursors, simulant materials, and related industrial chemicals[77]
Response times	Less than 2 s[71]	2 s[78]	3–10 min[19]
Limits of detection	GB-10 μg/m^3 and HD-400 μg/m^3[3,71]	G agents-20 μg/m^3 and HD-600 μg/m^3. Liquid VX- 3 μg/cm^2[78]	GA & GB-0.1 μg/m^3; GD- 0.03 μg/m^3; VX- 0.01 μg/m^3; blister agents-3 μg/m^3[3,77]
Simultaneous detection	Yes[74]	Yes[78]	
Portability and weight	Handheld, weighs <2.5 kg with battery and hydrogen storage device[74]	Handheld, weighs 2 kg with battery and hydrogen storage device[78]	Portable, weighs 9 kg[77]
Power requirements	7.3 V Lithium battery pack containing 2 LSH20, liquid cathode and lithium thionyl chloride batteries[74]	Battery, external power supply of rechargeable battery[78]	110 (+/−10%) VAC 50/60 Hz or 600 watts[77].
Operational life	12 h[74]	Dependent on power supply	
Operational temperature range (°C)	−32 to +55[72]	−31 to +50[78]	0 to +40[77]

Assessment criteria for available IR-based detectors

Criteria	M21 RSCAAL	JSLSCAD	MIRAN SapphIRE	AN/ KAS/I-1A	TravelIR HCI	HazMat ID	IlluminatIR
Detected agents	Nerve, HD and L vapour[27]	Nerve (GA, G B, GD, GF) and Blister (HD, L)[97]	Nerve and blister agents[90]		Nerve agents and vesicants. TICs, white powders, forensic drug and clandestine lab precursors, explosives & common chemicals[92]	Nerve agents, vesicants, precursors. TICs, forensic drug and clandestine lab precursors, white powders, explosives, common chemicals & pesticides[92]	
Response times	Line of sight dependent[27]	Line of sight dependent	Approximately 18 s[56]		Less than 20 s[92]	Less than 20 s[93]	
Limits of detection	Nerve agent (GA, GB & GD) 90 mg/m^3; L 500 mg/m^3; HD 2300 mg/m^3[94]	Nerve agents 135 mg/m^2, blister 3300 mg/m^2; blood (AC) 6600 mg/m^2; CK 6000 mg/m^2[92]	GA – 1.30 mg/m^3; GB- 0.83 mg/m^3 & HD 2.54 mg/m^3 [80]		% in isopropyl alcohol: GA 0.625, CB 5.0, VX 0.016, HD 0.25[90]		
Simultaneous detection	Yes[98]	Yes	Yes	Yes	Ye*		
Portability and weight	Two man portable, detector weighs 23.6 kg; & the tripod weighs 6.8 kg[99]	Vehicle mountable, standoff. 360° scanner weighs 18.6 kg, 60° scanner weighs 19.5 kg, Power adapter is 6.8 kg & operator display unit weighs 5 kg[97]	Man portable. Unit weighs approximately 10 kg[56]	standoff, weighs <12.5 kg[68]	Man-portable, weighing <12 kg[90]	Man-portable, weighs <10.5 kg[94]	

Power requirements	Batteries or by standard military generators[99]	28 Vdc vehicle power or 155 Vac[97]	Internal, rechargeable Nickel- Cadmium battery or 110 V AC adapters[50,56]	Powered by 115 V AC[55]	Powered by 110 V outlet, 12 Vvehicle power or battery pack[92]	Powered by an internal battery, mams or cigarette lighter[94]	
Operational life	277 h[99]					Battery runs from 2 h and charge time la 3 h[94]	
Operational temperature range (°C)	−32 to +48[99]	02 to 49[97]	5 to 40[59]		−7 to 50[92]	−7 to 50 0–100% humidity[96]	

Assessment criteria for available Raman-based detectors

Assessment criteria	Ahura's firstdefender	Ahura's firstdefender XL	JCSD
Detected agents	CAs, toxic chemicals, explosives, white-powders, narcotics, contraband and forensics[104]	CAs, toxic chemicals, white powders, narcotics, contraband and forensics[105]	20 CAs, 30 TICs, 9 Lnterferents[107]
Response time	Analysis time <15 s[104]		Approximately 40 s
Limits of detection			
Simultaneous detection	Yes		
Portability and weight	Hand held, weighs <1.5 kg[104]	Handheld, weighs <2 kg[105]	
Power requirements	Rechargeable lithium ion batteries[100]	Rechargeable lithium ion batteries[100]	
Operational life	5 h[104]	5 h[105]	
Operational temperature range (°C)	−20 and +40[100]		

Assessment criteria for available SAW-based detectors

Criteria	HAZMATCAD	JCAD chemsentry	CW sentry plus	SAW MINICAD mk II
Detected agents	Nerve – GA, GB, GD. GF and VX Blister – HD, HN3 Blood – AC, CK Choking – CG[112] TICs: Hydride gases – Arsine, Diborane, Silane, Halogen gases- Chlorine, Fluorine, Bromine Acidic gases Sulfur dioxide[112]	Nerve (GA, GB, GD, GF and VX), blister (HD, HN3 and L), and blood (AC and CK) agents[123]	Nerve – VX, GA, GB, GD and GF. Blister – HD, HN3 Blood – AC Choking – CG TICs: Hydride gases – Arsine, Diborane, Silane, Halogen gases- Chlorine, Fluorine, Bromine Acidic gases Sulfur dioxide[125]	GA- GB, GD and HD[119]
Response times	Responds between 20 and 120 s[112]	Responds between 10 and 90 s, depending on agent concentration[120]	Response time 20 s[115]	Analysis time 60 s[119]
Limits of detection	GD and HD in High Sensitivity Mode at dose to IDLJ-f limits in up to 4 minule[50]	GA: 100 mg/m^3; GB: 30 mg/m^3 and GD 50 mg/ m^3 within 12–13 s[28]. HD 40 mg/ m^3 within 8 s; Lewisite: 300–10000 mg/m^3 within 13 s; HCN: 30 mg/m^3 CNCI:1000 mg/m^3 within 2 min[28]	Nerve – 0.04–0.16 ppm Blister – 0.14 ppm Blood – 5 ppm Choking – 0.3 ppm Hydride – 0.5 ppm Halogen – 10 ppm[115]	GA: 0.2 mg/ m^3; G B: 0.5 mg/ m^3; GD: 0.1 mg/ m^3; HD: 1 mg/m^3

Criteria	HAZMATCAD	JCAD chemsentry	CW sentry plus	SAW MINICAD mk II
Simultaneous detection	Yes.	Yes.	Yes.	Yes.
Portability and weight	0.63 kg[555]	Handheld, weighing <1 kg[213]	Designed to be permanently installed, weighs 18.2 kg[135]	Light, handheld weighing 0.5 kg, including batteries[119]
Power requirements	Rechargeable Li-ion batteries[122]	Will operate on either platform power or an internal battery (BA-5800)[123].		Lithium cells[119]
Operational life	8 h in fast mode and 12 h in sensitive mode[112]	Battery life is greater than 18 h on the primary battery or approximately 12 h on a rechargeable battery[113]		5 year shelf life[119] Mission life of 6–8 h[119]
Operational temperature range <°C)	−10 to 50[112]	−32 to 49[113,114,120]	−20°C to 50[115]	5–40°C[119]

Assessment criteria for available colorimetric detectors

Criteria	M8 paper	M9 paper	3 Way paper	Tubes	Simultest sets	M256A1 Kit	M18A3 Kit	M272 Kit
Detected agents	G- nerve agents, VX, H[133]	G,VX,H-liquids[133]	G- nerve agents, VX, H[127]	Different tubes available for detection of specific TICs, TIMs or CAs[12]	Different sets available for detection of specific TICs, TIMs or CAs[12]	G- nerve agents, VX, HD,L,CX Blood- AC & CK in vapour or liquid[133]	G- nerve agents, VX, HD, HNs, L, ED, MD, CG, AC in vapour, liquid or aerosol form[133]	G- nerve agents, VX HD, L, AC in 6–7 min[133]
Response times	<30 s[133]	<20 s[133]		Variable[12]	Variable[12]	15 min or 25 min for AC[133]	2–3 min[133]	6–7 min[133]
Limits of detection	100 μ drops[133]	100 μ drops[133]			Detects below IDLH levels or close to AEL[12]	Nerve- 0.005 mg/ m^3, Mustard – 0.02 mg/m^3, L – 2 mg/ m^3, CX – 9mg/m^3, AC & CK – 3 mg/ m^3[133]	Nerve- 0.1 mg/m^3, Mustard – 0.5 mg/ m^3, L, ED, MD – 10mg/m^3, CG – 12 mg/m^3, AC – 80 mg/m^3[133]	Nerve- 002 mg/ m^3, Mustard – 2 mg/ m^3, L – 2 mg/ m^3, AC – 20 mg/m^3[133]
Simultaneous detection	Yes.	Yes.	Yes.					
Portability and weight	Handheld[133]	Handheld[133]	Handheld[133]	Handheld[133]	Handheld[133]	Handheld[133]	Handheld[133]	Portable/Han dheld[133]
Power requirements	NA	NA	NA	NA	NA	NA	NA	NA
Operational life		3 years shelf life[133]				5 years shelf life[133]		5 years shelf life[133]
Operational temperature range (°C)						−32 to 49[134]		

Assessment criteria for available PID-based detectors

Criteria	MiniRAE 2000	MiniRAE 3000	ppbRAE	ppbRAE 3000	ppbRAE plus	MultiRAE plus	ToxiRAE plus PID	TVA 1000B
Detected agents	VOCs[143]	VOCs[146]	VOCs[148]	VOCs[149]	VOCs including TICs and CWAs[150]	VOCs, combustible gases, O_2, CO, H_2S, SO_2, NO, NO_2, Cl, HCN, NH_3 and PH_3[152]	VOCs[153]	Organic and Inorganic vapours[154]
Response limes	< 3 s[143]	<3 s[146]	< 5 s[148]	< 3 s[149]	<5 s[150]		5 s[153]	3.5 s[154]
Limits of detection	0.1–10000 ppm[143]	0.1–15000 ppm[146]	1–200 ppm[148]	1–10000 ppm[149]	1–2000 ppm[150]	0.1–2000 ppm[152]	0.1–2000 ppm[153]	FID – 300 ppb hexane. PID – 100 ppb benzene[154]
Portability and weight	Handheld, weighs 553 grams with battery pack[143]	Handheld, weighing 738 grams[146]	Handheld, weighing 553 grams with battery pack[148]	Handheld, weighing 738 grams[149]	Handheld weighing 553 grams with battery pack[150]	Handheld, weighing 454 grams with battery[152]	Handheld, weighing 180 grams with battery[153]	Weighs 5.8 kg[154]
Power requirements	Rechargeable, external, field replaceable Nickel Metal Hydride battery pack or 4 AA batteries[143]	Interchangeable, drop in, rechargeable lithium ion and alkaline battery pack[146]	Rechargeable, external, field replaceable Nickel Metal Hydride battery pack or 4 AA batteries for alkaline battery pack[148]	Interchangeable, drop in, rechargeable lithium ion and alkaline battery packs[149]	Rechargeable, external, field replaceable Nickel-Metal-Hydride battery pack or alkaline bait cry holder for 4 AA batteries[150]	Interchangeable Lithium ion and alkaline battery pack[152]	Rechargeable 2.4V, 1110 mAh, nickel-cadmium battery pack or 2 AA alkaline battery adapter[153]	Rechargeable Nickel-Cadmium Battery[154]
Operational life	10 h continuous operation[143]	16 h continuous operation[146]	10 h, continuous operation[148]	16 h continuous operation[149]	10 h continuous operation[150]	14 h continuous operation with Li-ion battery. Can run and charge simultaneously[152]	10 h of continuous and instantaneous monitoring of VOCs[153]	8 h[154]

References

Abingdonhealth. n.d. What is a lateral flow immunoassay? https://www.abingdonhealth.com/contract-services/what-is-a-lateral-flow-immunoassay/.

Adamovica, M. 2017. Gas detection: How catalytic bead sensors work and why O2 sensors are omportant. PK Safety. https://pksafety.com/blog/gas-detection-how-catalytic-bead-sensors-work-and-why-o2-sensors-are-important. Accessed 8 Apr 2018.

Agilent Technologies. 2011. Webinar, microwave plasma – Atomic emission spectroscopy, a revolutionary new technique and plasma source that increases performance while eliminating expensive gas requirements. spectroscopyNOW.com, http://www.spectroscopynow.com/details/webinar/136bfa120bc/Microwave-Plasma%2D%2D-Atomic-Emission-Spectroscopy-a-Revolutionary-New-Technique-an.html. Accessed 8 Apr 2019.

AJP- 3.15. 2018. *Allied Joint Doctrine for countering improvised explosive devices.* Edition C Version 1, February 2018. Available at https://assets.publishing.service.gov.uk/government/uploads/system/uploads/attachment_data/file/686715/doctrine_nato_countering_ied_ajp_3_15.pdf.

Anderson, G. P, and C.A. Rowe-Taitt. 2000. *Water quality monitoring using an automated portable fiber optic biosensor: RAPTOR*. Photonic detection and intervention technologies for safe food. Proc. SPIE 2000, 4206, 418742.

Antec Scientific. n.d. Electrochemical detection. https://antecscientific.com/support/tutorials/electrochemical-detection. Accessed 8 Apr 2019.

Bakour, S., J. Rathored, S. Sankar, P. Biagini. 2016. *Identification of virulence factors and antibiotic resistance markers using bacterial genomics.* Future Microbiology, March 2016. DOI: https://doi.org/10.2217/fmb.15.149.

Beck, K. 2018. *The types of electrophoresis.* Sciencing (25 July 2018). https://sciencing.com/types-electrophoresis-5569711.html.

Bergström, T. 2016. *Protein identification and characterization through peptide mass spectrometry. Method development for improved ricin and botulinum neurotoxin analysis*. Doctoral Thesis, Swedish University of Agricultural Sciences, Umeå.

Carey, J.R., K.S. Suslick, K.I. Hulkower, et al. 2011. Rapid identification of bacteria with a disposable colorimetric sensing array. *Journal of the American Chemical Society* 133: 7571–7576.

CDC. 2012. Possession, use, and transfer of select agents and toxins, centers for disease control and prevention, department of health and human services. *Biennial Review; Final Rule Federal Register* 77 (194): 61083–61115. https://www.federalregister.gov/documents/2012/10/05/2012-24389/possessionuse-and-transfer-of-select-agents-and-toxins-biennial-review.

CERN. n.d. *Solid state detectors.* Available at http://riegler.home.cern.ch/riegler/lectures/lecture5.pdf.

Chalmers, J. M. 2013. *Infrared spectroscopy, reference module in chemistry, molecular sciences and chemical engineering.* Elsevier pp. 402–415. Available at https://www.sciencedirect.com/science/article/pii/B9780124095472002547. Accessed 8 Apr 2019.

Chasteen, T.G. 2009a. *Flame photometric GC detector.* The Chemiluminescence Home Page. https://www.shsu.edu/~chm_tgc/FPD/FPD.html. Accessed 7 Apr 2019.

———. 2009b. *The flame ionization detector.* Sam Houston State University. https://www.shsu.edu/~chm_tgc/primers/FID.html.

Clarke, W. 2017. Chapter 1 – Mass spectrometry in the clinical laboratory: Determining the need and avoiding pitfalls. In *Mass spectrometry for the clinical laboratory*, vol. 2017. Elsevier. https://doi.org/10.1016/B978-0-12-800871-3.00001-8.

Davies, A.M.C. n.d. An introduction to near infrared (NIR) spectroscopy, IMPublications website. http://www.impublications.com/content/introduction-near-infrared-nir-spectroscopy. Accessed 8 Apr 2019.

Dembek, Z.F., ed. 2011. *USAMRIID's medical management of biological casualties handbook.* 7th ed. Fort Detrick: U.S. Army Medical Research Institute of Infectious Diseases.

Dutrow, B.L, and C.M. Clark. n.d. X-ray powder diffraction (XRD). The Science Education Research Center at Carleton College. https://serc.carleton.edu/research_education/geochemsheets/techniques/XRD.html. Accessed 8 Apr 2019.

El Sohly, M.A., W. Gul, and M. Salem. 2008. Chapter 5 – Cannabinoids analysis: analytical methods for different biological specimens. In *Handbook of analytical separations*, vol. 6, 203–241. Elsevier. https://doi.org/10.1016/S1567-7192(06)06005-0. Accessed 7 Apr 2019.

European Nuclear Society (ENS). n.d. *Monitoring area.* https://www.euronuclear.org/info/encyclopedia/m/monitoringaerea.htm. Accessed 23 Mar 2019.

FAIMS website. n.d. *What is high-field asymmetric waveform ion mobility spectrometry?* http://www.faims.com/what.htm.

Favrot, C. 2015. Polymerase chain reaction: Advantages and drawbacks. In *3. Congresso Latinoamericano de Dermatologia Veterinaria*, 26–27. Argentina: Buenos Aires.

Feltis, B.N., B.A. Sexton, F.L. Glenn, et al. 2008. A Hand-held surface plasmon resonance biosensor for the detection of ricin and other biological agents. *Biosensors and Bioelectronics* 23: 1131–1136.

Feng, S., R. Caire, B. Cortazar, et al. 2014. Immunochromatographic diagnotic test analysis using google glass. *ACS Nano* 8: 3069–3079.

Fornara, A., P. Johansson, K. Petersson, et al. 2008. Tailored magnetic nanoparticles for direct and sensitive detection of biomolecules in biological samples. *Nano Letters* 2008 (8): 3423–3428.

Gannon, M. 2018. *An unknown 'disease X' could become an epidemic. Can we find it before it's too late?* LifeScience (18 Oct. 2018). https://www.livescience.com/63862-disease-x-animal-source.html.

Ghann, W., and J. Uddin. 2017. *Terahertz (THz) spectroscopy: a cutting-edge technology*. DOI 10.5772/67031. https://www.intechopen.com/books/terahertz-spectroscopy-a-cutting-edge-technology/terahertz-thz-spectroscopy-a-cutting-edge-technology. Accessed 8 Apr 2019.

Hao, R.Z., H.B. Song, G.M. Zuo, et al. 2011. Probe functionalized QCM biosensor based on gold nanoparticle amplification for bacillus anthracis detection. *Biosensors and Bioelectronics* (26): 3398–3404.

Hitoshi, K., F. Yosuke, D.M. Joel, et al. 2018. Vapor detection and discrimination with a panel of odorant receptors. *Nature Communications* 9 (1). https://doi.org/10.1038/s41467-018-06806-w.

Ho, Y.P., and P.M. Reddy. 2010. Identification of pathogens by mass spectrometry. *Clinical Chemistry* 56: 525–536.

Jamshaid, T., E.T.T. Neto, M.M. Eissa, et al. 2016. Magnetic particles: From preparation to lab-on-a-chip, biosensors, microsystems and microfluidics applications. *Trends in Analytical Chemistry* 79: 344–362.

Jasco website. n.d. *Principles of FTIR spectroscopy.* https://www2.chemistry.msu.edu/courses/cem434/Principles%20of%20IR%20spectroscopy.pdf.

JoVE. 2019. *Gas chromatography (GC) with flame-ionization detection.* Science Education Database. Analytical Chemistry. Cambridge. https://www.jove.com/science-education/10187/gas-chromatography-gc-with-flame-ionization-detection. Accessed 7 Apr 2019.

Krammer, M. n.d. *Scintillators.* Vienna: Institute of High Energy Physics. Available at https://www.hephy.at/fileadmin/user_upload/VO-5-Scintillators.pdf.

KROMEK website. n.d. *About CZT: cadmium zinc telluride.* Available at https://www.kromek.com/cadmium-zinc-telluride-czt/.

Kwak, S.-W., S.-S. Chang, and H.-S. Yoo. 2010. Radiation detection system for prevention of illicit trafficking of nuclear and radioactive materials. *Journal of Radiological Protection* 35 (4): 167–171. https://www.researchgate.net/publication/264192590_Radiation_Detection_System_for_Prevention_of_Illicit_Trafficking_of_Nuclear_and_Radioactive_Materials.

Lamont-Doherty Earth Observatory. n.d. *Ion chromatograph to detect major anions in precipitation (snow), groundwaters and drinking waters from New York.* Earth Institute, Columbia University. https://www.ldeo.columbia.edu/~martins/eda/Ic_lec.html. Accessed 7 Apr 2019.

LCGC. n.d. Quadrupole mass analyzers: An introduction. LCGC Europe 25(11).

Li, D., Y. Feng, L. Zhou, et al. 2011. Label-Free capacitive immunosensor based on quartz crystal Au electrode for rapid and sensitive detection of Escherichia coli O157: H7. *Analytica Chimica Acta* 687: 89–96.

Long, M. Scott. 2019. *Basic principles of LC, HPLC, MS, & MS*. Chemix website. https://www.chemyx.com/support/knowledge-base/applications/basic-principles-hplc-ms-lc-ms/. Accessed 8 Apr 2019.

Lonsdale, C.L., B. Taba, N. Queralto, et al. 2013. The use of colorimetric sensor arrays to discriminate between pathogenic bacteria. *PLoS One* 2013 (8): e62726.

MacDonald, J., et al. 2003. *Alternatives for landmine detection*. RAND.

Mariani, S., and M. Minunni. 2014. Surface plasmon resonance applications in clinical analysis. *Analytical and Bioanalytical Chemistry* 406: 2303–2323.

Maziejuk, M., W. Lisowski, M. Szyposzynska, T. Sikora, and A. Zalewska. 2015. *Differential ion mobility spectrometry in application to the analysis of gases and vapors. Solid State Phenomena* 223.

Mcdermott, J. 2016. *Professor designs explosives detector to rival a dog's nose.* Phys.org (16 February 2016). https://phys.org/news/2016-02-professor-explosives-detector-rival-dog.html#jCp.

Med.Navy.Mil. n.d. *Information paper: Hand held assay (HHA).* https://www.med.navy.mil/sites/nepmu2/Documents/lab/Hand%20Held%20Assays%20(HHA).pdf.

MedPhys. n.d. Chapter 4: Scintillation detectors. In *Radioisotopes and radiation methodology*. Med Phys 4R06/6R03, pp. 4-1 to 10. https://www.science.mcmaster.ca/radgrad/images/6R06CourseResources/4R6Notes4_ScintillationDetectors.pdf.

Meenakshi, A. 2018. Cell culture media: A review. *Mater Methods* 3:175. (Revised 11 Nov 2018). https://www.labome.com/method/Cell-Culture-Media-A-Review.html.

Mellon, F.A. 2003. Mass spectrometry – Principles and instrumentation. In *Encyclopedia of food sciences and nutrition*, 2nd ed. Elsevier.

Miglierini, M. 2004. *Detectors of radiation.* E. Wigner Course on Reactor Physics Experiments, April 27 – May 15, 2004, at http://oldweb.reak.bme.hu/Wigner_Course/2004/WignerManuals/Bratislava/detectors.pdf.

Miles, S. 1997. Rapid purification of small molecule libraries by ion exchange chromatography. *Tetrahedron Letters*. 38:3357–3358. Via Elsevier Science Direct.

MSA. n.d. *Photo ionization detectors (PIDs). Theory, uses and applications*, MSA. http://s7d9.scene7.com/is/content/minesafetyappliances/0803-11-MC%20Photoionization%20Detector%20Tech%20Brief%20-%20EN.

Nanophoton. n.d. *What is Raman spectroscopy?* https://www.nanophoton.net/raman/raman-spectroscopy.html. Accessed 8 Apr 2019.

Nationalmaglab. n.d.-a. *Tandem mass spectrometry (MS/MS).* National High Magnetic Field Laboratory. https://nationalmaglab.org/user-facilities/icr/techniques/tandem-ms.

———. n.d.-b. *Matrix-assisted laser desorption ionization (MALDI).* National High Magnetic Field Laboratory at https://nationalmaglab.org/user-facilities/icr/techniques/maldi.

NCRJS. n.d. 5.1.4.1 Immunoassay technologies, pp. 29–30. At https://www.ncjrs.gov/pdffiles1/nij/190747-c.pdf.

Neubauer, K. 2009. *Advantages and disadvantages of different column types for speciation analysis by LC-ICP-MS.* Spectroscopy Online. http://www.spectroscopyonline.com/advantages-and-disadvantages-different-column-types-speciation-analysis-lc-icp-ms. Accessed 8 Apr 2019.

Ong, T.-H., T. Mendum, G. Geurtsen, et al. 2017. Use of mass spectrometric vapor analysis to improve canine explosive detection efficiency. *Analytical Chemistry* 89 (12): 6482–6490.

Priest, J.R. 2017. A primer to clinical genome sequencing. *Current Opinion in Pediatrics* (29): 513–519.

Prime Faraday Technology Watch. 2002. *An introduction to MEMS (Micro-electromechanical Systems).* https://wwwe.lboro.ac.uk/microsites/mechman/research/ipm-ktn/pdf/Technology_review/an-introduction-to-mems.pdf. Accessed 8 Apr 2019.

Quan, K., and U. C. Davis. 2017. *Photoacoustic spectroscopy.* https://chem.libretexts.org/Bookshelves/Physical_and_Theoretical_Chemistry_Textbook_Maps/Supplemental_Modules_(Physical_and_Theoretical_Chemistry)/Spectroscopy/Photoacoustic_Spectroscopy. Accessed 8 Apr 2019.

Radboud University, Faculty of Science. n.d. *About NMR (nuclear magnetic resonance)*. https://www.ru.nl/systemschemistry/equipment/nmr-0/about-nmr/. Accessed 8 Apr 2019.

Rapiscan Systems. n.d. *Trace detection*. https://www.rapiscansystems.com/en/products/category/trace-detection.

Rebmann, A., M.H. Sorg, and E. David. 2000. *Cadaver dog handbook*. Boca Raton, London, New York, Washington D.C.: CRC Press LLC.

Reusch, W. 2013. *Visible and ultraviolet spectroscopy*. Michigan State University, Department of Chemistry. https://www2.chemistry.msu.edu/faculty/reusch/virttxtjml/spectrpy/uv-vis/spectrum.htm. Accessed 8 Apr 2019.

Ridha, A.-A. n.d. Chapter 7 – Nuclear detectors. Nuclear Physics 80–92. https://www.researchgate.net/publication/307593954_Chapter_Seven_Nuclear_Detectors.

Robertson, S. 2018. *What is flow cytometry?* https://www.news-medical.net/life-sciences/What-is-Flow-Cytometry.aspx.

Sanchez-Rodas, D., W.T. Corns, B. Chen, and P.B. Stockwell. 2010. Atomic fluorescence spectrometry: A suitable detection technique in speciation studies for arsenic, selenium, antimony and mercury. *Journal of Analytical Atomic Spectrometry* 25: 933–946.

SANDIA. 2011. *Rapid, automated point-of-care system (RapiDx)*. https://ip.sandia.gov/technology.do/techID=81. Accessed 7 Apr 2019.

———. 2013. *SpinDx™: Point-of-care diagnostics using centrifugal microfluidics*. https://ip.sandia.gov/technology.do/techID=82. Accessed 7 Apr 2019.

SAVER. 2005. *Guide for the selection of chemical agent and toxic industrial material detection equipment for emergency first responders*, Guide 100-04, Volume I and II: Summary; pp. 1–5.

Schelkanova, I., A. Pandya, A. Muhaseen, G. Saiko, and A. Douplik. 2015. Chapter 13 – Early optical diagnosis of pressure ulcers. In *Biophotonics for medical applications*, vol. 2015, 347–375. Woodhead Publication Series Biomaterials Elsevier.

Semrock website. n.d. *Surface-enhanced Raman scattering (SERS)*. https://www.semrock.com/surface-enhanced-raman-scattering-sers.aspx. Accessed 8 Apr 2019.

Senseor. n.d. *How SAW sensors operate?* https://www.senseor.com/saw-technology/saw-sensors-operation.

Seo, Y., J.-e Kim, Y. Jeong, et al. 2016. Engineered nanoconstructs for the multiplexed and sensitive detection of high-risk pathogens. *Nanoscale* 8: 1944–1951.

Sevostianova, E. n.d. *Atomic absorption spectroscopy*. https://web.nmsu.edu/~esevosti/report.htm. Accessed 8 Apr 2019.

Sferopoulos, R. 2008. *A review of chemical warfare agent (CWA) detector technologies and commercial-off-the-shelf items*. Human Protection and Performance Division, Defense Science and Technology Organization, Australian Government. DSTO-GD-0570.

Shimadzu. n.d. *What is TOF MS?* At https://www.shimadzu.com/an/lifescience/maldi/princpl1.html.

Short, M. 2015. *Geiger tube theory*. MIT lecture at https://ocw.mit.edu/courses/nuclear-engineering/22-s902-do-it-yourself-diy-geiger-counters-january-iap-2015/lecture-notes/MIT22_S902IAP15_lec02.pdf.

Siebach, J. 2010. *Characterization of He-3 detectors typically used in international safeguards monitoring*, Thesis, Brigham Young University. http://citeseerx.ist.psu.edu/viewdoc/download?doi=10.1.1.307.3890&rep=rep1&type=pdf.

Stauffer, E., J.A. Dolan, and R. Newman. 2008. Chapter 5 – Detection of ignitable liquid residues at fire scenes. In *Fire debris analysis*, vol. 2008, 131–161. Elsevier. https://doi.org/10.1016/B978-012663971-1.50009-9. Accesses 7 Apr 2019.

Sun, Y., and K.Y. Ong. 2005. *Detection technologies for chemical warfare agents and Toxic Vapors*. 1st ed, 272. Boca Raton: CRC Press.

Taitt, C.R., L.C. Shriver-Lake, M.M. Ngundi, and F.S. Ligler. 2008. Array biosensor for toxin detection: Continued advances. *Sensors* 8: 8361–8377.

Terelii, M., and A. Tüzüni. 2014. New molecular methods for detection of bioterrorism agents. *Türk Bilimsel Derlemeler Dergisi* 7 (1): 46–48.

Thakur, S.N. 2007. Chapter 2 – Atomic emission spectroscopy. In *Laser-induced breakdown spectroscopy*, 23–48. Elsevier.

The Sam Houston State University. n.d. *The pulsed flame photometric detector.* https://www.shsu.edu/~chm_tgc/primers/pfpd.html.

ThermoFisher Scientific. n.d.-a. *Inductively coupled plasma mass spectrometry (ICP-MS) information.* https://www.thermofisher.com/us/en/home/industrial/spectroscopy-elemental-isotope-analysis/spectroscopy-elemental-isotope-analysis-learning-center/trace-elemental-analysis-tea-information/inductively-coupled-plasma-mass-spectrometry-icp-ms-information.html. Accessed 8 Apr 2019.

———. n.d.-b. *Inductively coupled plasma optical emission spectroscopy (ICP-OES) information.* https://www.thermofisher.com/gr/en/home/industrial/spectroscopy-elemental-isotope-analysis/spectroscopy-elemental-isotope-analysis-learning-center/trace-elemental-analysis-tea-information/icp-oes-information.html. Accessed 8 Apr 2019.

Thomas III, S.W., Joly, G.D. and T.M. Swager. 2007. Chemical sensors based on amplifying fluorescent conjugated polymers, Chemistry Review vol. 107, 1339–1386. https://pdfs.semanticscholar.org/c2c3/3ef844107e028357cbdb49e98af137b107ef.pdf. Accessed 8 Apr 2019.

Tomšič, U. 2003. *Detection of explosives: Dogs vs. CMOS capacitive sensors.* University of Ljubljana, Faculty of Mathematics and Physics, Department of Physics. Seminars (March 2003). Available at http://mafija.fmf.uni-lj.si/seminar/files/2012_2013/DETECTION_OF_EXPLOSIVES_-_Dogs_vs._CMOS_Capacitive_Sensors.pdf.

University of Massachusetts Amherst UMASS. n.d. *The flame photometric detector.* http://www.ecs.umass.edu/eve/facilities/equipment/Agilent6890/The%20Flame%20Photometric%20Detector.pdf. Accessed 8 Apr 2019.

Vekey, K. 2001. Mass spectrometry and mass-selective detection in chromatography. *Journal of Chromatography A* 921: 227–236.

Vincent, A.T., N. Derome, B. Boyle, A.I. Culley, and S.J. Charette. 2017. Next-generation sequencing (NGS) in the microbiological world: How to make the most of your money. *Journal of Microbiological Methods* 138: 60–71.

Visser, M., R. Bester, J.T. Burger, and H.J. Maree. 2016. Next-generation sequencing for virus detection: Covering all the bases. *Virology Journal* 13: 85.

Vivekanand, L. S. n.d. *Radiation surveys, Instrumentation and Dosimetry. NXP Semiconductors.* Available at https://www.aiha.org/aihce07/handouts/rt240vivekanand.pdf.

Voborova, O., M. Bentahir, A.-S. Piette, and J.-L. Gala. 2015. CBRN: Detection and identification innovations. *Crisis Response Journal* 2015: 36–38.

Walper, S.A., G.L. Aragonés, K.E. Sapsford, et al. 2018. Detecting biothreat agents: From current diagnostics to developing sensor technologies. *ACS Sensing* 2018 (3): 1894–2024.

Warwick, B., and B. Dunn. 2011. Chapter 2 – Mass spectrometry in systems biology: An introduction. In *Methods in enzymology*, vol. 500. Elsevier.

Wirth, K., and A. Barth. n.d. *X-ray fluorescence (XRF).* The Science Education Research Center at Carleton College. https://serc.carleton.edu/research_education/geochemsheets/techniques/XRF.html. Accessed 8 Apr 2019.

Wunderlich, B. 2001. Thermal analysis. In *Encyclopedia of materials: Science and technology*, 2nd ed. Elsevier.

Zasada, A.A., K. Forminska, K. Zacharczuk, D. Jacob, and R. Grunow. 2015. Comparison of eleven commercially available rapid tests for detection of bacillus anthracis, francisella tularensis and yersinia pestis. *Letters in Applied Microbiology* 60: 409–413.

Zubarev, R.A., and A. Makarov. 2013. *Orbitrap mass spectrometry. Analytical Chemistry* 85 (11): 5288–5296.

Chemical Forensics

Paula Vanninen, Hanna Lignell, Harri A. Heikkinen, Harri Kiljunen, Oscar S. Silva, Sini A. Aalto, and Tiina J. Kauppila

1 Introduction

Chemical weapons (CW) have been used against civil populations numerous times since 2012, including attacks in the Syrian Arab Republic (Syria), the Republic of Iraq (Iraq), Malaysia, and the United Kingdom (UK). In Syria alone, 336 attacks involving chemical weapons have been confirmed (Schneider and Lütkefend 2019). The questions raised around the attacks include the identity of the threat chemicals, the responsible actor (terrorists or state party), and whether any new tools exist for investigating these crimes.

The Convention on the Prohibition of the Development, Production, Stockpiling and Use of Chemical Weapons and on their Destruction (the CWC) was opened for signature on January 13, 1993, and entered into force on April 29, 1997. Each State Party to the Convention undertakes never, under any circumstances, to use, develop, produce, otherwise acquire, stockpile, or retain chemical weapons, or transfer, directly or indirectly, chemical weapons to anyone (OPCW/PTS 1993). The CWC defines a toxic chemical as "any chemical which through its chemical action on life processes can cause death, temporary incapacitation, or permanent harm to humans or animals". This general-purpose criterion was included to protect the CWC from circumvention (CWC, Art. II para 1(a)). Chemical warfare agents (CWAs) are chemicals that act as chemical weapons. When the CWC entered into force, the Organization for the Prohibition of Chemical Weapons (OPCW) was established in The Hague, the Netherlands, to achieve the objectives and purposes of the CWC and to ensure the implementation of its provisions, including those for international verification of compliance with it, and to provide a forum for consultation and cooperation among the State Parties. The above-mentioned CWA attacks have been a clear and serious

P. Vanninen (✉) · H. Lignell · H. A. Heikkinen · H. Kiljunen · O. S. Silva
S. A. Aalto · T. J. Kauppila
VERIFIN, Department of Chemistry, University of Helsinki, Helsinki, Finland
e-mail: paula.vanninen@helsinki.fi

M. Martellini, R. Trapp (eds.), *21st Century Prometheus*,
https://doi.org/10.1007/978-3-030-28285-1_12

breach of the Convention. In most cases, the mandate of the OPCW has been to verify the use of chemical weapons. The next chapter describes the current procedures used by the OPCW in the investigations of alleged use of chemical weapons. Traditionally, the main focus of the OPCW investigations has been confirmation of CW use, the identification of the used CWA, and collection of additional evidence regarding the alleged attack. A wide variety of forensic evidence has been collected during the OPCW's Fact Finding Missions (FFM) and the Joint Investigative Mechanism (JIM) of the United Nations (UN) and the OPCW, and these are reviewed in this article. After the attacks in Syria and Iraq, the inspectors of the OPCW and local officials collected evidence in a hostile environment under tight time constraints. In these attacks, the use of sarin or sarin-like nerve agent, mustard gas and/or chlorine (OPCW S/2016/738; OPCW S/2017b/904) was reported. In the UK and Malaysia, evidence was collected as part of crime scene investigation.

The term "chemical forensics" is defined as application of chemistry and forensic toxicology in a legal framework. The following literature review concentrates on techniques and research methods applicable to CWC-related forensic crime scene investigations and their potential in future investigations. The focus is on recent studies related to CWA, illicit drug profiling, poisons, and incapacitating chemicals. A lot has been achieved in the field of chemical forensics during the past years, including a new emerging approach called attribution analysis. Attribution analysis is aimed to aid the OPCW in forensic crime scene investigations involving CWAs. An overview of the recent analysis techniques applied in chemical forensics together with an assessment of the status quo and the future benefits for the OPCW are discussed in this review.

2 Current Procedures of the OPCW

The OPCW has a worldwide network of designated laboratories for environmental, material, and biomedical sample analysis. These laboratories are required to have international accreditation and participate in annual proficiency tests to keep the designation. The OPCW is also developing capacity for analysis of biological toxins, like ricin, and organizes so-called confidence building exercises for laboratories pursuing designation for biological toxins. The samples collected by the OPCW inspectors are analyzed at the OPCW's designated laboratories under a strict scope of analysis. The Standard Operating Procedures (SOPs) of the OPCW for collecting and splitting samples under toxic and hazardous conditions on-site, handling of inspection samples, and chain-of-custody and documentation during the inspection period describe the work of the inspection team in detail. Many different sample types can be collected at the site. These include environmental samples such as wipe, liquid, and solid samples, as well as biomedical samples from the attack victims. The OPCW SOPs also describe how to prepare the environmental samples on-site for portable gas chromatography-mass spectrometry (GC-MS) analysis. During the sampling in Syria and Iraq, no samples were analyzed on-site. Instead, the collected samples were sent to the OPCW laboratory for recoding. When samples are distributed for analyzing laboratories, a strict chain of custody is followed every step of the way. Samples are split into subsamples, which are then sent to off-site laboratories for analysis. Currently (2018), the OPCW has 19 laboratories designated for environmental and 16 laboratories designated for biomedical/clinical samples worldwide, from which the Director General selects the laboratory for off-site sample analysis. The analyzing laboratories, anonymous to one another, receive the scope of analysis defining the focus of the analysis. The OPCW designated laboratories have constant readiness to analyze environmental and material samples from an investigated site, as well as clinical samples collected from the victims of an attack.

The analysis of samples is performed in two independent designated laboratories in two different countries. The verification protocol for determining the presence or absence of chemicals related to the CWC includes sample preparation followed by analysis by different chromatographic and spectrometric techniques. According to the analysis rules of the OPCW, the identification is considered unambiguous if two different spectrometric techniques giving consistent results confirm the presence of the same chemical. An example of the analysis strategy of VERIFIN is presented in Fig. 1.

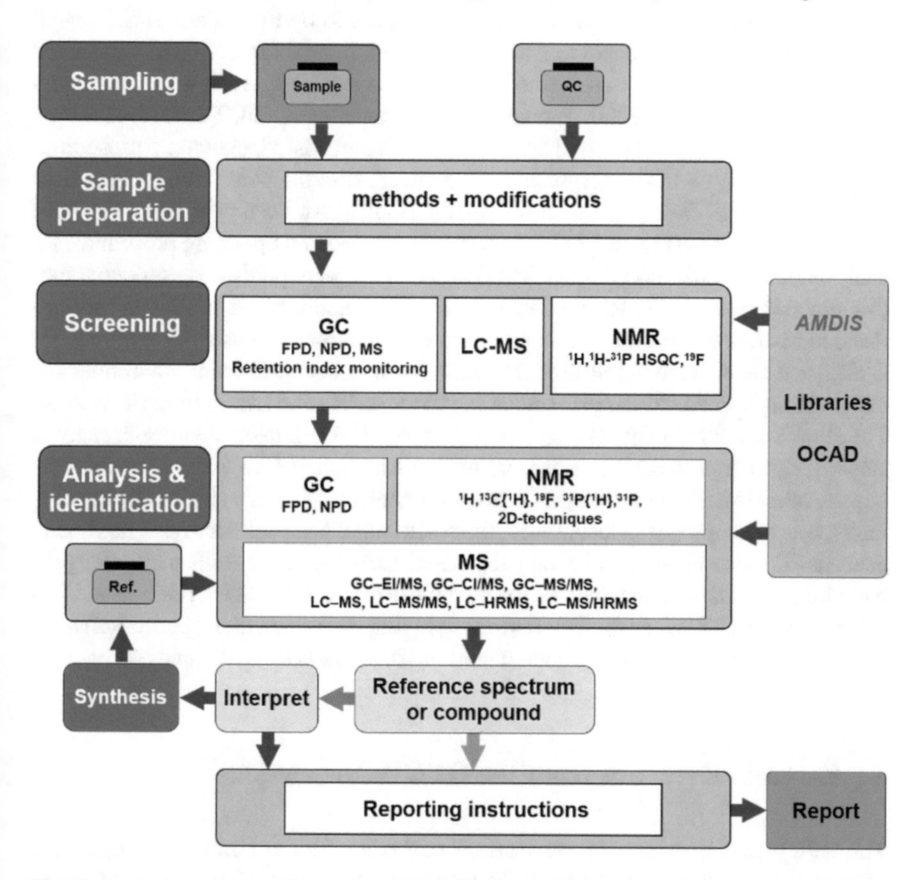

Fig. 1 Analysis strategy employed at VERIFIN for verification of CWC-related chemicals. (Reproduced by permission from Tiina Kauppila and Paula Vanninen, Working Instruction, Strategy for Analysis of CWC-related Chemicals, Quality Manual of VERIFIN, 2019)

A typical designated laboratory is equipped with versatile instrumentation, commonly including mass spectrometry (MS) techniques combined with gas chromatography (GC) and liquid chromatography (LC), and often also nuclear magnetic resonance (NMR) instruments. The measured data are searched against spectral libraries like the National Institute of Standards and Technology (NIST) MS library. The OPCW also has its own analytical database, OCAD (OPCW Central Analytical Database) containing a multitude of MS, NMR and infrared (IR) spectra, and retention indices of scheduled chemicals of the CWC (OCAD v.20_2018). In most cases, the identification is verified by comparison to a reference chemical analyzed under identical conditions. If reference data for the chemical of interest is not found in the available databases and no reference chemical is available, spectral interpretation or spectrum comparison to a measured standard of an alike chemical can be applied. However, synthesis and analysis of a reference chemical and the comparison of the analytical data is always preferred.

The verification methods for CWAs have been developed as a result of international collaboration. Initially they were tested by round robin tests among several laboratories, and later in the proficiency tests organized by the Technical Secretariat (TS) of the OPCW. The history of the International Interlaboratory Comparison and the proficiency tests are described in Encyclopedia of Analytical Chemistry (Mesilaakso and Rautio 2000; Vanninen 2011a, Vanninen 2019). The method testing eventually resulted in the publication of the recommended operation procedures (ROPs) for analysis in the verification of chemical disarmament ("the Finnish Blue Book"), and in 1977–2017, a total of 24 Blue Books have been published. The Blue Book 1993, 1994, 2011 and 2017 Editions, "Recommended operating procedures for analysis in the verification of chemical disarmament" were published as monographs (Vanninen 2011b, 2017; Rautio 1994, 1993). In general, the Blue Books' recommended operation procedures include sample preparation and analysis methods for CWAs and their degradation products and starting chemicals from environmental, material and biomedical samples (including blood, urine, and tissue) from the victims. The Blue Book 2017 (Fig. 2) includes chapters on CWC related chemicals, analysis strategy, sample preparation, analytical methods including GC- and LC-based techniques, other separation techniques like capillary electrophoresis and ion chromatography, spectroscopy-based techniques like NMR, other analyses (ricin and saxitoxin), analysis of biomedical samples and reporting. Utilizing these ROP methods is the basis for accreditation, and they are the guidelines for analyses of the designated laboratories of the OPCW, or the laboratories applying for designation. These ROPs are also used to train laboratory personnel working in the field of CWA analysis.

3 Forensic Evidence from the OPCW Inspections

The OPCW has done several inspections to Syria with the aim to investigate the alleged use of CWAs. The FFM of the OPCW and the JIM of the UN were established for investigation of individuals or organizations that were involved in the

Fig. 2 "Recommended operating procedures for analysis in the verification of chemical disarmament", i.e. "the Finnish Blue Books" are the basis of the analysis of CWC related chemicals, and guidelines for the OPCW designated laboratories. The 1993, 1994, 2011, and 2017 editions were published as monographs

use of chemical weapons, including chlorine and other toxic industrial gases (OPCW-UN FACT SHEET 2017a). The JIM was reporting to the UN Security Council about these investigations. The work of JIM was broadened to investigate other events, for example the actions of the Islamic State of Iraq and Syria (ISIS). The investigations of the JIM were performed in two phases: first, reports and materials produced by the FFM were reviewed and analyzed, and secondly the in-depth analysis of cases was continued until sufficient information was formed for the Security Council.

In April 2017, the JIM released the report of the 2016 Khan Shaykhun chemical attack, where the Government of Syria was attributed as the responsible party (OPCW S/2017b/904).

In addition to the identification of the actual CWA, there has been interest towards additional chemicals, such as stabilizers, synthesis by-products, and impurities. This chemical forensics type of investigation aims at discovering evidence to point out the origin of the CWAs, and eventually the party responsible for the attack.

The inspections are based on the mandates for the FFM and the JIM given by the OPCW and the UN, respectively. The safety situation and access to places to be inspected are cleared for the inspection team, and a predefined inspection plan is followed to the extent possible under the actually encountered circumstances.

The forensic evidence from an alleged chemical attack inspection consists of written documents, electronic data, sampling for chemical analysis, technical exploitation, and data analysis, which are available at the OPCW website (www.opcw.org). Below a review of the report of the FFM on Al-Hamanidiyah, Syria in 2016 is given as an example (OPCW S/1642/2018).

Written Documents Syria provided medical information of patients treated after the incident in Al-Hamanidiyah, 2016. The information was collected from casualties, treating personnel, and facilities. Plenty of medical records were provided, and some reports from forensic doctors.

The incident reports of the Technical Committee of the authorities or the Syrian Armed Forces were also provided as well as technical reports related to laboratory analysis done in the inspected country. These laboratory analysis reports were often detailed, containing information about equipment, working instructions, SOPs, and quality documents.

The inspection team made inspection notes and reports of the meetings between the inspection team and local authorities.

Electronic Data Relevant locations, personnel, and objects were photographed and carefully documented and analyzed. Possible videos and other electronic data were collected. The data was provided by the Syrian Arab Republic, or collected by the inspection team. Maps in electronic format were used for determining the coordinates and locations of incidents.

Interviews Interviews of local citizens and authorities were done using audio and video recordings. In some cases, written statements, including drawings etc., were also collected.

Samples Both environmental and biomedical samples were collected. Environmental samples were taken and secured by the inspection team, or received from the local authorities. Typical sample types included soil, water, metal pieces, such as outer shells from ammunition, bottles containing chemicals, gas cylinders, and clothing. The original samples were sent to the OPCW Laboratory following the OPCW SOPs for sample splitting, packaging, transporting, and handling. A representative of the Syrian authorities witnessed the procedures.

Biomedical samples were collected by the local medical staff and made available for the inspection team via the Syrian Arab Republic authorities. In some cases the samples could not be split because of the low volume, and thus the joint custody was not applicable. The analysis of biomedical samples was targeted towards the presence of specific chemicals and their markers, because of their low concentrations in blood and plasma.

There is limited scientific literature on these analyses due to the confidential nature of reports of the OPCW designated laboratories. A scientific article on fatal sarin poisoning in Syrian Arab Republic 2013 was published in 2017 with permission from the UN. It describes in detail the studies in two specialized laboratories in the Netherlands and Germany for forensic analysis of collected samples (John et al. 2018).

Technical Exploitation An unexploded munition was inspected and documented by the inspection team. Chemical detectors, physical measurements, and nondestructive evaluation techniques were used. An attempt was made to identify the munition type. The caliber, size, model, external and internal structures, and the fill were determined.

Data Analysis The data was analyzed following the succeeding principles: To collate facts in relation to the reported incidents, medical information, locations, and samples especially related to the use of toxic chemicals as a weapon. Further analysis of the data was performed to better understand the event, and to identify the potential sources of information like interviewees. A large part of the data was not available in English. Interpreters were used for translation and transcribing of the interviews and documents, and the interpretation was reviewed by a minimum of two inspection team members. The interviews were analyzed regarding to the respondent's descriptions of locations, sights, sounds, smells, symptoms, and actions. The inspection team tried to match the interviewees' responses for cross-checking. Gaps and discrepancies were analyzed to find out their significance. Some of the discrepancies were explained by the long time that had lapsed between the reported incidents. If the discrepancies in these cases were minor, they were discarded.

4 Techniques in Chemical Forensics

Based on the above-mentioned reports, it has become evident that new forensic tools are needed for investigating the alleged use of chemical weapons. In this chapter, the methods and techniques applied to chemical forensics are reviewed, with the aim to evaluate their usability for chemical forensics on CWA related events. The main focus is on applications related to CWAs, illegal drugs, poisons, and incapacitating agents.

4.1 Investigations of Synthetic Routes

4.1.1 Impurity Profiling

Synthetic chemical batches have certain impurities that are batch or stock specific, and these impurities can be used as forensic signatures (Fraga et al. 2011b). In many cases, most of these impurities are also present in the final product mixture. Some important impurities can be lost during the synthetic process by reactions with the reagents used or by purification processes, such as distillation. In addition, the sample preparation prior to analysis can modify or discriminate the impurities. E.g., the extraction solvent used in sample preparation can be unsuitable for certain impurity, or it can produce impurities that were not present in the original mixture.

The identification of impurities will become challenging if the concentrations are very low. The identification of impurities requires a comprehensive classification model. The impurity peaks are typically analyzed by statistical software that groups the resulting peaks and can be used to make dendrograms showing impurity profiles. These profiles are easier to analyze than individual peaks. An example dendrogram is shown in Fig. 3.

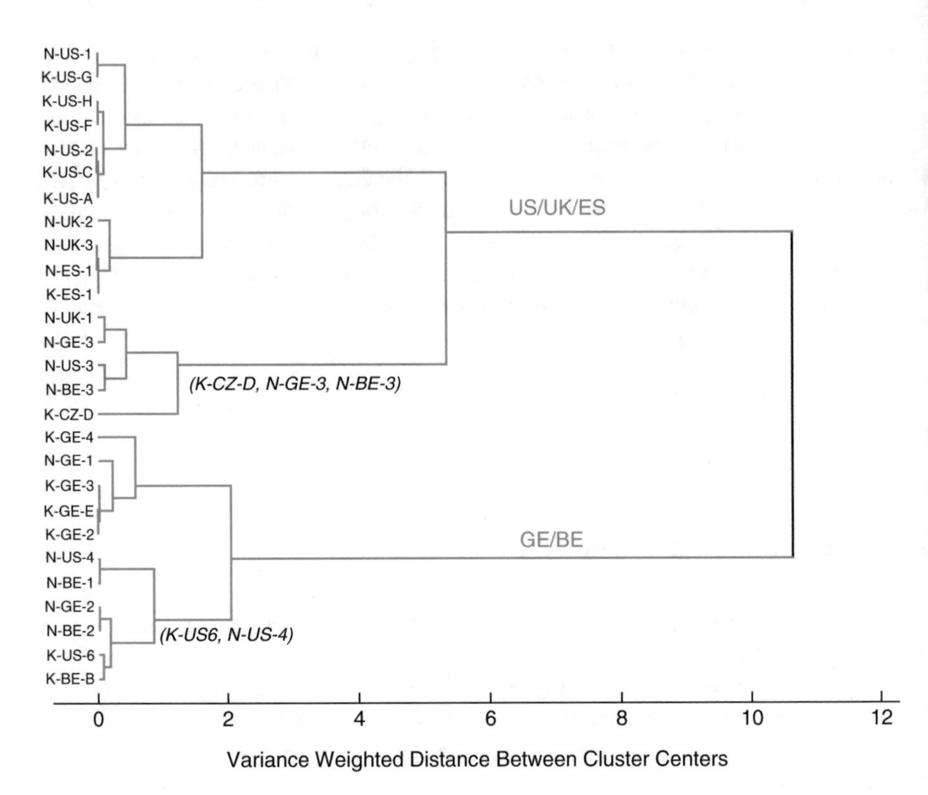

Fig. 3 Grouping of KCN and NaCN stocks from different origins using a hierarchical cluster analysis dendrogram. KCN and NaCN stocks cluster into two main groups: US/UK/ES (red) and GE/BE (green). Two stock samples reportedly from US (K-US-6 and N-US-4) cluster with the GE/BE group. Stock samples reportedly from Germany (N-GE-3), Belgium (N-BE-3), and Czech Republic (K-CZ-D) cluster with the US/UK/ES group. (Reprinted with permission by Mirjankar et al. (2016). Copyright (2016) American Chemical Society)

4.1.2 Side Product and Starting Material Analysis

"Chemical memory", i.e. the source attribution profile (SAP), can be utilized to discover the origin, history, or processing of a substance (i.e. synthesis route, event, manufacturer, batch, person, or country of origin). Starting materials that have not completely reacted during the synthesis can reveal the applied reaction pathway, and can provide valuable forensic information. In addition, side products can confirm the reaction route, and provide information about the reaction conditions and the applied methods and techniques during the synthesis process. This, in turn, can give information about the equipment type used, and limit the number of possible manufacturers.

The side products and impurities formed in the synthetic processes of fentanyl (Mayer et al. 2016), bicyclophosphate (Mazzitelli et al. 2012) and Russian VX (VR)

(Holmgren et al. 2018) have been studied. In most cases, the synthetic routes could be confirmed, and the impurities and side products identified.

In case of several synthesis steps, the mixture of impurities can easily get very complicated making it practically impossible to connect the impurities to a specific synthesis route. A relatively high threshold of chemical attribution signatures (CAS) can be used to reduce the number of CASs to manageable quantities.

In scientific papers, the detailed synthetic routes are typically not shown to prevent their unlawful employment and to hinder proliferation of CWAs (Holmgren et al. 2018).

4.2 Current Analysis Techniques

In this chapter, analytical techniques that are potentially applicable to CWA related attribution analysis are briefly reviewed. On-site analysis techniques include portable Raman spectroscopy and GC-MS instruments. For off-site analysis, various chromatographic, spectrometric, and spectroscopic techniques have been used, including GC-MS, LC-MS, inductively-coupled plasma-mass spectrometry (ICP-MS), NMR, Raman and IR spectroscopy, and ion chromatography.

4.2.1 Gas Chromatography-Mass Spectrometry

GC-MS is a robust and powerful method in detecting CAS. GC-MS techniques are used to analyze and identify any relevant chemicals which can be volatilized without thermal decomposition. Several studies exist, where different CASs, such as trace impurities that match a chemical or chemical mixture of interest to its source have been determined utilizing different variations of GC-MS. GC-MS methods can be applied to both untargeted and targeted analyses, depending on the instrumentation and mode of analysis. In addition to laboratory equipment, portable GC-MS instruments can be utilized in attribution-type analysis (Groenewold et al. 2013), as described in more detail later. In 2011, Fraga et al. demonstrated how synthetic chemicals could be matched to their precursors by precursor impurity profiling (Fraga et al. 2011b). GC-MS measurements in both full scan and selected ion monitoring modes were carried out for untargeted analysis of impurities in dimethyl methylphosphonate (DMMP) and methylphosphonic dichloride (DC), followed by peak detection and grouping by the marker-discovery tool XCMS. The data was processed using cluster analysis, including principal component analysis (PCA) and hierarchal cluster analysis (HCA). This work, that simulates a real-world situation where a CWA could be tracked back to its source, demonstrates the importance and usefulness of chemical forensics in order to aid in the forensic process.

In 2012, Fraga et al. used DMMP as a chemical threat agent (CTA) simulant to map the effects of real-world factors on impurity profiling and source matching

(Fraga et al. 2012). Four different stocks of DMMP with different impurity profiles were deposited as aerosols onto three different matrices representative of real-world situations (cotton, painted wall board, and nylon coupons) and analyzed using comprehensive two-dimensional chromatography-mass spectrometry (GC×GC-MS) after solvent extraction. The data was analyzed utilizing PCA and PARAFAC (generalization of PCA). It was observed that inappropriate choice of the extraction solvent could result in the same impurities being present in both the solvent and the CTA, which reduces the usefulness of the CTA's impurity profile. In addition, volatilization of forensic impurities should be minimized, when possible, by a short elapsed time between CTA release and sampling, as impurity profiles weighted toward higher volatility are more likely to be altered by evaporation after CTA exposure.

Another paper by Fraga et al. 2016 demonstrated trace impurities from the synthesis of nitrogen mustard gas (HN3) that point to the reagent and the specific reagent stocks used in the synthesis of this CW agent (Fraga et al. 2016). They synthesized thirty batches of HN3 from different reagent stocks (using different combinations of commercial stocks of triethanolamine, thionyl chloride, chloroform, and acetone) and analyzed both reagent and product stocks for impurities utilizing GC-MS with electron ionization (EI) (full scan and SIM). MetAlign 3.0 was used to generate peak tables from the GC-MS full scan data, and Automated Mass Spectral Deconvolution and Identification System (AMDIS) and the NIST library were used to tentatively identify the impurity compounds. An in-house software tool was utilized to determine the peak areas and determine the presence or absence of the selected impurities. HCA was used to determine if the impurities for each reagent type resulted in clustering according to stock. Data analysis by Partial Least-Squares Discriminant Analysis (PLS-DA) and K Nearest Neighbors (KNN) indicated that all reagent stocks had differing impurity profiles. They also noticed that some impurities found from the synthesized HN3 stocks had prevailing traces of certain reagent impurities, suggesting that those impurities are not altered during the synthesis process. Another important finding was the discovery of impurities in HN3 stocks, which were produced by impurity side reactions that are specific to triethanolamine and chloroform, this way potentially revealing the use of these chemicals in the synthesis. While this work is an important demonstration of the capabilities to point back to specific reagent stocks used in the synthesis of the CW agent in question, the applicability to real-world samples needs more work in order to determine the operational usability and limits for source attribution using specific impurities and impurity profiles. Another challenge for impurity profiling arises from the possibility of samples being adulterated or altered by environmental factors.

In 2016, Strozier et al. used comprehensive two-dimensional gas chromatography with time-of-flight mass spectrometric detection (GC×GC-TOFMS) and random forest pattern recognition techniques for classifying CTAs and detecting CAS (Strozier et al. 2016). Organophosphate pesticides were selected as examples (dichlorvos, dicrotophos, and chlorpyrifos) as they are highly toxic and could potentially be used as chemical threat agents. GC×GC-TOFMS technique has great potential for forensic analysis of CTAs, as the separation of chemical components by two orthogonal properties enhances the chromatographic separation of compounds from one another. TOFMS, on the other hand, is very sensitive, and provides spectra in full-scan mode (unlike SIM-techniques in quadrupole instruments), so it can be used in nontargeted analysis of chemical mixtures. Another advantage of the full-scan mass spectral data is the use of spectral deconvolution software, which allows further separation of overlapping peaks. The problem with GC×GC-TOFMS is the large amount of data it produces per sample, and finding the appropriate features that can be compared over the sample set. Strozier et al. used modified pattern recognition algorithm called balanced random forest (BRF), which consider all peaks present in the data without the need for dimension reduction (Strozier et al. 2016). This data can then be used to identify CASs. The classification success rate was 97−100% in three tests using Oval Area variables and 87–100% in three tests using In/Out variables. Four blind samples were correctly identified with high confidence using either variable type. This proof-of-concept study that illustrates that GC×GC-TOFMS analytical data can be used with random forest classification techniques to classify CTAs from different sources, was carried out in a best-case scenario, which means that a real life situation will likely be more complicated and the technique may be less effective in attributing samples from an actual attack site. Nonetheless, the models that were developed can be applied to any sample, and the technique seems to outperform more traditional statistical evaluation techniques. The data produced by this powerful technique can be used to create lists of chemicals that are potential CASs.

Isotopes are defined as atoms of the same element that differ in the number of neutrons present in their nuclei, thus they have different mass numbers. Hydrogen, carbon, oxygen, and nitrogen are fundamental components of multitude of toxic compounds, such as toxic industrial chemicals or CWAs. In nature, isotopic abundancies of elements vary because they are fractionated during chemical, physical, and biological processes. The isotope fractionation reflects isotope exchange reactions and mass-dependent differences in chemical reaction rates and physical processes such as evaporation and diffusion, and causes measurable effects on the isotopic composition of samples, characteristic of their chemical, biological, or physical history. The stable isotope content of different molecules has been used in source matching of forensically important species, as stable isotopic composition of e.g. different manufacturing batches of the same chemical can vary. This provides an additional level for sample matching which goes beyond chemical identity. Stable isotopic content can be used as a signature for sample matching and can be applied to variety of forensic applications. Compounds synthesized from isotopically distinct reagents or by a different process will yield a different stable isotopic content. Isotope ratio mass spectrometry (IRMS) is a specialization of mass spectrometry, where mass spectrometric methods are used to measure the relative abundance of isotopes in a given sample. IRMS technique encompasses variety of different mass spectrometric techniques, which differ in the type of mass separation, ionization, and inlet systems. These techniques include elemental analyzer IRMS (EA-IRMS), hyphenated gas- and liquid chromatography IRMS (gas chromatography-combustion-IRMS,

LC/IRMS), ultrahigh-resolution Fourier transform-ion cyclotron resonance mass spectrometry (FT-ICR MS), ICP-MS, nanoscale secondary ion mass spectrometry (Nano-SIMS), and ultra-high resolution IRMS (Ultra HR-IRMS).

In forensic analytics, IRMS technique has been utilized in stable isotopic characterization of toxic industrial chemicals, ammonium metavanadate (Volpe and Singleton 2011). Volve and Singleton showed, utilizing thermal conversion elemental analyzer IRMS, that stable isotope characterization is a valuable forensic tool, discriminating between different ammonium metavanadate samples due to differences in reagent materials, production routes, and production facility location.

In 2012, Moran et al. published a study, where multiple stable isotope characterization was used as a forensic tool to distinguish acid scavenger samples (Moran et al. 2012). They utilized GC-IRMS to report carbon, nitrogen, and hydrogen isotope content. In the study, 33 acid scavengers were selected as the target of the analysis, as they are known stabilizers in nerve agents and other CTAs. The study showed that stable isotope analysis can provide basis for distinguishing between different acid scavengers and thus could be utilized for sample matching between an inventory or database of these compounds and a threat chemical. The technique could also be used to link CTAs used at different occasions or supply facility to CTA event.
Another forensic application of IRMS was published in 2012 by Kreuzer et al., where stable carbon and nitrogen isotope ratios of sodium and potassium cyanide were used as forensic signatures (Kreuzer et al. 2012). Solid cyanides are highly toxic compounds, and probable agents to be used in a chemical terrorism event. EA-IRMS technique was utilized to analyze 65 cyanide samples, demonstrating that significant variation exists in both the C and N stable isotope ratios among commercially available cyanide samples.

4.2.2 Liquid Chromatography-Mass Spectrometry

Liquid chromatography-mass spectrometry (LC-MS) is a complementary technique to GC-MS, as it can be used to analyze polar and non-volatile compounds without the need to derivatize the analytes. However, LC-MS is not as reproducible as GC-MS, and this adds uncertainty to the statistical handling of the analytical data. LC-MS can be used for both untargeted and targeted analyses. Untargeted analysis is used for screening of unknowns from the sample. It is especially efficient if high resolution mass spectrometry (HRMS) is available, as it can be used to calculate the elemental composition of the detected species. In targeted analysis, the species of interest are already known, and high sensitivity can be achieved by e.g. multiple reaction monitoring (MRM) analysis.

Fraga et al. used high resolution LC-MS (Orbitrap), and bioinformatics and chemometric tools to trace the forensic signatures of ten stocks of methylphosphonic dichloride (dichlor), which is a commercially available, toxic organophosphorus compound and a nerve agent precursor (Fraga et al. 2010). Sample preparation and the sensitivity of the LC-MS method permitted the detection of dichlor impurities quantitatively estimated to be in the parts-per-trillion concentration range. In total, 34 impurity peaks were found with the bioinformatics software. Eight out of ten analyzed dichlor stocks were shown to have unique impurity profiles. The profiles could be successfully used to match dichlor samples from a test set back to their respective sources.

Jansson et al. chose LC-MS/MS for the analysis of attribution signals of different Russian VX (VR) synthesis routes in different food matrices (Jansson et al. 2018). The VR synthesis products are in hydrolyzed form in aqueous matrices, such as most food products, and LC-MS/MS is sensitive towards polar, hydrophilic compounds, such as the hydrolysis products. Seven different foods (water, orange juice, apple purée, baby food, pea purée, liquid eggs, and hot dog) were spiked with VR at lethal levels (32–50 ppm). Sample extraction and protein precipitation with acetonitrile was chosen as the sample preparation method. Some ion suppression was caused by matrix-related chemicals from some of the food matrices, which lead to the disappearance of some of the marker candidate signals. The method allowed the classification of 94% of test set samples to the correct synthesis route.

Mayer et al. used LC-MS/MS together with complementary techniques like GC-MS and ICP-MS for the chemical attribution of six different synthesis routes of fentanyl (Mayer et al. 2016). The highest amount of unique compounds was identified by LC-MS/MS. The identification was aided by tandem mass spectrometry (MS/MS) and the exact mass capability of the time-of-flight (TOF) instrument used. All in all, the combination of data from the three separate techniques was concluded to provide better discriminatory ability than that from individual analyses alone. LC-MS/MS in MRM mode has also been used for the source attribution analysis of acephate (an organophosphate pesticide) and several VX-and G-series agents (Favela et al. 2012).

4.2.3 Inductively Coupled Plasma-Mass Spectrometry

Inductively coupled plasma (ICP-MS) is an analytical technique that can be used for elemental analysis. ICP uses inductively heated, ionized argon plasma to ionize the atoms of the elements in the sample. The ions are then detected by MS or optical emission spectroscopy (OES) (Mirjankar et al. 2016). ICP-MS has been used for the quantitation of elemental impurities in sulfamide from seven common vendors, with the aim to differentiate the sulfamide samples and link the sulfamide back to its vendor (Hondrogiannis et al. 2013). Sulfamide is a precursor to the deadly neurotoxin, tetramethylenedisulfotetramine (TETS), which makes it a forensically relevant compound. Metal isotopes (iron, copper, nickel and zinc) were quantitated by ICP-MS, and the concentrations were normalized to the mass of sulfamide. Statistically significant differences in the concentrations of the elements existed among many of the vendors, and inorganic impurities were suggested to be feasible for vendor lot attribution. ICP-MS (along with GC-MS and LC-MS/MS) has also been used for the chemical attribution of fentanyl (Mayer et al. 2016). Six synthesis methods were studied, with the effort to find route specific signatures. GC-MS and LC-MS were used to identify the organic species, and ICP-MS to identify the inorganic species. For ICP-MS, sodium, potassium, iodine, cesium, and barium were found the most relevant elements for classification, and were chosen for quantitation. In this particular study, the quantitative ICP-MS data had little effect on the PLS-DA data analysis, and was therefore not found useful.

In a study by Fraga et al., ICP-MS was used to find elemental attribution signals of calcium ammonium nitrate (CAN), which is fertilizer frequently used to make homemade explosives (Fraga et al. 2017). ICP-MS was used to determine the concentrations of 64 elements in 125 samples from 11 CAN stocks from 6 different CAN factories. Na, V, Mn, Cu, Ga, Sr, Ba and U were selected for the classification of CAN samples. PLS-DA was used to develop a classification model, which was tested for a separate set of test samples. The model worked well for unadulterated samples, and samples adulterated with powdered sugar, but adulteration with aluminum powder confounded the source matching.

4.2.4 Nuclear Magnetic Resonance Spectroscopy

Nuclear magnetic resonance (NMR) spectroscopy is a technique that can probe diverse materials at the atomic level, and in principle is able to detect most stable nuclei found in the periodic table (with spin I $\geq 1/2$). Accordingly, the studied materials define the nuclei to be observed with NMR, and thus ^{1}H, ^{13}C, ^{19}F, and ^{31}P NMR spectra are typically measured from CWAs. Especially ^{1}H-^{31}P NMR correlation techniques are widely used for the screening and identification of organophosphorus chemicals related to the CWC. Comparing all the available analytical techniques available for CWA compound identification, NMR requires very little sample preparation. However, the inherent low sensitivity makes the NMR technique impractical for situations, where a limit of detection below 1 ppm is required. NMR hardware improvements (stronger external magnetic fields (>20 T) in combination with cryogenically cooled probe heads) have been introduced (Kovacs et al. 2005) to tackle the issue, yet the sensitivity of NMR is roughly in the order of one magnitude smaller compared to e.g. MS techniques.

Despite the acknowledged applicability of NMR spectroscopy e.g. in metabolite profiling from biofluids and tissue samples (Beckonert et al. 2007), NMR is not widely employed in CWA forensics. According to recent literature, NMR applications in the field of chemical forensics are mainly linked to identifying exact molecular structures encompassing materials, such as poisons (e.g. strychnine and brodifacoum) (Cort and Cho 2009; Metaxas and Cort 2013) and illicit drugs (e. g. cocaine and metamphetamine) (Benedito et al. 2018; Hays 2005). In drug analysis NMR has been applied to impurity profiling alongside with GC-MS e.g. during synthesis route and precursor marker analysis (Blachut et al. 2002; Heather et al. 2015; Doughty et al. 2016), and alone as ^{1}H NMR e.g. to analysis of organic impurities and cutting agents in heroin (Balayssac et al. 2014), as well as in combinatorial NMR study with multivariate analysis of cocaine samples seized at different times and locations (Pagano et al. 2013).

In 2013, Metaxas and Cort showed that NMR is able to distinguish and separate different strychnine salts based on chemical shift perturbation that is observed in the

^{1}H NMR spectrum compared to their free base forms (Metaxas and Cort 2013). Strychnine is a common poison, often used as a rodenticide, and the recognized strychnine salt form could serve as a link to its source. This kind of information could be utilized, if needed, in source attribution analysis and would also act as a complementary technique together with other techniques used for analysis of organic salts such as capillary electrophoresis, ion chromatography, and ion-pair LC. In another study, NMR spectroscopy was utilized in identification of different stereoisomers of brodifacoum, also used as a rodenticide, potential chemical for poisoning. Full assignment of ^{1}H and ^{13}C chemical shifts together with J-coupling values and 2D NMR correlation data (COrrelation SpectroscopY (COSY), heteronuclear multiple-quantum correlation (HMQC), heteronuclear multiple-bond correlation (HMBC)), revealed different diastereomers of brodifacoum, and thus data that could be used for source identification and analysis for brodifacoum (Cort and Cho 2009; Beckonert et al. 2007).

To this date, NMR spectroscopy is merely an off-site analysis tool for CWAs and their precursors. Superconductive magnets have high expenses connected to cryogen- and other maintenance costs, which are compulsory for the system to work properly and therefore the cost stops many forensic laboratories from holding such instrumentation. Recent developments in utilization of low static magnetic field NMR instruments and efforts to develop portable magnetic resonance detection devices hold promise that NMR could be more widely operated analytical technique in CWA forensics in the future (Duffy et al. 2019; Yoder et al. 2018).

4.2.5 Infrared Spectroscopy

Infrared (IR) spectroscopy is a non-destructive analysis technique that exploits the vibration of molecular bonds when they are being excited with electromagnetic radiation on a specific wavelength. Based on the absorption frequency, different functional groups can be distinguished. In most modern IR instruments, the data is Fourier transformed after analysis to produce a spectrum, and IR spectroscopic data is often treated with different computational programs, such as multivariate modelling and classification algorithms. This treatment is required, because the spectra may be visually similar to each other and background adsorption bands may overlap preventing unambiguous analysis. In CWA related analysis, it is also common to make comparisons between experimental and calculated data produced by e.g. density functional theory (DFT) or *ab initio* calculations. In chemical forensics, IR is mostly used in combination with other techniques, such as GC-MS, LC-MS, NMR, or Raman spectroscopy. IR can also be combined with GC-analysis in gas chromatography-Fourier transform infrared (GC-FTIR) spectroscopy, but then the advantage of minimal sample preparation is lost.

IR spectroscopy has mostly been used in the analysis of nerve agents (Mott and Rez 2012), mustard gas (Wiktelius et al. 2018), hydrogen cyanide (HCN) (Magnusson et al. 2012), fentanyl (Asadi et al. 2017; Liu et al. 2018), different

CWA simulants (Salter et al. 2009), and CWA hydrolysis products and precursors (Garg et al. 2009; Lee et al. 2005; Rummel et al. 2011). IR is a less common analysis technique for ricin, and only few studies on the subject can be found from the literature (Picquart et al. 1989; Riaz and Farrukh 2014; Rodriguez-Saona et al. 2000).

The IR spectra of nerve agents are well known, but still the separation between individual agents may prove to be challenging due to matrix effects, or other characteristics of the sample that lead to line broadening in the spectra. Separation between V- and G-series is easily done based on a spectral line at ~500 cm^{-1} caused by the stretching of the phosphorous-sulfur bond in the excited molecule (Mott and Rez 2012). Although spectral predictions may be helpful when differentiating separate agents from each other, caution should be taken due to the intrinsic distortion of experimental spectra compared to the calculated one (Mott and Rez 2012). IR spectroscopy is applicable also in synthesis route determinations and kinetic studies. Synthetic route determination is usually based on attribution analysis, and generally requires the use of a parallel technique, such as Raman spectroscopy, since replicate samples from the same route may produce inconsistent data (Wiktelius et al. 2018). IR spectroscopy has also been used in kinetic studies, for monitoring of processes, such as decontamination, and degradation of the nerve agent simulant DMMP (Lee et al. 2005; Salter et al. 2009),

The major disadvantages of IR spectroscopy are instrument influence (e.g. light source intensity) (Scafi and Pasquini 2001) and physical properties (e.g. water content (Magnusson et al. 2012), solid sample tablet geometry, granule size etc. (Scafi and Pasquini 2001; Liu et al. 2018)) of the sample that easily cause distortion in the IR spectra.

The recent literature on fentanyl analysis by IR spectroscopy has mostly concentrated on structural assignment with the help of DFT based calculations (Asadi et al. 2017), but various other analysis techniques have also been used in characterization (Liu et al. 2018). The most common technique among these studies has been FTIR spectroscopy combined with NMR, GC-MS, and sometimes with different forms of LC-MS.

GC-FTIR can be used parallel to GC-MS in analysis of CWC related chemicals. Similarly to GC-MS, GC-FTIR is a separation technique, where the GC effluent is monitored using IR spectrum. GC-FTIR is less sensitive than GC-MS, as the compound identification is carried out by measuring IR spectra, but this can be overcome by concentrating the sample, lowering the temperature, or by fluorinating the analytes to increase absorptivity prior to analysis. Sample concentration typically increases the background and requires additional sample preparation. According to Garg et al., heptafluorobutyrylimidazole increases the sensitivity by 60–125 fold depending on the analyte. Trimethylsilyl iodide in n-hexane is also a good option, while N,O-Bis(trimethylsilyl)trifluoroacetamide produces a huge background (Garg et al. 2009). Another issue in GC-FTIR technique is the relatively small number of vapor phase reference spectra in the OCAD.

4.2.6 Raman Spectroscopy

Raman spectroscopy is an analysis technique that is based on inelastic scattering of photons during sample radiation, usually in the range of near IR or near ultraviolet. Like with IR radiation techniques, Raman spectra are nowadays most commonly produced by Fourier transform (FT) technique by leading the scattered light through an interferometer at some point after detection. Unlike in IR, a change in the dipole moment is not needed in Raman spectroscopy. In CWA related analysis, the applicability of the Raman technique has increased, mostly because of its flexibility in sample composition, its non-destructiveness, and its suitability for portable devices. Raman spectroscopy has mainly been used by itself, or in combination with its new variants, but when used with other analysis techniques it is usually paired with IR spectroscopic methods.

Unambiguous Raman spectroscopic structural assignments have not been made for different nerve agents and other CWAs, but relatively comprehensive suggestions have been proposed (Inscore et al. 2004; Kondo et al. 2018). Nerve agents all have unique fingerprint profiles at approx. 400–1800 cm^{-1}, but due to their closely related chemical structures their Raman spectra are otherwise so similar that identifying individual CWAs based on only their Raman spectra is usually not possible. However, the G- and V-series (sarin and VX related compounds, respectively) can be easily differentiated from each other (Kondo et al. 2018; Wu et al. 2017).

In addition to detecting single agents, Raman spectroscopy can be used to differentiate between synthetic routes based on impurities and side products specific to each route. This has been demonstrated in a study of aged mustard gas samples produced via eight different synthetic routes (Wiktelius et al. 2018).

The main advantage of Raman spectroscopy in comparison to other techniques, is the general lack of required sample preparation. Raman spectrum can be obtained from a sample that is in any phase or solution, whether it is aqueous or organic. This trait has been exploited when using traditional Raman and its variations in the analysis of ricin from biomedical samples (Tang et al. 2016; Notingher et al. 2004; Campos et al. 2016) and foodstuff (Tang et al. 2016; He et al. 2011). Usually, Raman analysis can be executed through the sample container and even the spectra of a sample in a closed and paper- or cloth-covered container has been successfully produced (Kondo et al. 2018; Cletus et al. 2013). This makes the technique ideal for in situ applications. Some restrictions of Raman spectroscopy are related to imperfect sample homogeneity and colored samples (Wiktelius et al. 2018) or containers, all of which may cause distortion of spectra (Hopkins et al. 2017) or a complete failure to obtain it.

One of the biggest disadvantages of Raman spectroscopy is the effect of background light on spectra, especially when the interfering light is fluorescent. In a worst case scenario fluorescent light may produce background, which causes the signals of certain substances to disappear even when others can be easily detected. E.g. chlorpyrifos in acetonitrile solution has been shown to be undetectable in artificial fluorescent light, while acquiring the spectra of the same sample in darkness poses no problem (Cletus et al. 2013). Similar results have been achieved with nitrogen mustard gas (HN3) and Adamsite (Kondo et al. 2018). Other disturbances of the spectrum may be caused by background chemicals as they create a competitive environment for absorption, which can be seen e.g. for mixtures of soman and sulfur-containing chemicals in comparison with pure soman samples (Wu et al. 2017).

Several different methods for overcoming the background light effect have been presented in literature, such as short-wave infrared excited Raman spectroscopy (SWIR), time-gating using electronical methods (Ariese et al. 2009; Martyshkin et al. 2004; Petterson et al. 2010), optical Kerr gating, and multi-excitation techniques such as shifted excitation Raman difference spectroscopy (McCain et al. 2008). Out of these methods the most used in CWA related analysis is SWIR, in which a 1064 nm excitation frequency is used instead of the more conventional 785 nm (Kondo et al. 2018; Wiktelius et al. 2018; Wu et al. 2017).

Traditional Raman spectroscopy does not always provide sufficient signal intensity, and especially for nerve agents, there have traditionally been problems with high limits of detection (Inscore et al. 2004). Many techniques have been developed to help with this issue, out of which especially spatially offset Raman spectroscopy (SORS) and surface-enhanced Raman spectroscopy (SERS) have been used in the field of chemical forensics with good results. SORS has been applied e.g. to the analysis of chlorpyrifos and various precursors of explosives (Cletus et al. 2013), and SERS has been used for increasing the signal intensity of organophosphorus nerve agents (Wu et al. 2017), hydrolysis products of G-series agents, and fentanyl and its analogues (Haddad et al. 2018; Leonard et al. 2017).

Some helpful applications of Raman spectroscopy from the forensic point of view are portable Raman instruments. The results produced by these instruments are fairly consistent when compared to e.g. GC analysis of the same samples (Wiktelius et al. 2018; Cletus et al. 2013; Christensen and Morris 1999). It should be noted, however, that their sensitivity is markedly lower than that of the benchtop Raman instruments (Zheng et al. 2017).

4.2.7 Ion Chromatography

Ion chromatography (or ion exchange chromatography) is a powerful tool for separating mixtures based on their charged groups. Advantages of the technique are its good matrix tolerance and elution patterns based on the presence of ionizable groups. Two types of ion exchange chromatography exist: anion exchange and cation exchange chromatography. Anion exchange chromatography is typically used in the field of chemical forensics, as it may provide information about different inorganic chemicals that could be difficult to observe with other analytical techniques. Ion chromatography is usually combined with conductivity or ultraviolet/visible detection. This analytical technique can be used with almost any kind of charged molecules, even amino acids and proteins.

High performance ion chromatography (HPIC) with conductivity detection has been used for analysis of anionic impurity profiles for sample matching of potassium cyanide (KCN) stocks (Fraga et al. 2011a). It was hypothesized that these anionic impurities are dependent on the geographic location of manufacture. The selected anions after PCA were sulfate, oxalate, phosphate, and one unknown anion. Anion exchange column and the conductivity detector were used. The samples were from different manufacturers from four countries. Statistical methods were used for classifying the analysis results. The concentrations of selected anions were used for matching a test set of KCN to their origin. Also, each stock had a unique anionic impurity profile.

Sodium and potassium cyanides were also studied by Mirjankar et al. (2016). This study involved analysis of 27 stocks of KCN and NaCN from six individual countries, utilizing HPIC for trace anions, IRMS for carbon stable isotope ratios, and inductively coupled plasma optical emission spectroscopy (ICP-OES) for trace elements. Various statistical techniques were used to evaluate the data. The paper demonstrated the potential for cyanide source attribution to a production factory through combination of anionic, elemental, and isotopic profiling, though future work is needed to understand the limits and realistic error rates for the complete integration of the techniques.

4.3 Sampling and Sample Preparation

The ROPs for sampling and sample preparation for a wide variety of sample types (organic and aqueous liquids, soil, air, solid material etc.) are described in the Blue Book (Vanninen 2017). However, there are scientific papers that report sample preparation methods for sample types not represented in the Blue Book, or methods that otherwise complement the established methods. The methods described in scientific literature are shortly discussed here, and summarized in Table 1.

The suitable sample preparation method depends on the sample matrix, as well as the analysis technique. GC-MS typically requires more exhaustive sample preparation than LC-MS, and sometimes in addition derivatization of non-volatile analytes. The sample preparation method should preserve the analytes of interest to the largest possible degree. Whereas traditional sample preparation aims to extract the target compounds as efficiently as possible and remove the interfering impurities and background components, attribution analysis focuses on the impurity profiles. Therefore, the CWA specific sample preparation method is not necessarily optimum for its attribution signals. It should also be noted that in the case of attribution analysis, some extraction solvents can contain the same impurities as the studied CTA (Fraga et al. 2012). Long storage times may also result in loss of volatile impurity components. Combining sampling with extraction and/or derivatization steps speeds up the analysis procedure, and may help in preservation of the components of interest.

Table 1 Summary of the sample preparation methods used in the analysis of CWA from different sample types

Sample type	Analytes	Sample preparation	Analysis	References
Painted wallboard	Organophosphorus compounds	ASE, solvent extraction (methylene chloride:acetone (1:1))	GCxGC-MS	Wahl and Colburn (2010)
Glass, painted wallboard, concrete and carpet	Nonvolatile nerve agent hydrolysis products, amines, thiodiglycol	On-matrix derivatization and dynamic headspace sampling by field and laboratory emission cell	GC-MS	Harvey and Wahl (2012)
Cotton, painted wall-board, nylon coupons	DMMP impurities	Solvent extraction (acetone)	GCxGC-MS	Fraga et al. (2012)
Vinyl tiles	Organophosphonates	Field-vacuum extractor (FVE)	portable GC-MS	Groenewold et al. (2013)
Painted wall board	DMMP impurities	Solvent extraction (acetone) or accelerated diffusion sampler-solid-phase microextraction (ADS-SPME)	GC-MS	Mo et al. (2017)
Food	Russian VX chemical attribution signals	Solvent extraction (acetonitrile)	LC-MS/MS	Jansson et al. (2018)
Food	Russian VX chemical attribution signals	Solvent extraction (methanol), solid-phase extraction and/or QuEChERS	GC-MS, LC-MS/MS	Williams et al. (2018)
Dust	Acephate, VX- and G-series agents	Solvent extraction (dichloromethane or acetone)	LC-MS/MS	Favela et al. (2012)

ADS-SPME = accelerated diffusion sampler with solid-phase microextraction, *ASE* = accelerated solvent extraction, *DMMP* = dimethyl methylphosphonate, *FVE* = field-vacuum extractor, *GC-MS* = gas chromatography-mass spectrometry, *GCxGC-MS* = two-dimensional gas chromatography-mass spectrometry, *LC-MS/MS* = liquid chromatography-tandem mass spectrometry, *QuEChERS* = quick, easy, cheap, effective, rugged, and safe

During a chemical attack, airborne chemicals can be absorbed by indoor materials, which can be analyzed to get evidence for the criminal investigation. Pressurized fluid extraction using an accelerated solvent extraction (ASE) system was used for targeted organophosphorus compounds and sarin simulants from painted wallboard (Wahl and Colburn 2010). In ASE, the painted wallboard sample was ground to pieces and placed in an extraction cell filled with solvent, after which the isolated cell was brought to an elevated temperature and pressure for extraction. Subsequently, the extraction solvent was removed and analysed, in this case with GCxGC-MS. Methylene chloride:acetone (1:1) was found to be the most robust and consistent extraction solvent and it could also extract known impurities. Harvey and Wahl used dynamic headspace sampling with field and laboratory emission cell (FLEC) with on-matrix derivatization of forensically important residues having low vapour pressures or low volatilities (Harvey and Wahl 2012). The test compounds were nerve agent hydrolysis products on glass, painted wallboard, concrete and carpet. Volatile alkylated esters were formed by derivatization, sampled by FLEC, and analyzed by GCxGC-MS. Sampling by FLEC was found especially important for collection of samples from difficult substrates, such as concrete and carpet matrices that cannot be sampled using wipe approach.

Food is a complicated matrix, and different food types may require very different sample preparation procedures. Different analytical techniques have also different requirements for sample purity. The analysis of Russian VX chemical attribution signatures from food has been studied in two publications (Jansson et al. 2018; Williams et al. 2018). The studied matrices were bottled water, orange juice, apple purée, baby food, pea purée, liquid eggs, and hot dog. In the publication by Jansson et al., a simple solvent extraction with acetonitrile was used for all the food matrices (except homogenization of the hot dog, followed by dilution with water), after which the samples were analysed by LC-MS/MS (Jansson et al. 2018). Some ion suppression due to matrix-related chemicals that could have been prevented by more thorough sample preparation was observed in the LC-MS/MS analysis. A more thorough sample preparation protocol for the same sample types was developed in a publication by Williams et al. (Williams et al. 2018). Solid-phase extraction (SPE) with polymeric reversed-phase cartridges was used for all matrices; in addition, the different matrices were processed differently before the SPE. Baby food was solvent extracted with methanol, and milk was diluted. For liquid eggs and hot dogs QuEChERS (quick, easy, cheap, effective, rugged, and safe) (Schenck and Hobbs 2004) protocol was used. The extraction efficiencies for all matrices were reported to be reasonable (>75%).

Dust has the ability to collect, concentrate, and retain toxic chemicals in indoor environments, and it may therefore give valuable evidence of a chemical attack. Favela et al. chose dust as the collection media, and acephate, VX- and G-series agents as test compounds (Favela et al. 2012). Dust was collected from storage sheds and loaded with different chemical profiles using a custom-made exposure chamber. The chemicals were extracted with dichloromethane (acephate) or acetone (VX- and G-agents) and analyzed with GC-MS and LC-MS. The data was subjected to PCA for distinguishing the chemical profiles. For most of the studied compounds, the exposed dust could be correctly grouped with the source. It was suggested that the matching accuracy could be improved by the development of a universal extraction method for dust that could capture compounds with a range of polarities. In a field situation, collection of replicate samples at different time points was suggested for studying the variability as a function of time.

In some cases, it may be necessary or advantageous to do the sample processing and analysis in addition to the sampling already on-site. Field vacuum extractor (FVE) can be used for the non-destructive sampling of compounds from surfaces. Groenewold et al. combined FVE with a portable, fast-duty cycle GC-MS for the rapid analysis of organophosphonate compounds from vinyl floor tile (Groenewold et al. 2013). The combination enabled the analysis of a surface sample approximately every 3 min. The approach was capable of a semi-quantitative measurement of a surface concentration. The combination of FVE with the portable GC-MS was able to identify the organophosphonate compounds from exposure quantities of 10–40 ng, which is about one molecular layer. It was concluded that if multiple sampling units combined with the portable analyzers were deployed onto an accident site, near real-time information regarding the threat chemical and its location could be achieved. Another technology that has been recommended for on-site analysis, is accelerated diffusion sampler with solid-phase microextraction (ADS-SPME) (Mo et al. 2017). Mo et al. used ADS-SPME to capture CAS impurities from painted wallboard. An advantage of ADS-SPME is that it does not require removing or cutting the material, like traditional solvent extraction. ADS-SPME with a portable detector (e.g. GC-MS) was recommended for use in field screening for the rough determination of which surfaces have the highest CTA or CAS content, and could therefore be selected for sample collection and laboratory analysis.

4.4 Multivariate Statistical Analysis

Analyzing chemical data using mathematical and statistical methods is a study field generally known as chemometrics. In CWA forensics, analytical data are processed chemometrically to find patterns in the sample data sets. Chemometrics uses mathematical and statistical procedures to process high-dimensional data, from where it identifies statistical regularities and patterns. Multivariate data-analysis can be divided into *data preprocessing* and *pattern recognition*.

Prior to multivariate analysis or other chemometric analysis, the acquired data are cleaned by *data preprocessing*. The goal of the preprocessing is to produce noise-free and reliable data, maintaining all relevant information. Preprocessing is carried out at early stages of the analysis and may include several procedures such as binning, smoothing, baseline corrections, peak alignment, and scaling.

The second part in data analysis includes *pattern recognition*, where software and algorithms are used to analyze the data by making use of its high-dimensionality. Pattern recognition is further divided into unsupervised and supervised techniques. In unsupervised techniques, only plain data about the samples are used, whereas in supervised techniques additional prior knowledge is introduced to the data. This knowledge can be used as a training set to "teach" a model. The prior knowledge can be e.g. the class membership (i.e. synthesis route or manufacturer) in the classification setting, or dependent variable in regression (e.g. peak area). The trained model can be used to predict the class, or the dependent variable of data for unknown samples. In chemical forensics, the interest has mainly been in the classification of the samples according to tabular peak data, derived from chromatographic or spectroscopic data. The most popular pattern recognition techniques include PCA and HCA, which are examples of unsupervised techniques (Table 2). Supervised pattern recognition techniques used in CWA forensics include PLS-DA and KNN. HCA can be used to cluster the data, and visualize the clustering through dendrograms. PCA and PLS-DA, on the other hand, can be used to identify peaks of interest, to compress the original data to a lower-dimensional form, and to visualize statistical regularities, PLS-DA can also be used for classification. KNN samples are classified

Table 2 List of selected publications, where multivariate statistical analysis has been utilized in CWA forensic studies

Compound	Technique	Preprocessing/variable-selection		Pattern recognition		Validation	References
		Normalized to unit area (×)		Unsupervised	Supervised		
DC, impurities	LC-MS	Autoscaled/PCA, NMF	×	HCA, PCA, NMF	KNN	Hold-out	Fraga et al. (2010)
GB, DF, DC, impurities	GC-MS	Autoscaled/PCA	×	HCA			Fraga et al. (2011a)
DMMP, impurities	GCxGC-MS	–/target PARAFAC	×	PCA			Fraga et al. (2012)
Acephate, VX and G-series agents related compounds	GC-MS, LC-MS/MS	Centered/–	×	PCA-varimax			Favela et al. (2012)
DMMP, impurities	GC-MS	–/–			GLM, CT	LOO	Mo et al. (2017)
Model pesticide compound impurities	GCxGC-TOFMS	Norm. to standard, indicator variable/custom selection	×		PCA-LDA, PLS-LDA, BRF	Hold-out	Strozier et al. (2016)
Cyanide: trace anions/ elements, isotope ratio	HPIC, ($\delta^{13}C$) IRMS, ICP-OES	Autoscaled/Fisher-ratio, iPLS, GAPLS	×	HCA	PLS-DA, kernel-SVM-DA, KNN	Hold-out	Mirjankar et al. (2016)
Cyanide trace anions	HPIC	–/Fisher-ratio – DCS		HCA, PCA	KNN	Hold-out	Fraga et al. (2011a)
VX synthesis impurities and by-products	GC-MS (EI, CI), EI-MS, GC/HR-MS	Centered, log-transformed, (pareto/variance scaled) / AMDIS	×	PCA	PLS-DA	CV-R^2/Q^2, hold-out	Holmgren et al. (2018)
Russian VX chemical attribution signals	LC-MS/MS	Log-transformed, autoscaled			Sequential PLS-DA	CV-R^2/Q^2, hold-out	Jansson et al. (2018)
Russian VX chemical attribution signals	GC-MS, LC-MS	Log-transformed, autoscaled			PLS-DA, boosted CT	Bootstrap	Williams et al. (2018)
Mustard gas	GC-MS	Autoscaled / Fisher-ratio – DCS	×	HCA	KNN, PLS-DA	Hold-out, CV	Fraga et al. (2016)
Cocaine impurities & adulterants	NMR	Norm. to analyte signal, binned, pareto scaled	×	PCA, HCA	PLS-DA	Permutation test	Pagano et al. (2013)
Fentanyl organic and inorganic synthesis impurities	GC-MS (EI, CI), LC-TOFMS/MS, ICP-MS	Mean centered by sample, mean normalized			PLS-DA	CV	Mayer et al. (2016)

(continued)

Table 2 (continued)

Compound	Technique	Preprocessing/variable-selection		Pattern recognition		Validation	References
		Normalized to unit area (×)		Unsupervised	Supervised		
3-methylfentanyl by-products, reagents	LC-QTOF, GC-QTOF	–/Boruta			Boosted CT	Bootstrap	Mayer et al. (2018)
Elemental impurities (Zn, Fe, Ni, Cu) in sulfamide	ICP-MS	Normalized to analyte signal /–	×		LDA		Hondrogiannis et al. (2013)
Elemental impurities in calcium ammonium nitrate	ICP-MS	Normalized to sum of element concentration / Fisher-ratio – DCS	×	PCA, HCA	PLS-DA	Hold-out	Fraga et al. (2017)
Ricin preparation biomarkers	GC-MS, LC-MS	Normalized to standard, log-transformed, variance scaled/–	×		OPLS-DA	Hold-out, CV-Q^2	Fredriksson et al. (2018)
Ricin cultivar biomarkers	LC-MS	Log-transformed, center scaled/–	×	PCA	PLS-DA, OPLS-DA	Q^2, permutation test	Ovenden et al. (2014)
Ricin trace elements & isotopes	LA-ICP-MS	Log-transformed, pareto scaled/–	×	PCA	OPLS-DA	CV-Q^2	Bagas et al. (2017)

($\delta^{13}C$) IRMS = carbon stable isotope ratio by isotope ratio mass spectrometry, *BRF* = balanced random forest, *CI* = chemical ionization, *CT* = classification trees, *CV* = cross-validation, *CWA* = Chemical warfare agents, *DC* = methylphosphonic dichloride, *DCS* = degree-of-class separation, *DF* = methylphosphonic difluoride, *DMMP* = dimethyl methylphosphonate, *EI* = electron ionization, *GAPLS* = genetic algorithm-based partial least squares, *GC-MS* = gas chromatography-mass spectrometry, *GCxGC-MS* = two-dimensional gas chromatography-mass spectrometry, *GB* = isopropyl methylphosphonofluoridate, *GLM* = generalized linear model, *HCA* = hierarchical cluster analysis, *hold-out* = part of data held out for model testing/validation, *HPIC* = high performance ion chromatography, *HR-MS* = high resolution-mass spectrometry, *ICP-MS* = inductively coupled plasma-mass spectrometry, *ICP-OES* = inductively coupled plasma optical emission spectroscopy, *IPBCP* = isopropyl bicyclophosphate, *iPLS* = interval partial least-squares, *kernel-SVM-DA* = (radial) Kernel support vector machine discriminant analysis, *KNN* = K-nearest neighbors classification, *LA-ICP-MS* = laser ablation inductively coupled plasma mass spectrometry, *LC-MS* = liquid chromatography-mass spectrometry, *LC-QTOF* = liquid chromatography-quadrupole time-of-flight, *LDA* = linear discriminant analysis, *LOO* = leave-one-out cross-validation, *MS/MS* = tandem mass spectrometry, *NMF* = non-negative matrix factorization, *NMR* = nuclear magnetic resonance, *OPLS-DA* = orthogonal partial least squares discriminant analysis, *PARAFAC* = parallel factor analysis, *PCA* = principal component analysis, *PLS-DA* = partial least squares discriminant analysis, *PLS-LDA* = partial least squares combined with LDA, Q^2 R^2/Q^2 = coefficient of determination of fit/prediction, *TOFMS* = time-of-flight mass spectrometry, *varimax* Varimax rotation of basis, *CT* = classification trees

based on similarity to a preselected number of most similar training samples. Chemical forensic data sets typically contain a relatively low number of samples, which makes the statistical analysis vulnerable to over-fitting and limits the types of applicable models. Typically, in CWA chemical forensics studies, chemometrics is used for e.g. source matching or classification into different synthetic routes (Beebe et al. 1998).

As multivariate statistical analysis is a valuable and growing tool in the attribution analysis of CWAs, in the future attention must be paid on drawing and creating guidelines for model validation.

5 Conclusions and Future Considerations

Until recently, the focus of the OPCW and its designated laboratories has been on the development of screening and identification methods of CWC-related chemicals, rather than analysis of impurities. However, in attribution analysis, the impurity profiles or side-product chemicals are used to get information about the origin of the chemicals and the synthesis route. The value and usefulness of attribution analysis in the framework of forensic investigations remains to be seen, but certainly more research and collaboration are required before it can reach its full potential. It should also be remembered, that chemical forensics offers only one method among others in the toolbox of criminal investigation. A suggested workflow for attribution analysis is illustrated in Fig. 4.

In attribution analysis, initial sampling procedure plays a critical role and production of representative samples is crucial. During sample preparation, general techniques should be favoured over analyte specific methods in order to avoid discrimination of impurity or side-product chemicals that could be helpful for attribution purposes. Currently used sampling and sample preparation methods mostly focus on the identification of an alleged chemical like sarin, mostly ignoring the impurities in the sample. Since attribution analysis is entering the OPCW's mandate for investigations, the feasibility of the sampling and sample preparation procedures will need to be evaluated and procedures modified and developed accordingly.

Various analytical techniques have been applied to attribution analysis. Whereas on-site analysis can be achieved by e.g. portable Raman spectroscopy and GC-MS instruments, off-site techniques include various spectrometric and spectroscopic techniques, including GC-MS, LC-MS, ICP-MS, IRMS, NMR, HPIC and Raman and IR spectroscopy. These techniques have different advantages and disadvantages, sensitivities, and specificities to different compound properties (e.g. volatility and polarity). It is apparent that techniques applicable for attribution analysis are already available, and better source matching can be achieved by combining data collected by different analytical techniques like NMR, GC-EI/MS, and LC-MS. In the future, detailed guidelines for sampling requirements, analytical conditions, dataset types of complimentary techniques, and a theoretical framework will need to be defined to allow for reliable data production and data comparisons between laboratories.

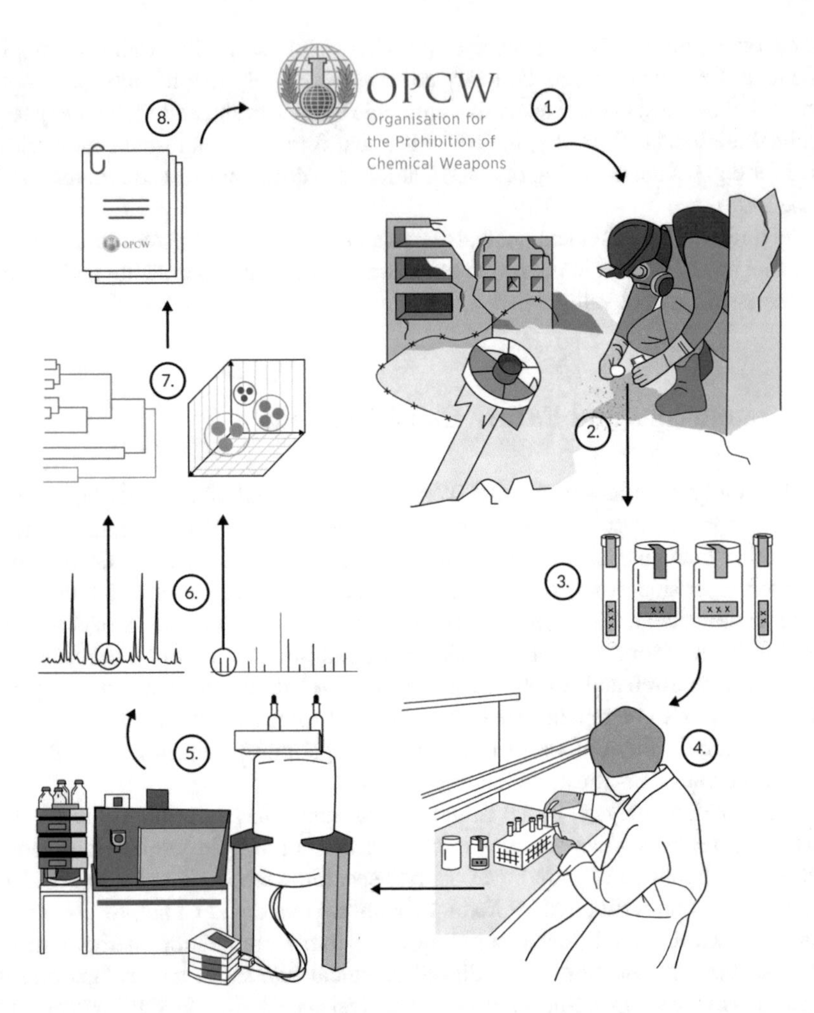

Fig. 4 Illustration of chemical attribution analysis workflow. 1. Mandate from the OPCW, 2. On-site sampling, 3. Sample splitting and coding, 4. Sample preparation in off site laboratory, 5. Instrumental analysis, 6. Data processing, 7. Data analysis, 8. Reporting. (Illustration: Mene Creative / Kiira Koivunen)

As already pointed out, complementary analytical techniques will be required for successful attribution analysis. Impending techniques such as LC-ICP/MS, LC- and GC-HRMS, and elemental (metal) analyses, as well as data fusion of large datasets from different techniques require further studies.

In the OPCW framework, quality control procedures for GC-MS, LC-MS, and NMR have already been established and tested in numerous proficiency tests among laboratories worldwide. For attribution analysis, new types of quality control measures for different techniques will be needed for both on-site and off-site applications. Currently, blank samples (matrix, solvent, and system blanks) are used in the analysis of CWC-related chemicals, but for attribution analysis normal background, from which the chemical attribution signals differ, is harder to define. Common quality control procedures should be established at an early stage of the development of attribution analysis methods in order to enable the comparison of data produced in different laboratories, as well as to enable combining data from various sources to build larger databases.

For data analysis and interpretation, knowhow on chemometric tools is of increasing importance. Numerous statistical methods are currently available, but their validity and benchmarking for available data in the area should be evaluated by establishing a benchmarking dataset. In addition to the statistical analysis methods presented so far (PCA, PLS-DA, HCA, KNN), the choice of pattern recognition techniques in chemical forensics has been on the conservative side. This can be partly attributed to the fact that state-of-the art algorithms sacrifice model interpretability over performance. Care should always be taken with data preprocessing, as it influences the outcome of the data analysis. Even with the same statistical methods, application of different data preprocessing tools, such as smoothing of signals, could lead to different outcomes. Therefore, guidelines for data treatment and processing should also be defined. Data interpretability is of crucial importance in chemical forensics research problems, as decision making has to be justifiable.

Despite the challenges, chemical forensics as a new field has many positive features that can assist in crime investigation when applied in a controlled manner. According to the research reviewed in this article, attribution analysis shows promise for CW related investigations, such as source matching and impurity profiling. However, the determination of the origin of real world samples may turn out to be implausible due to the real world factors that affect the reliability of the results. In the long term, new ROPs for attribution analysis should be created and tested in International Interlaboratory Comparison tests by using both on-site and off-site techniques to reveal possible false positive and negative results. The emphasis of these exercises should be on unprecedented scopes and scenarios that will widen the current focus of the OPCW proficiency tests from challenge inspections to inspections of allegations of use. In the future, designated laboratories should also be tested on their abilities to analyze impurities, synthesis side-products, and stabilizers in addition to verification of scheduled chemicals or closely related compounds. This also brings forth the requirement to create a usable database of the impurities with spectra and CAS registry number information. Reporting criteria should also be developed at the early stages to benchmark them during progress.

In conclusion, development of chemical forensics and attribution analysis requires systematic high quality research and collaboration between laboratories for benefit of the whole CWC community.

References

Ariese, F., H. Meuzelaar, M.M. Kerssens, J.B. Buijs, and C. Gooijer. 2009. Picosecond Raman spectroscopy with a fast intensified CCD camera for depth analysis of diffusely scattering media. *Analyst* 134 (6): 1192–1197.

Asadi, Z., M.D. Esrafili, E. Vessally, M. Asnaashariisfahani, S. Yahyaei, and A. Khani. 2017. Structural study of fentanyl by DFT calculations, NMR and IR spectroscopy. *Journal of Molecular Structure* 1128: 552–562.

Bagas, C.K., R.L. Scadding, C.J. Scadding, R.J. Watling, W. Roberts, and S.P.B. Ovenden. 2017. Trace isotope analysis of Ricinus communis seed core for provenance determination by laser ablation-ICP-MS. *Forensic Science International* 270: 46–54.

Balayssac, S., E. Retailleau, G. Bertrand, M.P. Escot, R. Martino, M. Malet-Martino, and V. Gilard. 2014. Characterization of heroin samples by H-1 NMR and 2D DOSY H-1 NMR. *Forensic Science International* 234: 29–38.

Beckonert, O., H.C. Keun, T.M.D. Ebbels, J.G. Bundy, E. Holmes, J.C. Lindon, and J.K. Nicholson. 2007. Metabolic profiling, metabolomic and metabonomic procedures for NMR spectroscopy of urine, plasma, serum and tissue extracts. *Nature Protocols* 2 (11): 2692–2703.

Beebe, K., R.J. Pell, and M.B. Seasholtz. 1998. *Chemometrics: A practical guide.* New York: John Wiley & Sons.

Benedito, L.E.C., A.O. Maldaner, and A.L. Oliveira. 2018. An external reference 1H qNMR method (PULCON) for characterization of high purity cocaine samples. *Analytical Methods* 10: 12.

Blachut, D., K. Wojtasiewicz, and Z. Czarnocki. 2002. Identification and synthesis of some contaminants present in 4-methoxyamphetamine (PMA) prepared by the Leuckart method. *Forensic Science International* 127 (1-2): 45–62.

Campos, A.R., Z. Gao, M.G. Blaber, R. Huang, G.C. Schatz, R.P. Van Duyne, and C.L. Haynes. 2016. Surface-enhanced Raman spectroscopy detection of ricin B chain in human blood. *Journal of Physical Chemistry C* 120 (37): 20961–20969.

Christensen, K.A., and M.D. Morris. 1999. Hyperspectral Raman microscopic imaging using powell lens line illumination. *Applied Spectroscopy* 52 (9): 1145–1147.

Cletus, B., W. Olds, P.M. Fredericks, E. Jaatinen, and E.L. Izake. 2013. Real-time detection of concealed chemical hazards under ambient light conditions using Raman spectroscopy. *Journal of Forensic Sciences* 58 (4): 1008–1014.

Cort, J.R., and H. Cho. 2009. ^{1}H and ^{13}C NMR chemical shift assignments and conformational analysis for the two diastereomers of the vitamin K epoxide reductase inhibitor brodifacoum. *Magnetic Resonance in Chemistry* 47 (10): 897–901.

Doughty, D., B. Painter, P.E. Pigou, and M.R. Johnston. 2016. The synthesis and investigation of impurities found in Clandestine Laboratories: Baeyer-Villiger Route Part I; Synthesis of P2P from benzaldehyde and methyl ethyl ketone. *Forensic Science International* 263: 55–66.

Duffy, J., A. Urbas, M. Niemitz, K. Lippa, and I. Marginean. 2019. Differentiation of fentanyl analoques by low-field NMR spectroscopy. *Analytica Chimica Acta* 1049: 161–169.

Favela, K.H., J.A. Bohmann, and W.S. Williamson. 2012. Dust as a collection media for contaminant source attribution. *Forensic Science International* 217 (1–3): 39–49.

Fraga, C.G., B.H. Clowers, R.J. Moore, and E.M. Zink. 2010. Signature-Discovery approach for sample matching of a nerve-agent precursor using liquid Chromatography−Mass spectrometry, XCMS, and chemometrics. *Analytical Chemistry* 82 (10): 4165–4173.

Fraga, C.G., O.T. Farmer, and A.J. Carman. 2011a. Anionic forensic signatures for sample matching of potassium cyanide using high performance ion chromatography and chemometrics. *Talanta* 83 (4): 1166–1172.

Fraga, C.G., G.A. Pérez Acosta, M.D. Crenshaw, K. Wallace, G.M. Mong, and H.A. Colburn. 2011b. Impurity profiling to match a nerve agent to its precursor source for chemical forensics applications. *Analytical Chemistry* 83 (24): 9564–9572.

Fraga, C.G., L.H. Sego, J.C. Hoggard, G.A.P. Acosta, E.A. Viglino, J.H. Wahl, and R.E. Synovec. 2012. Preliminary effects of real-world factors on the recovery and exploitation of forensic impurity profiles of a nerve-agent simulant from office media. *Journal of Chromatography. A* 1270: 269–282.

Fraga, C.G., K. Bronk, B.P. Dockendorff, and A. Heredia-Langner. 2016. Organic chemical attribution signatures for the sourcing of a mustard agent and its starting materials. *Analytical Chemistry* 88 (10): 5406–5413.

Fraga, C.G., A.V. Mitroshkov, N.S. Mirjankar, B.P. Dockendorff, and A.M. Melville. 2017. Elemental source attribution signatures for calcium ammonium nitrate (CAN) fertilizers used in homemade explosives. *Talanta* 174: 131–138.

Fredriksson, S.A., D.S. Wunschel, S.W. Lindstrom, C. Nilsson, K. Wahl, and C. Astot. 2018. A ricin forensic profiling approach based on a complex set of biomarkers. *Talanta* 186: 628–635.

Garg, P., A. Purohit, V.K. Tak, and D.K. Dubey. 2009. Enhanced detectability of fluorinated derivatives of N,N-dialkylamino alcohols and precursors of nitrogen mustards by gas chromatography coupled to Fourier transform infrared spectroscopy analysis for verification of chemical weapons convention. *Journal of Chromatography. A* 1216: 7906–7914.

Groenewold, G.S., J.R. Scott, E.D. Lee, and S.A. Lammert. 2013. Rapid analysis of organophosphonate compounds recovered from vinyl floor tile using vacuum extraction coupled with a fast-duty cycle GC/MS. *Analytical Methods* 5 (9): 2227.

Haddad, A., M.A. Comanescu, O. Green, T.A. Kubic, and J.R. Lombardi. 2018. Detection and quantitation of trace fentanyl in heroin by surface-enhanced Raman spectroscopy. *Analytical Chemistry* 90 (21): 12678–12685.

Harvey, S.D., and J.H. Wahl. 2012. On-Matrix derivatization for dynamic headspace sampling of nonvolatile surface residues. *Journal of Chromatography. A* 1256: 58–66.

Hays, P.A. 2005. Proton nuclear magnetic resonance spectroscopy (NMR) methods for determining the purity of reference drug standards and illicit forensic drug seizures. *Journal of Forensic Sciences* 50 (6): 1342–1360.

He, L., B. Deen, T. Rodda, I. Ronningen, T. B, C. Haynes, F. Diez-Gonzalez, and T.P. Labuza. 2011. Rapid detection of ricin in milk using immunomagnetic separation combined with surface-enhanced Raman spectroscopy. *Journal of Food Science* 76: 49–53.

Heather, E., R. Shimmon, and A.M. McDonagh. 2015. Organic impurity profiling of 3,4-methylen edioxymethamphetamine (MDMA) synthesised from catechol. *Forensic Science International* 248: 140–147.

Holmgren, K.H., C.A. Valdez, R. Magnusson, A.K. Vu, S. Lindberg, A.M. Williams, A. Alcaraz, C. Åstot, S. Hok, and R. Norlin. 2018. Part 1: Tracing Russian VX to its synthetic routes by multivariate statistics of chemical attribution signatures. *Talanta* 186: 586–596.

Hondrogiannis, E., A. Schmidt, F. Iannaconi, and E. Ehrlinger. 2013. Feasibility study into use of elemental impurities of sulfamide for use in characterizing different vendors by inductively coupled Plasma/Mass spectrometry. *Forensic Science International* 232 (1-3): 56–59.

Hopkins, R.J., L. Lee, and N.C. Shand. 2017. Correcting transmission losses in short-wave infrared spatially offset Raman spectroscopy measurements to enable reduced fluorescence through-barrier detection. *Analyst* 142 (19): 3725–3732.

Inscore, F., A. Gift, P. Maksymiuk, and S. Farquharson. 2004. Characterization of chemical warfare G-agent hydrolysis products by surface-enhanced Raman spectroscopy. *Chemical and Biological Point Sensors for Homeland Defense II* 2004: 46–52.

Jansson, D., S.W. Lindström, R. Norlin, S. Hok, C.A. Valdez, A.M. Williams, A. Alcaraz, C. Nilsson, and C. Åstot. 2018. Part 2: Forensic attribution profiling of Russian VX in food using liquid chromatography-mass spectrometry. *Talanta* 186: 597–606.

John, H., M.J. van der Schans, M. Koller, H.E.T. Spruit, F. Worek, H. Thiermann, and D. Noort. 2018. Fatal sarin poisoning in Syria 2013: Forensic verification within an international laboratory network. *Forensic Toxicology* 36 (1): 61–71.

Kondo, T., R. Hashimoto, Y. Ohrui, R. Sekioka, T. Nogami, F. Muta, and Y. Seto. 2018. Analysis of chemical warfare agents by portable Raman spectrometer with both 785nm and 1064nm excitation. *Forensic Science International* 291: 23–38.

Kovacs, H., D. Moskau, and M. Spraul. 2005. Cryogenically cooled probes – A leap in NMR technology. *Progress in Nuclear Magnetic Resonance Spectroscopy* 46 (2-3): 131–155.

Kreuzer, H.W., J. Horita, J.J. Moran, B.A. Tomkins, D.B. Janszen, and A. Carman. 2012. Stable carbon and nitrogen isotope ratios of sodium and potassium cyanide as a forensic signature. *Journal of Forensic Sciences* 57 (1): 75–79.

Lee, B.M., B. Veriansyah, S.H. Kim, J.D. Kim, and Y.W. Lee. 2005. Total organic carbon disappearance kinetics for supercritical water oxidation of dimethyl methylphospate used as a chemical agent simulant. *Korean Journal of Chemical Engineering* 22 (4): 579–584.

Leonard, J., A. Haddad, O. Green, R.L. Birke, T. Kubic, A. Kocak, and J.R. Lombardi. 2017. SERS, Raman, and DFT analyses of fentanyl and carfentanil: Toward detection of trace samples. *Journal of Raman Specroscopy* 48 (10): 1323–1329.

Liu, C., T. Li, Y. Han, Z. Hua, W. Jia, and Z. Qian. 2018. The identification and analytical characterization of 2,2′-difluorofentanyl. *Drug Testing and Analysis* 10 (4): 774–780.

Magnusson, R., S. Nyholm, and C. Astot. 2012. Analysis of hydrogen cyanide in air in a case of attempted cyanide poisoning. *Forensic Science International* 222 (1-3): e7–e12.

Martyshkin, D.B., R.C. Ahuja, A. Kudriavtsev, and S.B. Mirov. 2004. Effective suppression of fluorescence light in Raman measurements using ultrafast time gated charge coupled device camera. *The Review of Scientific Instruments* 75 (3): 630–635.

Mayer, B.P., A.J. DeHope, D.A. Mew, P.E. Spackman, and A.M. Williams. 2016. Chemical attribution of fentanyl using multivariate statistical analysis of orthogonal mass spectral data. *Analytical Chemistry* 88 (8): 4303–4310.

Mayer, B.P., C.A. Valdez, A.J. DeHope, P.E. Spackman, and A.M. Williams. 2018. Statistical analysis of the chemical attribution signatures of 3-methylfentanyl and its methods of production. *Talanta* 186: 645–654.

Mazzitelli, C.L., M.A. Re, M.A. Reaves, C.A. Acevedo, S.D. Straight, and J.E. Chipuk. 2012. A systematic method for the targeted discovery of chemical attribution signatures: Application to isopropyl bicyclophosphate production. *Analytical Chemistry* 84 (15): 6661–6671.

McCain, S.T., R.M. Willett, and D.J. Brady. 2008. Multi-excitation Raman spectroscopy technique for fluorescence rejection. *Optics Express* 16 (15): 10975–10991.

Mesilaakso, M., and M. Rautio. 2000. Verification of chemicals related to the chemical weapons convention. In *Encyclopedia of analytical chemistry: Applications, theory and instrumentation*. Chichester: John Wiley.

Metaxas, A.E., and J.R. Cort. 2013. Counterion influence on chemical shifts in strychnine salts. *Magnetic Resonance in Chemistry* 51 (5): 292–298.

Mirjankar, N.S., C.G. Fraga, A.J. Carman, and J.J. Moran. 2016. Source attribution of cyanides using anionic impurity profiling, stable isotope ratios, trace elemental analysis and chemometrics. *Analytical Chemistry* 88 (3): 1827–1834.

Mo, K.-F., A. Heredia-Langner, and C.G. Fraga. 2017. Evaluating and modeling the effects of surface sampling factors on the recovery of organic chemical attribution signatures using the accelerated diffusion sampler and solvent extraction. *Talanta* 164: 92–99.

Moran, J.J., H.W. Kreuzer, A.J. Carman, J.H. Wahl, and D.C. Duckworth. 2012. Multiple stable isotope characterization as a forensic tool to distinguish acid scavenger samples. *Journal of Forensic Sciences* 57 (1): 60–63.

Mott, A.J., and P. Rez. 2012. Calculated infrared spectra of nerve agents and simulants. *Spectrochimica Acta. Part A, Molecular and Biomolecular Spectroscopy* 91: 256–260.

Notingher, I., C. Green, C. Dyer, E. Perkins, N. Hopkins, C. Lindsay, and L.L. Hench. 2004. Discrimination between ricin and sulphur mustard toxicity in vitro using Raman spectroscopy. *Journal of the Royal Society Interface* 1 (1): 79–90.

OPCW – Note by the technical secretariat. 2018. *Report of the OPCW fact-finding mission in Syria regarding the incidents in Al-hamadaniyah on 30 October 2016 and in Karm Al-Tarrab on 13 November 2016*, S/1642/2018 (2 Jul 2018). Available from https://reliefweb.int/report/syrian-arab-republic/note-technical-secretariat-report-opcw-fact-finding-mission-syria-1.

OPCW – UN Joint Investigative Mechanism. 2016. *Third report of the Organization for the Prohibition of Chemical Weapons - United Nations Joint Investigative Mechanism*, S/2016/738 (24 Aug 2016). Available from https://undocs.org/S/2016/738.

OPCW - UN Joint Investigative Mechanism. 2017a. "*FACT SHEET OPCW – UN JOINT INVESTIGATIVE MECHANISM*" *(Organization for the Prohibition of Chemical Weapons-United Nations Joint Investigative Mechanism, June 2017)*. https://s3.amazonaws.com/unoda-web/wp-content/uploads/2017/07/JIM-Fact-Sheet-Jul2017.pdf.

OPCW - UN Joint Investigative Mechanism. 2017b. *Seventh report of the Organisation for the Prohibition of Chemical Weapons - United Nations Joint Investigative Mechanism*, S/2017/904 (26 Oct 2017). Available from https://undocs.org/S/2017/904.

OPCW/PTS. 1993. *Convention on the prohibition of the development, production, stockpiling and use of chemical weapons and on their destruction*. Signed in January 1993. Printed and Distributed by the OPCW/PTS. The Depositary of this Convention is the Secretary-General of the United Nations, from whom a Certified True Copy can be Obtained.

Ovenden, S.P., E.J. Pigott, S. Rochfort, and D.J. Bourne. 2014. Liquid chromatography-mass spectrometry and chemometric analysis of Ricinus communis extracts for cultivar identification. *Phytochemical Analysis* 25 (5): 476–484.

Pagano, B., I. Lauri, S. De Tito, G. Persico, M.G. Chini, A. Malmendal, E. Novellino, and A. Randazzo. 2013. Use of NMR in profiling of cocaine seizures. *Forensic Science International* 231 (1–3): 120–124.

Petterson, I.E.I., P. Dvorak, J.B. Buijs, C. Gooijer, and F. Ariese. 2010. Time-resolved spatially offset Raman spectroscopy for depth analysis of diffusely scattering layers. *Analyst* 135 (12): 3255–3259.

Picquart, M., E. Nicolas, and F. Lavialle. 1989. Membrane-damaging action of ricin on DPPC and DPPC-cerebrosides assemblies. A Raman and FTIR analysis. *European Biophysics Journal* 17 (3): 143–149.

Rautio, M. 1993. *Recommended operating procedures for sampling and analysis in the verification of chemical disarmament*. Helsinki: The Ministry for Foreign Affairs of Finland.

Rautio, M. 1994. *Recommended operation procedures for sampling and analysis in the verification of chemical disarmament*. Helsinki: The Ministry for Foreign Affairs of Finland.

Riaz, S., and M.A. Farrukh. 2014. Toxicological analysis of ricin in medicinal castor oil with evaluation of health hazards. *Asian Journal of Chemistry* 26 (2): 499–503.

Rodriguez-Saona, L.E., F.S. Fry, and E.M. Calvey. 2000. Use of Fourier transform near-infrared reflectance spectroscopy for rapid quantification of castor bean meal in a selection of flour-based products. *Journal of Agricultural and Food Chemistry* 48 (11): 5169–5177.

Rummel, J.L., J.D. Steill, J. Oomens, C.S. Contreras, W.L. Pearson, J. Szczepanski, D.H. Powell, and J.R. Eyler. 2011. Structural elucidation of direct analysis in real time ionized nerve agent simulants with infrared multiple photon dissociation spectroscopy. *Analytical Chemistry* 83 (11): 4045–4052.

Salter, B., J. Owens, R. Hayn, R. McDonald, and E. Shannon. 2009. N-chloramide modified Nomex as a regenerable self-decontaminating material for protection against chemical warfare agents. *Journal of Materials Science* 44: 2069–2078.

Scafi, S.H., and C. Pasquini. 2001. Identification of counterfeit drugs using near-infrared spectroscopy. *Analyst* 126: 2218–2224.

Schenck, F.J., and J.E. Hobbs. 2004. Evaluation of the Quick, Easy, Cheap, Effective, Rugged, and Safe (QuEChERS) approach to pesticide residue analysis. *Bulletin of Environmental Contamination and Toxicology* 73 (1): 24–30.

Schneider, T., and Lütkefend, T.. 2019. Global Public Policy Institute. *Nowhere to hide: The logic of chemical weapons use in Syria*. Available from https://www.gppi.net/2019/02/17/the-logic-of-chemical-weapons-use-in-syria (accessed Jan 13, 2020).

Strozier, E.D., D.D. Mooney, D.A. Friedenberg, T.P. Klupinski, and C.A. Triplett. 2016. Use of comprehensive two-dimensional gas chromatography with time-of-flight mass spectrometric detection and random forest pattern recognition techniques for classifying chemical threat agents and detecting chemical attribution signatures. *Analytical Chemistry* 88 (14): 7068–7075.

Tang, J.J., J.F. Sun, R. Lui, Z.M. Zhang, J.F. Liu, and J.W. Xie. 2016. New surface-enhanced Raman sensing chip designed for on-site detection of active ricin in complex matrices based on specific depurination. *ACS Applied Materials and Interfaces* 8 (3): 2449–2455.

Vanninen, P. accepted 2019. Verification of chemicals related to the chemical weapons convention. *In Encyclopedia of analytical chemistry: Applications, theory and instrumentation*. John Wiley & Sons.

Vanninen, P. 2011a. Verification of chemicals related to the chemical weapons convention. In *Encyclopedia of analytical chemistry: Applications, theory and instrumentation*. Chichester: John Wiley & Sons.

Vanninen, P. 2011b. *Recommended operating procedures for analysis in the verification of chemical disarmament*, 2011 Edition. Helsinki: University of Helsinki.

Vanninen, P. 2017. *Recommended operating procedures for analysis in the verification of chemical disarmament*, 2017 Edition. Helsinki: University of Helsinki.

Volpe, A.M., and M.J. Singleton. 2011. Stable isotopic characterization of ammonium metavanadate (NH_4VO_3). *Forensic Science International* 209 (1-3): 96–101.

Wahl, J.H., and H.A. Colburn. 2010. Extraction of chemical impurities for forensic investigations: A case study for indoor releases of a sarin surrogate. *Building and Environment* 45 (5): 1339–1345.

Wiktelius, D., L. Ahlinder, A. Larsson, K. Hojer Holmgren, R. Norlin, and P.O. Andersson. 2018. On the use of spectra from portable Raman and ATR-IR instruments in synthesis route attribution of a chemical warfare agent by multivariate modeling. *Talanta* 186: 622–627.

Williams, A.M., A.K. Vu, B.P. Mayer, S. Hok, C.A. Valdez, and A. Alcaraz. 2018. Part 3: Solid phase extraction of Russian VX and its chemical attribution signatures in food matrices and their detection by GC-MS and LC-MS. *Talanta* 186: 607–614.

Wu, J., Y. Zhu, J. Gao, J. Chen, J. Feng, L. Guo, and J. Xie. 2017. A simple and sensitive surface-enhanced Raman spectroscopic discriminative detection of organophosphorus nerve agents. *Analytical and Bioanalytical Chemistry* 409: 5091–5099.

Yoder, J.L., P.E. Magnelind, M.A. Espy, and M.T. Janicke. 2018. Exploring the limits of overhauser dynamic nuclear polarization (O-DNP) for portable magnetic resonance detection of low gamma nuclei. *Applied Magnetic Resonance* 49 (7): 707–724.

Zheng, J.K., C.Y. Zhao, G.F. Tian, and L.L. He. 2017. Rapid screening for ricin toxin on letter papers using surface enhanced Raman spectroscopy. *Talanta* 162: 552–557.

Recent Developments in the Clinical Management of Weaponized Nerve Agent Toxicity

Alexander F. Barbuto and Peter R. Chai

1 Introduction

Nerve agents are a class of highly potent organophosphate cholinesterase inhibitors, weaponized with the intent to cause catastrophic human morbidity and mortality. Human exposure generally occurs in the form of military and terrorist attacks, assassinations, laboratory accidents, and suicide attempts. These agents irreversibly inhibit acetylcholinesterase, the enzyme that degrades acetylcholine, an important neurotransmitter with a wide scope of parasympathetic and neuromuscular effects.

Symptoms of nerve agent toxicity include excessive secretions (bronchorrhea, rhinorrhea, lacrimation, diaphoresis), neuromuscular effects (fasciculations, weakness, paralysis), and central nervous system effects (loss of consciousness, seizures). Weaponized nerve agents can be fatal within minutes due to respiratory failure from seizure, paralysis, and airway compromise from copious secretions. Survivors may have long-term neurologic symptoms that are postulated to be a result of multiple intracellular pathways and neurodegenerative effects. The mainstays of treatment are supportive care and antidotal therapy in the form of oximes. Supportive treatments are aimed at supporting normal respiration and minimizing the toxic effects of excessive cholinergic activation. Oximes treat organophosphate poisoned patients by regenerating functional cholinesterase enzyme. With the reemergence of nerve agents as a threat to the public, novel and adjunctive therapies are being explored.

A. F. Barbuto (✉)
Harvard Medical Toxicology Program at Boston Children's Hospital, Boston, MA, USA
e-mail: abarbuto@alumni.nd.edu

P. R. Chai
Harvard Medical School, Division of Medical Toxicology, Department of Emergency Medicine, Brigham and Women's Hospital, Adjunct Faculty, The Fenway Institute, Boston, MA, USA
e-mail: pchai@bwh.harvard.edu

M. Martellini, R. Trapp (eds.), *21st Century Prometheus*,
https://doi.org/10.1007/978-3-030-28285-1_13

2 Nerve Agent Exposures

2.1 *Classic V and G Nerve Agents*

The original nerve agents were developed in Germany in the 1930s–40s. These "G-series" agents which include sarin (GB), soman (GD), and tabun (GA) are volatile substances, therefore are a threat for vapor and inhalational exposure. The later "V-series," first developed in the United Kingdom following World War II, include VX, VM, VG, and VR. These are oily substances which can persist in the environment or the skin. Despite many global wars after their development, the first weaponized use of nerve agents took place during the Iran-Iraq conflict in the 1980s when the Iraqi military used them against Iranian military and civilians.

Sarin was used in a terrorist attack by the cult Aum Shinrikyo in the Tokyo subway system in 1995. Over 1000 patients were evaluated at nearby hospitals and 12 died. The cult also used sarin in Matsumoto, Japan, one year prior in 1994; in this smaller attack, over 500 patients were exposed and 8 died. These agents continue to be used in the twenty-first century, most recently in the cities of Damascus (Rosman et al. 2014), Douma and Ghouta during the Syrian civil war in 2013–2017. Nerve agent exposures have led to estimated death tolls between 200 and 1729. Lessons learned by hospital systems in Syria can be helpful in the treatment of mass casualty nerve agent events. VX (2-diisopropylamino-O-ethylmethylphospho-nothioate) was used in the assassination of Kim Jong-nam, half-brother of North Korean leader Kim Jong-un in 2017. It is believed that VX was created by mixing two less toxic substances together, generating VX upon their interaction.

2.2 *Novichok and Novel Organophosphates*

During the 1970s, Russian chemists began to develop the Novichok agents, or "newcomers," under a secret project *Foliant*. The suspected aims of this project were to create new nerve agents that were not detectable by current field detection kits and were more potent than the traditional V and G series agents (Vaserhelyi and Foldi 2007). Little is known about their pharmacokinetics or pharmacodynamics, in part because their structures are not well known. Early agents in this project included substance 33, A-230, and A-232. Next, binary agents were developed, including Novichok-5 and Novichok-7 (United States Senate 1995). Much of the information known about the Novichok program is from Dr. Vil Mirzayanov, a Russian scientist who gave testimony about chemical weapons developed in the Soviet Union (Chai et al. 2018). Scant evidence is available regarding human exposure to Novichok agents. The deliberate poisoning in 2018 of Sergei Skripal, a Russian expatriate and his daughter Yulia Skripal, near their home in Salisbury, England using a Novichok agent (OPCW 2018; UK Government 2018) has led to continued interest in specific toxicities pertaining to these agents. Both patients required lengthy intensive care unit stays and ultimately survived their poisoning.

2.3 Unitary Agents vs Binary Agents

Many nerve agents exist in a unitary, stable, ready-for-use form. In contrast, binary agents exist as two (or more) non-toxic or less-toxic agents that can be mixed to form nerve agents. Binary agents are used for safer storage and handling. An example of this is the US Army M687 munition, in which methylphosphonyl difluoride was held in one reservoir and a mixture of isopropyl amine and isopropyl alcohol in another. A disk separating these canisters ruptures upon firing the artillery shell, allowing the reagents to mix, thereby forming sarin by the time the munition reached its target. Non-munition binary nerve agents were suspected in the assassination of Kim Jong-nam. Binary agents could also be used to circumvent the Organisation for the prohibition of chemical weapons, or OPCW, which does not ban inert precursor agents.

2.4 Route of Exposures

In recent years, significant nerve agent poisonings have occurred in a variety of settings. Exposures have occurred by chemical munitions, smearing nerve agent on objects that will be handled by their intended assassination targets, dispersal in a closed environment, and attacking a target by splashing or wiping nerve agent directly to the victim's face. As observed in the Salisbury, England event in 2018, secondary exposures can occur weeks to months later if the nerve agent persists in the environment or on an object. Two bystanders were subsequently exposed to Novichok in Salisbury almost 4 months later, after one of them picked up what appeared to be a small perfume vial and shared it with the other, demonstrating the persistence in the environment and the potential for delayed or unintended victim exposures. Ultimately, one of these victims did not survive the poisoning (BBC 2018). First responders and medical personnel can also be affected by victims or environments which are not adequately decontaminated. In the Tokyo sarin subway attacks, 135 of the 1364 firefighters who responded to the incident had a secondary exposure, though none died (Okumura et al. 2005).

Routes of exposure depend, at least in some degree, on the volatility of the substance. G-series agents are volatile compounds and exert severe toxicity when inhaled. Sarin is the most volatile nerve agent, with vapor pressure nearly four orders of magnitude greater than that of VX (Buchanan et al. 2009). As seen in the sarin attacks in Tokyo, the vapor can also make contact with the eyes, typically causing miosis without significant systemic effects. In contrast, VX and other oily, or "persistent" nerve agents can be readily absorbed through the skin or mucus membranes but do not pose significant vapor threat.

Less sinister exposures can happen in research and development settings. Although not intended for weaponized use, organophosphate pesticides are used in agricultural settings and are also implicated in suicides. Examples of organophosphate pesticides include malathion, parathion, dichlorvos, phosmet, and others. The treatment of organophosphate agents is similar, whether weaponized or agricultural. Organophosphate pesticide toxicity is observed much more frequently, and therefore, most of the clinical trials in treatment of cholinesterase inhibitor toxicity involve pesticides.

2.5 *Pediatric Considerations (Exposure)*

There are several key considerations in nerve agent exposures in children. First, children have a higher respiratory rate and minute ventilation per kilogram when compared to adults (Miller et al. 2016); therefore, they will experience greater exposure to inhaled agents. Additionally, vapors that are heavier than air concentrate nearer the ground and closer to a child's mouth and nose than an adult's, leading to increased exposure. Second, children are likely to explore the world around them with their hands and mouth, and may be at a heightened risk of exposure from a contaminated environment or object. Finally, dermally absorbed agents may pose higher risk to children due to a higher body surface area to body weight ratio, however the rate of dermal absorption is not well studied in children.

3 Nerve Agent Toxicology

3.1 *Pharmacokinetics of Nerve Agents*

Nerve agents are extremely potent organic phosphorous cholinesterase inhibitors. Human LD_{50} (the dose required to kill 50% of exposed patients) varies by route and agent. Volatile agents, like sarin or soman, exhibit toxicity almost immediately following inhalation or mucus membrane exposure. Dermal exposures may have somewhat delayed effects, though the lipophilicity of the particular nerve agents impacts time-to-toxicity. Although toxicity will most likely occur within minutes to hours following a skin exposure to liquid nerve agent, symptoms have been delayed up to 18 h (Miller 2004).

Potency of nerve agents varies by structure and is stereo-specific, with the toxicity dependent on the stereochemistry at the phosphorous center (Benschop and DeJong 1988). Elimination from the body is largely by hydrolysis or oxidation, and occurs at varying rates. Metabolites may exert their own toxicities as well, but are typically irritants and do not have the morbidity and mortality associated with the parent compound (Moshiri et al. 2012) (Table 1).

Table 1 Nerve agent properties

Agent	LD50 (mg)	Lct50 (mg-min/m^3)	Hydrolysis $t_{1/2}$	Rate of action	Modes	Aging $t_{1/2}$
GA (tabun)	1500	70 (inhalation)	8.5 h in water; rapid in strong acids/alkalis	Rapid	Inhalation > eyes > skin	46 h
GB (sarin)	1700	35	80 h in water; 27 mins in strong acid; <1 min in strong alkali	Rapid	Inhalation > eyes > skin	5 h
GD (soman)	350	35 (inhalation)	Varies with pH. Complete hydrolysis occurs in <5 min in a 5% NaOH sol'n	Rapid	Inhalation > eyes > skin	2 mins
GF (cyclosarin)	350	35 (inhalation/ ocular)	42 h in water.	Rapid	Inhalation > eyes > skin	40 h
VX	5	15 (inhalation/ ocular)	60 h in water; 1.8 min in 1.25 M NaOH; 31 min in 0.10 M NaOH	Rapid	Skin=inhalation	48 h

Adapted from Potential military chemical/biological agents and compounds. Chapter 2 Chemical Warfare agents and their properties (US Army 2005)

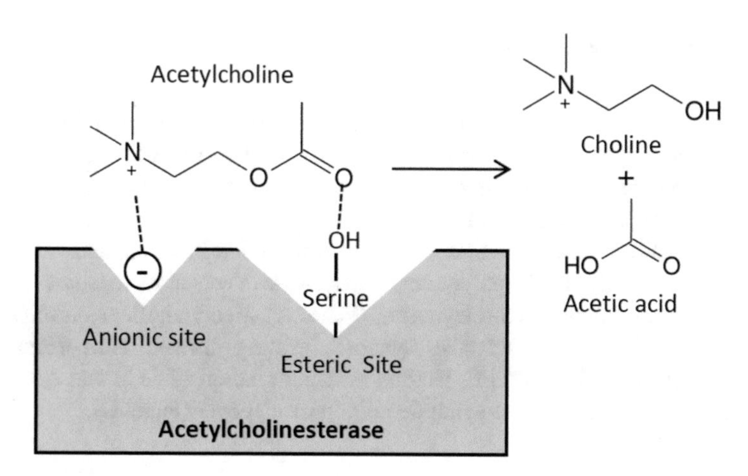

Fig. 1 Acetylcholinesterase

3.2 Cholinesterase Enzymes

Nerve agents exert their toxicity by inhibiting the enzyme acetylcholinesterase (AChE) – see Fig. 1. Cholinesterases are a family of enzymes that hydrolyze esters of choline, most notably acetylcholine (ACh). AChE is found at the synapses of the cholinergic nervous system. When a neuron releases ACh at a receptor site, AChE begins rapidly hydrolyzing the neurotransmitter, regulating its targeted effects. Two other relevant cholinesterases are found in the blood: butyrylcholinesterase (BuChE) and red blood cell cholinesterase (RBC-cholinesterase). Butyrylcholinesterase, found in the plasma, is also known as plasma cholinesterase or pseudocholinesterase. RBC-cholinesterase is isoenzyme of AChE that is found on the membranes of red blood cells. Blood cholinesterases are of clinical significance, as their activity levels can be measured and monitored as an approximation of tissue or nervous system AChE activity following a known or suspected organophosphate poisoning. Point-of-care laboratory assays to detect RBC-cholinesterase activity level have been developed. In the Matsumoto attack, BuChE was not inhibited as strongly as red blood cell AChE by sarin (OPCW 2001). While BuChE activity levels may be more practical to measure, one should not use this as sole evidence of ongoing poisoning.

The acetylcholinesterase enzyme rapidly hydrolyzes the bond in acetylcholine, producing acetic acid and choline. The serine residue is the active site of the enzyme, while a pocket of negative charge in the protein helps align acetylcholine by attracting the positive charge of the quaternary amine. This is a highly efficient enzyme, capable of degrading 25,000 molecules of acetylcholine per second.

3.3 Mechanism of Action

Organophosphate nerve agents inhibit AChE, leading to a buildup of acetylcholine at nerve synapses and at the neuromuscular junction. This leads to excessive stimulation post-synaptically and the muscarinic and nicotinic effects described below. The mechanism is illustrated in Fig. 2.

At synapses between neurons or at the neuromuscular junction, acetylcholine is released from vesicles in the pre-synaptic neuron. ACh acts at its receptors, leading to transmission of downstream signaling (Panel A). Acetylcholinesterase (AChE) cleaves acetylcholine within the synapse, ending transmission (Panel B). Organophosphates inhibit AChE, leading to accumulation of ACh and continuous signal transmission at the post-synaptic neuron or myocyte (Panel C).

3.4 Enzymatic "Aging"

After binding to the AChE or BuChE, nerve agents can undergo an enzyme-independent molecular rearrangement, leading to the removal of one alkyl group bound to the phosphorous atom, and the formation of a covalent bond to the serine residue. This covalent bond is permanent, rendering the enzyme inactive. The rate of aging varies by agent. The half-time to aging for soman is a few minutes, while sarin takes 5 h, and VX more than 40 h. As additional enzymes are irreversibly poisoned, normal physiologic function will not resume until new enzymes are created.

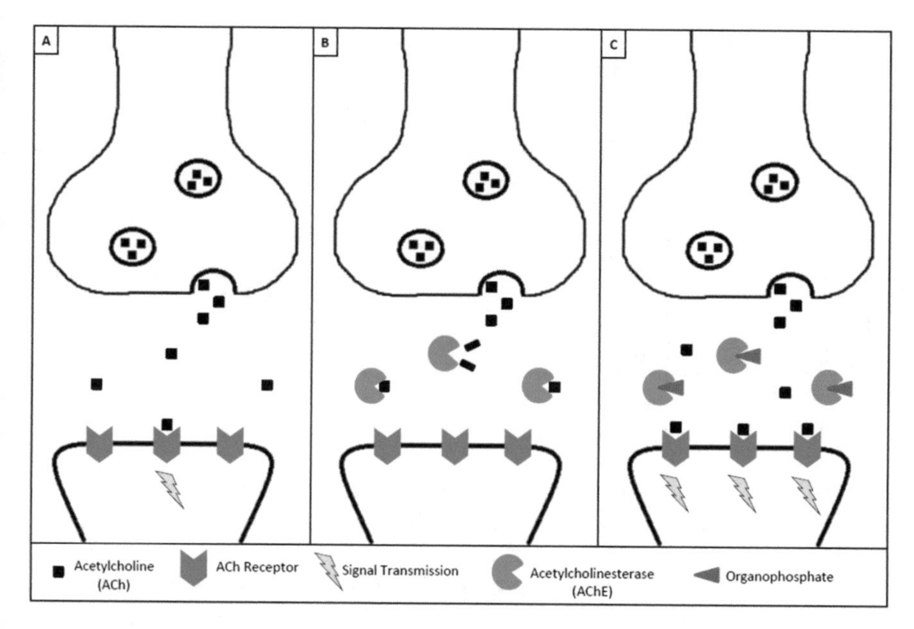

Fig. 2 Schematic illustration of synapse

Oxime therapy, described in detail later, aims to interrupt the aging process, and therefore is most effective if initiated early. The remaining side chain of the nerve agent shields the serine-agent bond from nucleophilic attack by oximes. The particular side chains of each nerve agent contribute to the varied efficacy of each oxime in treating each exposure.

3.5 *Clinical Effects*

The clinical manifestations of nerve agent poisoning are the result of excessive ACh activity at two major receptors, the nicotinic and the muscarinic receptor. Excessive muscarinic activity largely leads to increased secretory effects including lacrimation, diaphoresis, bronchorrhea, and rhinorrhea. Urinary incontinence, increased gastrointestinal motility, diarrhea, and emesis are all expected phenomena. Vapor exposure to the eyes can cause pupillary constriction (miosis), without systemic toxicity, which was observed in many patients in the Tokyo sarin attack. The absence or presence of miosis can be useful in triaging patients following vapor exposure (Ohbu et al. 1997). Nicotinic activity is manifested largely at the neuromuscular junctions; fasciculations leading to muscle weakness or paralysis can occur. Central nervous system (CNS) effects include seizure and loss of consciousness. Mnemonics to remember these constellations of symptoms include SLUDGEM or DUMBBB ELS for muscarinic effects and days-of-the-week for nicotinic effects (Table 2).

Table 2 Mnemonic devices to remember the expected signs and symptoms of the cholinergic toxidrome caused by inhibition of acetylcholinesterase by nerve agents

DUMBBBELS	SLUDGEM	MTWTF (days of the week)
(Muscarinic)	*(Muscarinic)*	*(Nicotinic)*
Diarrhea	**S**alivation	**M**ydriasis
Urination	**L**acrimation	**T**achycardia
Miosis	**U**rination	**W**eakness
Bradycardia	**D**iarrhea	**T**witching
Bronchorrea	**G**astrointestinal Distress	**F**asciculation
Bronchospasm	**E**mesis	
Emesis	**M**iosis	
Lacrimation		
Salivation		

Cardiac effects include bradycardia and hypotension. Patients may have a transient catecholamine surge as a response to their exposure, leading to initial tachycardia or hypertension. This misleading tachycardia could prevent accurate and prompt recognition of cholinergic toxicity. Excessive secretions, seizure, or miosis in the appropriate context should prompt treatment with anticholinergic therapy, regardless of heart rate.

The combination of bronchospasm, bronchorrhea, and paralysis of respiratory muscles ultimately progresses to respiratory failure, which is the main cause of mortality in nerve agent victims. Patients at the extremes of age appear to have the highest mortality from status epilepticus (Towne et al. 1994), and clinicians should be aggressive in treating seizures in these at-risk populations.

There are effects of nerve agents that are not mediated by the buildup of acetylcholine at synapses and neuromuscular junctions, specifically long term neurologic sequelae. Survivors of acute nerve agent poisoning have demonstrated neurodegenerative changes, including impaired memory, dementia, and ataxia. Duration of seizure has correlation with worse neuropathology in animal models (Carpentier et al. 2000). While the precise mechanism is unknown, there may be multiple intracellular pathways that are affected by nerve agents (RamaRao and Bhattacharya 2012), including induction of gene transcription that leads to neurodegeneration (RamaRao et al. 2014). Some treatments, which are described later, may have impact on decreasing these sequelae. Animal studies suggest early and aggressive treatment may reduce these long-term effects (Dillman III et al. 2009).

3.6 Diagnosis and Laboratory Findings

Immediate diagnosis is clinical. Organophosphate or nerve agent poisonings cause a cholinergic toxidrome, and the presence of excessive secretions, bradycardia, and miosis should prompt clinicians to consider and empirically treat affected patients.

Routine laboratory tests like blood counts, electrolytes, and liver or kidney function can be obtained, but are unlikely to be directly affected by nerve agent poisonings in the acute setting. Over time, critical illness can lead to renal insufficiency, fluid and electrolyte losses, or other metabolic derangements. Predicting severity of illness can be difficult. Absence of symptoms soon after exposure likely excludes serious toxicity. In organophosphate pesticide poisoning, the APACHE II, APACHE III, and SAPS scores have correlation with duration of atropine and pralidoxime treatment, length of stay, and mortality (Sungurtekin et al. 2008), though these are not studied in nerve agent toxicity.

Cholinesterase activity levels can be measured using plasma samples, and have been used to monitor exposures to both pesticides (Higgins et al. 2001) and nerve agents. There is substantial variability, up to 23%, in baseline ChE activity levels for un-poisoned patients (Lessenger and Reese 1999). Inhibition of 70% of RBC-AChE activity is generally indicative of significant poisoning (Knechtges 2008). While plasma BuChE can predict the severity and potential recovery from nerve agent exposure, these levels are frequently unavailable in real time to be of clinical relevance for treating physicians. It is important to note that there have been advances in point-of-care testing devices for plasma BuChE activity, originally used to identify severe inflammatory responses (Zivkovic et al. 2016), although these devices are not widely available.

Clinical laboratories should examine their resources and the extent of their capabilities to provide laboratory support to treating clinicians before a nerve agent exposure takes place (Jortani et al. 2000). Pre-established arrangements for using regional or government laboratories should be made in order to respond to mass casualty events. In addition to monitoring cholinesterase activity, samples taken from the environment can aid government agencies in identifying the precise agents of exposure. Biologic samples can be difficult to interpret, in part because the free, native compounds do not persist in the body for long periods of time as they become hydrolyzed, metabolized, or bound to proteins. Nerve gases and their hydrolysis products can be identified by liquid or gas chromatography with mass spectrometry.

Phosphorylated butyrylcholinesterases can be detected by tryptic digestion followed by liquid chromatography-tandem mass spectrometry and used to identify nerve agent exposure (Van Der Schans et al. 2008). The same reactive phosphate atom that creates covalent bonds to the serine residue of cholinesterase enzymes also reacts with sulfhydryl or oxygen moieties on a host of other proteins and enzymes (Black 2010). Assessing for these protein adducts is a promising method to identify past exposures or chronic low-dose exposure.

4 Treatment Strategies

4.1 Exposure Avoidance

The first step in the treatment of any poisoned patient is to limit exposure. Patients should be removed, if possible, from the contaminated area. If removal is not possible, the patient should be placed in a recovery position to avoid aspiration of expected emesis. Rescuers may also be exposed, so should wear appropriate protective equipment. Level A personal protective equipment is preferred, as nerve agents pose a vapor risk to skin and mucous membranes (CHEMM website 2019). Patients may have other significant traumatic injuries, therefore a rapid survey of life-threatening injuries should occur prior to transport. If possible, critical procedures should occur after decontamination.

4.2 Decontamination Procedures

Due to the rapid absorption of nerve agents on the skin, decontamination is unlikely to serve any significant benefit to the poisoned casualty. Instead, thorough decontamination is necessary prior to patients reaching a treatment area to avoid exposing healthcare workers. Volatile agents like sarin or soman are likely to have been absorbed or evaporated from skin exposures, leaving little agent left to affect others. Liquid or residues can remain on clothing or gear, so the patient should be undressed (Tang and Chan 2002). Clothing and exposed equipment should be sealed in biohazard bags and appropriately marked. In contrast, viscid nerve agents including Novichok and VX, remain on the skin and will continue to absorb. These should be washed from the skin to avoid secondary exposures. Adsorptive resins like Fuller's Earth can be used to adsorb nerve agents.

Copious water with soap or dilute bleach solution (0.5% sodium hypochlorite) can be used to wash the skin. Basic solutions increase the rate of hydrolysis of nerve agents and should be the mainstay of decontamination for sarin, soman, and VX. There is a theoretical risk that hydrolysis of tabun in bleach could generate cyanogen chloride (US Army 2005) and lead to cyanide toxicicty, though this has not been described in victims of tabun poisoning. Hydrolysis of the Novichok agents in dry bleach powder or in concentrated bleach can potentially unleash toxic metabolites like hydrofluoric acid, hydrochloric acid, or hydrogen cyanide (Ellison 2008). For this reason, any attempts at decontamination should be accompanied by copious irrigation. Equipment, including patient transport equipment, should be decontaminated between each patient contact. Stronger bleach solutions (5%) can be used on equipment. Victims should be washed from head to toe after removing clothing. Hair can carry nerve agents; showering with water alone can reduce contamination, though pre-treating the hair with Fuller's Earth may improve the decontamination rate (Josse et al. 2015). Ocular exposure to liquid nerve agent should prompt irrigation of the eyes with copious water or isotonic saline for 5–10 min. Eyes exposed to vapor only do not require irrigation (CDC 2014).

4.3 Pretreatment

In an effort to reduce the susceptibility of cholinesterase to nerve agents, pretreatments have been considered in at-risk populations, such as military personnel. Pyridostigmine is the most commonly used pretreatment agent. The intended effect is that the pretreatment agent will intermittently occupy the cholinesterase enzyme, competing with the nerve agent and displacing molecules before aging occurs. Pyridostigmine is a carbamate cholinesterase inhibitor that is used in the treatment of myasthenia gravis and to reverse the effects of non-depolarizing paralytics. Pretreatment with pyridostigmine appears to enhance the efficacy of atropine and pralidoxime in soman poisoning (US Army 2005, pp. 11–14), though does not appear to be effective VX or sarin poisoning (Balali-Mood and Balali-Mood 2008). Typical dosing is 30 mg by mouth every 8 h prior to exposure (Madsen n.d.).

Physostigmine has been examined as a pretreatment strategy in several animal studies. Unlike pyridostigmine, it crosses the blood brain barrier, causing some inherent toxicity and marked behavioral side effects. It is coupled with an anticholinergic such as trihexyphenidyl, scopolamine, benactyzine, or others (Lennox et al. 1992). Pretreatment along with scopolamine or trihexyphenidyl showed improved survival in non-human primates exposed to two-times the LD_{50} of soman. The same combination provided mortality benefit when non-human primates were exposed to five-times the LD_{50} of soman and were subsequently treated with atropine and 2-PAM (von Bredow et al. 1991). Combining physostigmine with trihexyphenidyl might be preferred over scopolamine due to larger therapeutic index (Solana et al. 1989). In guinea pigs, physostigmine coupled with an anticholinergic reduced time to recovery and frequency of convulsions following nerve agent exposure (Harris et al. 1991). Human efficacy is not well studied and there is insufficient evidence to recommend routine use of pretreatment with physostigmine at this time.

4.4 Supportive Care

As with any toxic exposure, excellent supportive care is critical. Given that the most immediate cause of mortality involving nerve agents is respiratory failure, maintaining a patent airway and supporting respiration should be the first priority. Nasopharyngeal airways or oropharyngeal airways can be temporizing measures prior to establishing a definitive airway, though will not assist with respiratory paralysis or bronchorrhea. Endotracheal intubation should be performed if a patient has respiratory paralysis or fails to improve after anticholinergic treatment. Endotracheal intubation with positive pressure ventilation is the preferred airway maneuver for those with respiratory failure and allows frequent deep suctioning to remove mucus plugging. In mass casualty events with a limited supply of ventilators, two patients of similar size can be ventilated with one machine, using a circuit splitter and doubling the typical tidal volume, a technique learned from experience in mass casualty events. Third-spacing and excessive excretion of fluids may lead to intravascular volume depletion, which should be replaced. Maintain adequate nutrition as in any critically ill patient. While nerve agents are not expected to cause renal failure, should renal insufficiency develop, hemodialysis or continuous renal replacement therapy may be needed.

4.5 Anticholinergics

Anticholinergic medicines are used to alleviate the symptoms produced by excessive acetylcholine. More precisely, they are anti-muscarinics, as they competitively antagonize acetylcholine at the muscarinic receptors and have no effect on nicotinic receptors or the cholinesterase enzymes. The therapeutic endpoint of these medications is to dry respiratory secretions, thereby improving ventilation and oxygenation. Atropine is the most commonly used anticholinergic and is used in many auto-injector drug combinations. Atropine and scopolamine act both centrally and peripherally. When given early, centrally-acting anticholinergic agents may provide some protection against nerve agent-induced seizures (McDonough et al. 2000). One potential complication of large doses of centrally-acting anticholinergics is anti-muscarinic delirium, which should be avoided when possible. Glycopyrrolate (Proakis and Harris 1978), ipratropium, and tiotropium are peripherally acting anticholinergic with poor central nervous system penetration, thus avoiding delirium.

Atropine Atropine is the most commonly used anticholinergic drug for organophosphate poisoning. It acts both centrally and peripherally. Many nations use 1–2 mg of atropine in auto-injectors for point-of-injury treatment. The recommended dose of atropine for severe poisoning is 2–5 mg IV or IM, with repeating doses every 5 min until bronchorrhea resolves. Patients will likely develop tachycardia as a result of atropine administration, although this is not a contraindication to continued dosing. Maintenance dosing can be started at 1–2 mg/h and titrated to adequate clinical effect. These requirements may rapidly consume a hospital's supply of atropine, which highlights the need for stockpiling this medication and evaluating alternative anticholinergic agents. Patients with isolated ocular vapor exposure to volatile nerve agents who have severe miosis, blurred vision, and eye pain due to ciliary spasm may be amenable to topical atropine alone since systemic treatment is not without side effects. In the event of mass exposure, providers should consider the use of atropine eye drops, veterinary atropine, and early outreach to government or regional stockpiles.

Adverse reactions to atropine are uncommon; however, serious sensitivity and allergic reactions have been reported. In military settings or in patient populations expected to be exposed to nerve agents, it is appropriate to identify patients with prior adverse reactions to atropine (for instance, systemic anti-muscarinic symptoms after atropine eye drops or low dose IV atropine) and perform sensitivity testing (Hague and Derr 2004). A suggested regimen is to perform hypersensitivity testing by subcutaneous injection of 1% atropine solution or skin patch testing (Eyal et al. 2002). We would not recommend hypersensitivity testing in patients without prior adverse reactions to atropine. If hypersensitivity is present, atropine should be avoided in the future. Alternative anticholinergics like scopolamine or glycopyrrolate may be more appropriate and are discussed below.

Sublingual (Raipal et al. 2010) and intranasal (Raipal et al. 2009) preparations of atropine sulfate have been developed, with favorable bioavailability and rapid absorption when compared with intramuscular administration, though they are not yet widely used. In settings with shortages of intravenous or intramuscular formulations, 1% atropine solutions (for example, ophthalmic preparations) can be given sublingually (Stolbach et al. 2018).

Scopolamine Scopolamine is a centrally- and peripherally-acting anticholinergic drug that crosses the blood brain barrier. It is often used in patch form for the prevention of post-anesthesia nausea and vomiting or motion sickness. It can be administered intravenously or intramuscularly. The initial recommended dose is 0.2–0.6 mg (Eyal et al. 2002), though higher doses have been used in animal studies (Anderson et al. 1994). It is typically given three to four times daily according to labeled dosing for reducing salivation or emesis (McEvoy 2012), though frequency in organophosphate exposure is not as well defined. Scopolamine's adverse effects include drowsiness, amnesia, and anti-muscarinic delirium. Because of these adverse effects, it should only be given if a patient displays clear CNS toxicity. If seizures are present, scopolamine should be given early (first 20 min), as its efficacy in aborting seizures diminishes the longer the seizure progresses (McDonough and Shin 1993).

Glycopyrrolate Glycopyrrolate is a peripherally-acting anticholinergic drug with very little CNS penetration (Proakis and Harris 1978). It is used to reduce respiratory or gastric secretions (Baraka et al. 1977), peri-operatively (Sacan et al. 2007), and in the treatment of abdominal pain (Guard and Wiltshire 1996). It can be administered orally, intramuscularly, or intravenously. It is effective at reducing the peripheral muscarinic effects of organophosphate poisoning, but does not treat seizures or central muscarinic effects.

Combining glycopyrrolate and atropine may have the benefit of reducing total atropine doses and the unintended central anti-muscarinic effects of higher doses of atropine (Tracey and Gallaghar 1990), and in one study, demonstrated mortality benefit when compared with atropine alone (Arendse and Irusen 2009). In another study of organophosphate poisoned patients, glycopyrrolate was equally as effective as atropine (Bardin and VanEeden 1990). A suggested treatment regimen is 1 mg of glycopyrrolate intravenously every 10–15 min until evidence of anti-muscarinic effects occur. If IV access is not available, intramuscular administration repeated every 30–40 min is appropriate. Time to peak effect is slower for intramuscular administration, about 20–45 min compared with 1–2 min for IV form (Ali-Melkkila et al. 1990). Continuous infusion could also be used, titrated to drying of respiratory secretions.

Other Anticholinergics Ipratropium bromide, often used in asthma or COPD (Rowe et al. 2001), can be nebulized and inhaled, and may have efficacy at reducing bronchial secretions (Peronne et al. 2003). Other anticholinergic agents such as diphenhydramine, trihexyphenidyl, and benztropine are not as well studied in nerve agent poisoning. If other supportive measures are failing to alleviate cholinergic toxicity, it may be reasonable to give one of these agents.

4.6 *Oximes*

While anticholinergics effectively relieve muscarinic effects, they offer no relief from nicotinic symptoms, which supports the use of oximes as antidotal therapy. Oximes work to regenerate active cholinesterase by binding to the phosphorous atom of the organophosphate-serine complex. This leads to cleavage of the covalent bond between the nerve agent and the cholinesterase enzymatic site, thereby regenerating the enzyme's function. Additionally, oximes are reversible ligands of the cholinesterase anionic site and enzymatic site, and may have some prophylactic benefit by shielding this site from further attack by nerve agents. Unfortunately, oxime's short half -lives make them impractical as pre-treatment agents. Each oxime varies in its safety profile, solubility, and ability to rescue AChE. Furthermore, oxime efficacy in reversing agent-specific AChE inhibition varies by nerve agent (Lundy et al. 1992). Unfortunately, there is no single all-encompassing oxime that effectively reactivates AChE in any organophosphate poisoning.

Pralidoxime (2-PAM) Pralidoxime (pyridinium-2-aldoxime or 2-PAM) was the first pyridinium oxime synthesized. In 1956, 2-PAM was used in parathion poisonings in Japan. Pralidoxime is soluble in aqueous solution and is shelf-stable in liquid ampules or in auto-injectors. In animal models, 2-PAM is efficient in reactivating AChE poisoned by sarin or VX, but not effective in tabun or soman poisonings. Definitive human data involving nerve agent poisoning is lacking.

An initial 2-PAM dosing regimen is 1–2 g IV over 15–30 min followed by 0.5 g/h. In pediatric patients, 25–50 mg/kg bolus followed by 10–20 mg/kg/h infusion. In prehospital settings, or if IV access cannot be obtained, 600 mg can be given intramuscularly for up to three doses. MARK-1, ATNAA (Antidote Treatment-Neve Agent Autoinjector), and DuoDote kits each contain 600 mg of pralidoxime. The UK's Combopen contains 500 mg of pralidoxime. Adverse effects of pralidoxime include injection site pain, nausea, tachycardia, hypertension, dizziness, weakness, and headache. Transient elevations in aminotransferases as well as creatine phosphokinase have been observed in healthy volunteers who received pralidoxime (Meridian Medical Technologies 2008).

Asoxime Chloride (HI-6) Asoxime chloride, also known as HI-6, is an oxime that is available for use in Canada, Sweden, and the Czech Republic. It is a fairly broad-spectrum AChE reactivator, effectively treating Russian VX, sarin, cyclosarin, and soman inhibited AChE (Kassa et al. 2007, Medical Countermeasures Database 2017). It also appears effective in treating percutaneous VX exposure in guinea pigs (Whitmore et al. 2018). Despite the poor ability of HI-6 to reactivate tabun-poisoned AChE, its otherwise broad efficacy has led it to be investigated by several other nations to possibly be incorporated in military autoinjectors. Pre-mixed HI-6 is not shelf stable, so it must be formulated in a dry/wet delivery device. Once dry HI-6 dichloride is dissolved in solution in a dry/wet auto-injector, it has a half-life on the order of 32–39 days (Schlager et al. 1991), though its stability is improved at lower temperatures (6–8 degrees Celsius) and an acidic environment (pH 2.5) (Eyer et al. 1988). HI-6 dimethanesulphonate is a salt with higher solubility and stability than usual the more commonly prepared HI-6 dichloride preparation, and appears to be a promising candidate for autoinjector use (Thiermann et al. 1996).

Obidoxime Obidoxime, or LüH-6, was developed in Germany and its pharmacology was first published by Erdman and Engelhard in 1964 (Erdmann and Engelhard 1964). Animal studies suggest it is particularly effective for tabun-inhibited AChE. Although it has some activity in regenerating sarin- and VX-poisoned AChE, it does not appear to be as effective as other oximes. In addition to reactivating AChE, obidoxime also possesses a small amount of anti-muscarinic activity, though the practical benefit of this is unknown (Soukup et al. 2011). One dosing strategy is a 250 mg intravenous bolus, followed by 750 mg given over a 24-h period (Thiermann et al. 1997). Dosing ranges from 4–8 mg/kg have been used in pediatric cases of carbamate and organophosphate poisoning (Lifshitz et al. 1994; El-Naggar et al. 2009).

HLo-7 HLo-7 has the benefit of successfully reversing tabun-inhibited AChE, so may provide the broadest spectrum ability to regenerate poisoned AChE (Worek et al. 1998), though human data is generally lacking. Tabun-inhibited AChE is difficult to reactivate, with only obidoxime and HLo-7 having significant efficacy. Pralidoxime and HI-6 only recover about 5% AChE after tabun poisoning in vitro. In one in-vitro study, HLo-7 seems more effective than other oximes in treating soman, sarin, cyclosarin, or VX poisonings: HI-6, obidoxime, and pralidoxime follow from most to least efficacy (Worek et al. 1998).

Other Oximes Several other oximes are in various stages of investigation or implementation. Methoxime, or MMB-4, is suggested to be superior to obidoxime and 2-PAM in animal studies, but human data is lacking (Lundy et al. 2011). Timedoxime bromide, or TMB4, has been employed in Israeli auto-injectors due to good shelf-stability, but has not been widely used in treatment of organophosphorus poisoned patients (Kozer et al. 2005). In organophosphorus insecticide poisoning, TMB4 was found to have good efficacy in regenerating AChE, but has significant adverse effect including hepatotoxicity, severe headaches and paresthesias that limit its use (Xue et al. 1985).

The newest oximes, so called K-Oximes, include K-27 and K-48. In rat models of organophosphate pesticide poisoning, they demonstrate improved AChE regeneration when compared with obidoxime, HI-6, and TMB-4 (Petroianu et al. 2007a, b). K-203 shows in-vitro efficacy in regenerating tabun-inhibited AChE, though is not as successful in animal studies (Gorecki et al. 2018). Human data is generally lacking at this time.

4.7 Benzodiazepines

Benzodiazepines are the first-line treatment for seizures caused by nerve agents. Benzodiazepines act on the GABA-A chloride channel and work by increasing the frequency that this channel is open. This leads to hyperpolarization of the neuron and prevents normal triggering of action potentials. Their action is independent of acetylcholine receptors and these medications are not metabolized by cholinesterases, making them ideal in treating nerve agent induced seizure. Midazolam, diazepam, or lorazepam are all appropriate treatments. Diazepam is the most familiar benzodiazepine, and diazepam auto-injectors are popular among military and national stockpiles. A growing body of literature suggests midazolam may be more effective at alleviating nerve agent-induced seizures, at least in part due to better absorption when given intramuscularly compared with diazepam (Reddy and Reddy 2015; McDonough et al. 1999; McDonough 2002).

4.8 Scavengers

Scavengers are used as alternate targets for nerve agents, with the intended goal of decreasing nerve agent binding to cholinesterase enzymes. While they sequester the nerve agent, they do not regenerate cholinesterase active sites. A potential advantage, however, is that bioscavengers will provide a longer duration of effect in treating poisoned patients. Human butyrylcholinesterase (HuBChE) has been demonstrated as an effective countermeasure in several species. Raveh et al. demonstrated that HuBChE injected intramuscularly in monkeys was able to protect them against both intravenous soman and VX toxicity, and that HuBChE remained in the blood an average of 34 h (Raveh et al. 1997). Prophylactic HuBChE has also been demonstrated to provide protection against inhalational soman exposure in guinea pigs (Allon et al. 1998) and sarin vapor in Göttingen minipigs (Saxena et al. 2015). HuBChE may prove to be a valuable pretreatment strategy in high risk populations, but human data is thus far lacking. Additionally, there are concerns regarding cost, storage, potential immunogenicity, and the difficulty in timing pre-exposure administration (Rice et al. 2016).

There have been several attempts at developing strategies to detoxifying nerve agents. Bacterial enzymes, like phosphotriesterase from the bacterium *pseudomonas diminuta* can detoxify an area by hydrolyzing or cleaving organophosphate insecticides and nerve agents (Ghanem and Raushe 2005).

4.9 Pediatric Considerations (Treatment)

Atropine and oxime auto-injectors can be used safely in pediatric patients (Baker 2007). A review by Kozer et al studied 142 pediatric patients who accidentally or unintentionally received auto-injector therapy in Israel. Auto-injectors with age-appropriate doses were distributed to families during the Persian Gulf War (children up to 3 years of age: 0.5 mg atropine and 20 mg TMB4 [trimedoxime bromide]; ages 3–8 and >60: 1 mg atropine and 40 mg TMB4; ages 8–60: 2 mg atropine and 80 mg TMB4). Fifteen percent of exposures involved children receiving an inappropriately high dose of atropine and TMB4. All of the toxic symptoms were due to

Table 3 Treatment recommendations and dosing strategies

Drug	Starting dose	Pediatric dosing	Route	Frequency	Maintenance dosing
Antimuscarinic					
Atropine	2–5 mg	0.05–0.1 mg/kg	IV or IM	q 5 min	initial 1–2 mg/h, titrated to clinical effect
Glycopyrrolate	1 mg	0.01 mg/kg	IV or IM	q 15 min (IV) or q 30 min (IM)	0.5–1 mg/h, titrated to clinical effect
Scopolamine	0.2–0.6 mg	0.1–0.15 mg	IV, SubQ, or IM		0.6 mg q6 h
Oximes					
Pralidoxime	1–2 g IV over 15 min or 600 mg IM	25–50 mg/kg	IV or IM	q 8–12 h	0.5 g/h or 10–20 mg/kg/h (pediatric)
Obidoxime	250 mg	6–8 mg/kg	IV	Repeat 2 h after initial, then q 6–12 h	750 mg/day
Asoxime (HI-6)	500 mg	10–15 mg/kg	IV	Repeat 2 h after initial (severe poisonings), then q 6–12 h	500 mg q6–12 h
Benzodiazepines					
Lorazepam	2–4 mg	0.1–0.2 mg/kg	IV or IM	PRN	Repeat as needed
Diazepam	5–10 mg	0.2–0.5 mg/kg	IV or IM	PRN	Repeat as needed
Midazolam	5–10 mg	0.15 mg/kg	IV or IM	PRN	Repeat as needed

atropine, without any observed toxicity from TMB4. Despite a range of clinical effects like mydriasis (26.7%), tachycardia (22.5%), and abnormal neurologic state (2.8%), no patients had serious adverse events or death and none required physostigmine (Kozer et al. 2005). Due to the time-sensitive nature of effective oxime therapy, it may be reasonable to distribute age and weight appropriate auto-injectors to at-risk populations. As with any medication in the home, unintentional exposures are possible. However, there do not appear to be serious negative outcomes of atropinization in this setting.

Table 3 below provides an overview on treatment recommendations and dosing strategies.

4.10 Possible Advances in Treatment

Intravenous Lipid Emulsion Intravenous lipid emulsion (ILE) therapy has gained recognition as a possible treatment for severe poisonings of pharmaceuticals, including local anesthetics (Hoegberg et al. 2016), tricyclic antidepressants, bupropion, calcium channel blockers (Meaney et al. 2013), and some beta-blockers (Cave and Harvey 2009; Ozcan and Weinberg 2014). There are several proposed mechanisms of action for ILE. The most accepted is a "lipid sink" model, in which the xenobiotic dissolves into lipid micelles in the bloodstream, thus decreasing agent availability at biologically active sites. Other possible mechanisms include augmenting fatty acid metabolism and providing energy substrate for poisoned myocardium, thereby increasing inotropy (ACMT 2017).

VX is the most lipophilic of the common nerve agents, as measured by its octanol-water partition coefficient. VX has Log K_{ow} of 2.09, while those of soman and sarin are 1.78 and 0.3, respectively (Eisenkraft and Falk 2016). Since organophosphates are generally lipid soluble, there is theoretical benefit to ILE therapy. Organophosphate insecticide poisonings have been treated successfully with ILE in vitro (Uysal and Karaman 2018). In rat models, ILE has prevented neurotoxicity (Basarslan et al. 2014) and delayed the onset of respiratory failure (Dunn et al. 2012). Human case reports of ILE to treat nerve agent toxicity are lacking, though in one case report, ILE reversed cardiotoxicity from parathion poisoning (Mir and Rasool 2014).

Typical dosing strategy for local anesthetic poisoning is 1.5 mL/kg of 20% lipid emulsion bolus over 2–3 min, followed by an infusion of 0.25 mL/kg/min. If a patient remains unstable, a repeat bolus dose can be given. A theoretical maximum of 12 mL/kg cumulative dose should be considered. Adverse effects of lipid emulsion therapy include hypertrigylceride-induced pancreatitis, acute respiratory distress syndrome (ARDS), and interference with laboratory studies due to grossly lipemic blood samples. This same dosing strategy was also used in severe oral medication overdoses until recently, when a revised recommendation was published with the goal of maintaining adequate vascular lipid compartment while lowering the incidence of adverse events. This updated regimen is 1.5 mL/kg loading bolus over 2–3 min, followed by 0.25 mL/kg/min for 3–5 min, then 0.025 mL/kg/min thereafter (ACMT 2017). These dosing strategies have not been studied-head-to-head in organophosphate poisoning.

Red Blood Cell Transfusion Fresh packed red blood cell (RBC) transfusion has been demonstrated to improve blood levels of cholinesterase activity following organophosphate pesticide overdose. Red blood cell membranes contain acetylcholinesterase linked to a glycosylphosphatidylinositol anchor (Prall et al. 1998). As RBCs age in-vitro and membranes deform, activity of acetylcholinesterase in stored blood drops precipitously after 10–14 days (Leal et al. 2017). A case report of a critically-ill 16 year old male who ingested parathion in a suicide attempt documented improvement in cholinesterase activity from below threshold for determination to 55% activity 6 h after receiving 500 mL of whole blood transfusion. This correlated with clinical improvement and he survived to discharge (Ryniak et al. 2011). In a study of organophosphate poisoning by Bao et al., 80 patients were randomly assigned to transfusion of fresh RBCs, longer-storage RBCs, or non-transfusion. The patients randomized to transfusion arms received 200–400 mL of packed RBCs up to 3 h after their exposure. Blood cholinesterase levels were measured 6 h after transfusion. The study authors found that patients who received either fresh or longer-storage RBCs had decreased required duration of atropine and pralidoxime therapy compared with controls. There was an observed benefit in hospitalized days and total amount of atropine required in patients who received fresher RBCs (10 days storage or fewer) compared with longer storage (11–35 days) (Bao et al. 2017). In an environment where atropine supplies are limited, one could consider treatment with fresh packed red blood cells, especially if transfusion early after exposure is feasible. However, exposure to weaponized organophosphates may be accompanied by trauma, and blood supplies should be given where they will have the greatest likelihood for benefit (i.e., in trauma resuscitation).

Melatonin Melatonin may be beneficial in reducing the duration of delirium (as measured by the Confusion Assessment Method for ICU) when given to organophosphate poisoned patients. A double-blind randomized placebo controlled trial of 56 organophosphate-poisoned patients showed a reduction in delirium from an average of 9–6 days when 3 mg of melatonin was given each night (Vijayakumar et al. 2016).

Magnesium Sulfate Magnesium sulfate ($MgSO_4$) inhibits presynaptic acetylcholine release by inhibiting calcium channels and has been investigated in OP pesticide poisonings over the past two decades (Eddleston and Chowdhury 2016; Jamshidi et al. 2018). It has also demonstrated bronchodilatory effects in treating acute asthma exacerbation (Okayama et al. 1987). An infusion of 4 g of 20% $MgSO_4$ solution within 24 h of admission has been demonstrated in a randomized controlled trial to be effective in decreasing atropine requirements, rates of intubation, and ICU days in OP pesticide poisoning (Vijayakumar et al. 2017). Another study demonstrated mortality benefit and reduced length of stay without significant differences in atropine requirements (Pajoumand et al. 2004). Research in nerve agent toxicity is sparse, showing reduced convulsions in rats poisoned with sarin (Katalan et al. 2013). Given its efficacy in OP poisoning, it would be reasonable to give $MgSO_4$ in as an adjunctive treatment NA toxicity, though should not replace usual anticholinergic and oxime therapy.

NMDA Receptor Antagonists NMDA receptor antagonists, such as gacyclidine (Lallement et al. 1999b) or ketamine (Barbier et al. 2015), have been proposed to offer neuroprotection in organophosphate and nerve agent poisoning (Lallement et al. 1999b). This is thought to be due to altering the effects of organophosphates on non-acetylcholine dependent processes and reducing seizure severity (Hirbec et al. 2001). Gacyclidine has been evaluated in soman poisoning in primates, and it appears to inhibit soman-induced seizures and improve neurologic recovery (Lallement et al. 1999a). Ketamine is not as thoroughly studied as gacyclidine, but has a similar mechanism of action, possesses antiepileptic effects, and is typically available in hospital formularies (Moeller et al. 2018). Treating nerve-agent induced seizures with ketamine should be considered (Mion et al. 2003).

5 Medical Resource Management

5.1 Mass Poisonings and Resource Demanding Exposures

Rapid identification of patients who need treatment or observation is crucial in the management of mass exposures. Patients seizing or in respiratory distress will be obvious to first responders, and should receive immediate anti-muscarinic treatment with or without airway intervention. If responders have adequate personal protective equipment, they should rapidly administer atropine prior to decontamination. More subtle findings, like miosis or diaphoresis, could be a rapid way to identify patients truly exposed to nerve agents, but may not accurately predict eventual severity of toxicity. While intramuscular and intravenous atropine are well described, intranasal or sublingual atropine may provide benefit that can be rapidly administered with minimal equipment (no auto-injectors or syringes and needles).

Hospitals should have planned and rehearsed decontamination and mass casualty processes. Patients may arrive from ambulance, personal vehicles, by foot, or other routes. Controlling crowds and identifying an appropriate flow from decontamination, emergency treatment, and triage is critical. Following a mass exposure, accessing additional stores of anti-muscarinic medications and antidotal therapy will likely be needed. Regional disaster planners should plan for redistribution of hospitalized patients to other local or regional hospitals. Ventilators may be a limiting resource. As described in other mass casualty events, using a single ventilator for two patients is acceptable by connecting two patients with similar ideal body weight, connecting their ventilator circuit with a Y-piece, and doubling the tidal volume (Menes et al. 2017).

5.2 Strategic Stockpiles

Several nations stockpile medications, personal protective equipment, ventilators, and other medical supplies needed in the treatment of chemical, biological, and radiological emergencies (CDC 2018). The United States maintains a stockpile of atropine, pralidoxime, and medical equipment to be used in the event of mass exposure. Pre-positioned CHEMPACKs contain medications to treat nerve agent exposures, and 90% of the US population lives within 1 h of a CHEMPACK location (PHE 2018). Other nations have similar storages. Practitioners and disaster planners should identify how to access these stockpiles, as in some locations, this is a federal resource that can only be released on order by certain government officials (Elbe et al. 2015). Local hospitals should identify strategies for treating causalities prior to arrival of stockpiled medications.

5.3 Medication Alternatives

When in-hospital stores of medications are depleted, alternative therapies can be considered. Anticipate that atropine will be the most limited resource, as dosing requirements per patient may exhaust a hospital's usual supply. Ophthalmic atropine solution can be given sublingually or intranasally. Other antimuscarinics could be used either in combination or instead of atropine, though atropine is the best studied and most effective anti-muscarinic. Expired medications may provide effective treatment, though as drug concentration is gradually reduced, higher doses are needed. Since the therapeutic endpoint is based on symptoms, and not on total dose administered, concerns about remaining effective drug should not hinder their use. Contact your local health department or drug administration to determine appropriateness of using expired medicines. In some hospitals, strict guidelines requiring on-label use of medications prevent appropriate treatment. To combat this, the United States introduced Emergency Use Authorization procedure (Institute of Medicine 2010), which is a FDA-issued authorization for using "unapproved medical products or unapproved uses of approved medical products to be used in an emergency to diagnose, treat, or prevent serious or life-threatening diseases or conditions caused by CBRN threat agents when there are no adequate, approved, and available alternatives (FDA n.d.; Schwartz et al. 2018).

6 Summary

Combined respiratory failure due to excessive secretions and ineffective muscles control is the leading cause of fatality in nerve agent poisoning. Early anti-muscarinic treatment with atropine along with supportive care is essential in preventing death. Antidotal treatment in the form of oximes should be given early to prevent enzymatic aging, though the speed with which aging occurs varies by agent. Seizures should be treated with benzodiazepines. Adjunctive therapies such as intravenous lipid emulsion, enzymatic bioscavengers, magnesium sulfate, or fresh red blood cell transfusion may prove to be beneficial, but more research is needed before comprehensive recommendations can be given.

References

ACMT. 2017. *American college of medical toxicology. Position statement: Guidance for the use of intravenous lipid emulsion. Journal of Medical Toxicology* 13 (1): 124–125.

Ali-Melkkila, T.M., T. Kaila, J. Kanto, and E. Iisalo. 1990. *Pharmacokinetics of I.M. glycopyrronium. British Journal of Anaesthesia* 64: 667–669.

Allon, N., L. Raveh, E. Gilat, E. Cohen, J. Grunwald, and Y. Ashani. 1998. *Prophylaxis against soman inhalation toxicity in guinea pigs by pretreatment alone with human serum butyrylcholinesterase. Toxicology Science* 43 (2): 121–128.

Anderson, D.R., L.W. Harris, S.L. Bowersox, W.J. Lennox, and J.C. Anders. 1994. *Efficacy of injectable anticholinergic drugs against soman-induced convulsive/subconvulsive activity. Drug and Chemical Toxicology* 17 (2): 139–148.

Arendse, R., and E. Irusen. 2009. *An atropine and glycopyrrolate combination reduces mortality in organophosphate poisoning. Human & Experimental Toxicology* 28: 715–720.

Baker, M.D. 2007. *Antidotes for nerve agent Poisoning: Should we differentiate children from adults? Current Opinion in Pediatrics* 19: 211–215.

Balali-Mood, M., and K. Balali-Mood. 2008. *Neurotoxic disorders of organophosphorus compounds and their managements. Archives of Iranian Medicine* 11: 65–89.

Bao, H.X., P.J. Tong, C.X. Li, J. Du, B.Y. Chen, Z.H. Huang, and Y. Wang. 2017. *Efficacy of fresh packed red blood transfusion in organophosphate poisoning. Medicine* 96 (11).

Baraka, A., M. Saab, M.R. Salem, and A.P. Winnie. 1977. *Control of gastric acidity by glycopyrrolate premedication in the parturient. Anesthesia and Analgesia* 56 (5): 642–645.

Barbier, L., et al. 2015. *Beneficial effects of a ketamine/atropine combination in soman-poisoned rats under a neutral thermal environment. Neurotoxicology* 50: 10–19.

Bardin, P., and S. VanEeden. 1990. *Organophosphate poisoning: grading the severity and comparing treatment between atropine and glycopyrrolate. Critical Care Medicine* 18 (9): 956–960.

Basarslan, S.K., H. Alp, S. Senol, O. Evliyaoglu, and U. Ozkan. 2014. *Is intralipid fat emulsion a promising therapeutic strategy on neurotoxicity induced by malathion in rats? European Review for Medical and Pharmacological Sciences* 18 (4): 471–476.

BBC. 2018. Russian spy: What happened to Sergei and Yulia Skripal. BBC News. 27 Sep 2018. https://www.bbc.com/news/uk-43643025. Accessed 1 Feb 2019.

Benschop, H.P., and L.P.A. DeJong. 1988. *Nerve agent stereoisomers: Analysis, isolation and toxicology. Accounts of Chemical Research* 21 (10): 368–374.

Black, R.M. 2010. *History and perspectives of bioanalytical methods for chemical warfare agent detection. Journal of Chromatography. B, Analytical Technologies in the Biomedical and Life Sciences* 878 (17-18): 1207–1215.

Buchanan J, Sumpter K, Abercrombie P, Tevault D. 2009. *Vapor pressure of GB*. ECBC-TR-686. Edgewood Chemical Biological Center, US Army Research, Development and Engineering Command, April, 2009.

Carpentier, P., A. Foquin, G. Rondouin, M. Lerner Natoli, De Groot DMG, and G. Lallement. 2000. *Effects of atropine sulphate on seizure activity and brain damage produced by soman in guinea-pigs: ECoG correlates of neuropathology. Neurotoxicology* 2: 521–540.

Cave, G., and M. Harvey. 2009. *Intravenous lipid emulsion as antidote beyond local anesthetic toxicity: A systematic review. Academic Emergency Medicine* 16: 815–824.

CDC. 2014. Toxic substances portal – Nerve agents (GA, GB, GD, VX). Last updated 21 Oct 2014. https://www.atsdr.cdc.gov/mmg/mmg.asp?id=523&tid=93. Accessed 17 Jan 2018.

———. 2018. Strategic National Stockpile. Last updated 12 Dec 2018. https://www.cdc.gov/cpr/stockpile/products.htm. Accessed 27 Dec 2018.

Chai, P.R., B.D. Hayes, T.B. Erickson, and E.W. Boyer. 2018. *Novichok agents: A historical, current, and toxicological perspective. Toxicology Communication* 2 (1): 45–48. Epub 2018 Jun 29.

CHEMM website. 2019. Nerve Agents –Emergency Department/Hospital management. Last updated 23 Jan 2019. https://chemm.nlm.nih.gov/na_hospital_mmg.htm. Accessed 1 Feb 2019.

Dillman, J.F., III, C.S. Phillips, D.M. Kniffin, C.P. Tompkins, T.A. Hamilton, and R.K. Kan. 2009. *Gene expression profile of rat hippocampus following exposure to the acetylcholinesterase inhibitor soman. Chemical Research in Toxicology* 22: 633–638.

Dunn, C., S.B. Bird, and R. Gaspari. 2012. *Intralipid fat emulsion decreases respiratory failure in a rat model of parathion exposure. Academic Emergency Medicine* 19 (5): 504–509.

Eddleston, M., and F.R. Chowdhury. 2016. *Pharmacological treatment of organophosphorus insecticide poisoning: The old and the (possible) new. British Journal of Clinical Pharmacology* 81: 462–470.

Eisenkraft, A., and A. Falk. 2016. *The possible role of intravenous lipid emulsion in the treatment of chemical warfare agent poisoning. Toxicology Reports* 3: 202–210.

Elbe, S., A. Roemer-Mahler, and C. Long. 2015. *Medical countermeasures for national security: A new government role in the pharmaceuticalization of society. Social Science & Medicine* 131: 263–271.

Ellison, D.H. 2008. *Handbook of chemical and biological warfare agents*. 2nd ed. Boca Raton, FL: Taylor and Francis Group.

El-Naggar, Ael-R, M.S. Abdalla, A.S. El-Sebaey, and S.M. Badawy. 2009. *Clinical findings and cholinesterase levels in children of organophosphates and carbamates poisoning. European Journal of Pediatrics* 168 (8): 951–956.

Erdmann, W.D., and H. Engelhard. 1964. *Pharmacological-toxicological investigations with the dichloride of the bis- [4-hydroxyiminomethyl-pyridinium (1) -methyl] -ether, a new esterase-reactivator. Drug Discovery* 14: 5–11.

Eyer, P., I. Hagedorn, and B. Ladstetter. 1988. *Study on the stability of the oxime HI 6 in aqueous solution. Archives of Toxicology* 62 (2-3): 224–226.

FDA. n.d. U.S. Food & Drug Administration, "Emergency use authorization" https://www.fda.gov/EmergencyPreparedness/Counterterrorism/MedicalCountermeasures/MCMLegalRegulatoryandPolicyFramework/ucm182568.htm#current. Accessed 27 Dec 2018.

Ghanem, E., and F.M. Raushe. 2005. *Detoxification of organophosphate nerve agents by bacterial phosphotriesterase. Toxicology and Applied Pharmacology* 207: S459–S470.

Gorecki L, Soukup O, Kucera T, Malinak D, Jun D, Kuca K, Musilek K, Korabecny J. (2018): *Oxime K203: a drug candidate for the treatment of tabun intoxication,* Arch Toxicol. Dec 18.

Guard, B.C., and S.J. Wiltshire. 1996. *The effect of glycopyrrolate on postoperative pain and analgesic requirements following laparoscopic sterilization. Anaesthesia* 51 (12): 1173–1175.

Hague, J., and J. Derr. 2004. *Military implications of atropine hypersensitivity. Military Medicine* 169 (5): 389–391.

Harris, L., B. Talbot, W. Lennox, D. Anderson, and R. Solana. 1991. *Physostigmine (alone and together with adjunct) pretreatment against soman, sarin, tabun and VX intoxication. Drug and Chemical Toxicology* 14 (3): 265–281.

Higgins, G.M., J.F. Muniz, and L.A. McCauley. 2001. *Monitoring acetylcholinesterase levels in migrant agricultural workers and their children using a portable test kit. Journal of Agricultural Safety and Health* 7 (1): 35–49.

Hirbec, H., M. Gaviria, and Vignon J. Gacyclidine. 2001. *A new neuroprotective agent acting at the N-methyl-D-aspartate receptor. CNS Drug Reviews* 7 (2): 172–198.

Hoegberg, L.C.G., T.C. Bania, V. Lavergne, et al. 2016. *Systematic review of the effect of intravenous lipid emulsion therapy for local anesthetic toxicity. Clinical Toxicology* 54: 167–193.

Institute of Medicine. 2010. Medical countermeasures dispensing: Emergency use authorization and the postal model, workshop summary. Institute of Medicine (US). Forum on Medical and Public Health Preparedness for Catastrophic Events. Washington, DC. National Academies Press. 2010. Available at www.ncbi.nlm.nih.gov/books/NBK53126/pdf/TOC.pdf. Accessed 27 Dec 2018.

Jamshidi, F., A. Yazdanbakhsh, M. Jamalian, et al. 2018. *Therapeutic effect of adding magnesium sulfate in treatment of organophosphorus poisoning. Open Access Macedonia Journal of Medical Science* 6 (11): 2051–2056.

Jortani, S., J. Snyder, and R. Valdes. 2000. *The role of the clinical laboratory in managing chemical or biological terrorism. Clinical Chemistry* 12: 11.

Josse, D., J. Wartelle, and C. Cruz. 2015. *Showering effectiveness for human hair decontamination of the nerve agent VX. Chemico-Biological Interactions* 232: 94–100.

Kassa, J., D. Jun, K. Kuca, and J. Bajgar. 2007. *Comparison of reactivating and therapeutic efficacy of two salts of the oxime HI-6 against tabun, soman and cyclosarin in rats. Basic & Clinical Pharmacology & Toxicology* 101 (5): 328–332.

Katalan, S., S. Lazar, R. Brandeis, I. Rabinovitz, I. Egoz, E. Grauer, E. Bloch-Shilderman, and L. Raveh. 2013. *Magnesium sulfate treatment against sarin poisoning: dissociation between overt convulsions and recorded cortical seizure activity. Archives of Toxicology* 87 (2): 347–360.

Knechtges P. (2008): *An evaluation of blood cholinesterase testing: methods for military health surveillance*, USACEHR Technical Report 0801. May, 2008. United States Army Center for Environmental Health Research. Fort Detrick.

Kozer, E., A. Mordel, S. Bar Haim, M. Bulkowstein, M. Berkovitch, and Bentur Y. Pediatric. 2005. *Poisoning from trimedoxime (TMB4) and atropine automatic injectors. The Journal of Pediatrics* 146: 41–44.

Lallement, G., D. Baubichon, D. Clarençon, M. Galonnier, M. Peoc'h, and P. Carpentier. 1999a. *Review of the value of gacyclidine (GK-11) as adjuvant medication to conventional treatments of organophosphate poisoning: primate experiments mimicking various scenarios of military or terrorist attack by soman. Neurotoxicology* 20 (4): 675–684.

Lallement, G., D. Clarençon, M. Galonnier, D. Baubichon, M.F. Burckhart, and M. Peoc'h. 1999b. *Acute soman poisoning in primates neither pretreated nor receiving immediate therapy: Value of gacyclidine (GK-11) in delayed medical support. Archives of Toxicology* 73 (2): 115–122.

Leal, J.K., M.J. Adjobo-Hermans, R. Brock, and G.J. Bosman. 2017. *Acetylcholinesterase provides new insights into red blood cell ageing in vivo and in vitro. Blood Transfusion* 15 (3): 232.

Lennox, W.J., L.W. Harris, D.R. Anderson, R.P. Solana, M.L. Murrow, and J.V. Wade. 1992. *Successful pretreatment/therapy of soman, sarin and VX intoxication. Drug and Chemical Toxicology* 15 (4): 271–283.

Lessenger, J.E., and B.E. Reese. 1999. *Rational use of cholinesterase activity testing in pesticide poisoning. The Journal of the American Board of Family Practice* 12 (4): 307–314.

Lifshitz, M., M. Rotenberg, S. Sofer, T. Tamiri, E. Shahak, and S. Almog. 1994. *Carbamate poisoning and oxime treatment in children: A clinical and laboratory study. Pediatrics* 93 (4): 652–655.

Lundy, P.M., A.S. Hansen, B.T. Hand, and C.A. Boulet. 1992. *Comparison of several oximes against poisoning by soman, tabun and GF. Toxicology* 72: 99–105.

Lundy, P., M. Hamilton, T. Sawyer, and J. Mikler. 2011. *Comparative protective effects of HI-6 and MMB-4 against organophosphorous nerve agent poisoning. Toxicology* 285 (3): 90–96.

Madsen, JM. (n.d.): USAMRICD special publication 98-01, pyridostigmine. United States Army Medical Research Institute of Chemical Defense. https://www.hsdl.org/?view&did=1081. Accessed 23 Jan 2018.

McDonough JH. 2002. *Midazolam: An improved anticonvulsant treatment for nerve agent-induced seizures,* Army Medical Research Inst of Chemical Defense. Aberdeen Proving Ground, MD. 2002.

McDonough, J.H., and T.M. Shin. 1993. *Pharmacological modulation of soman-induced seizures. Neuroscience and Biobehavioral Reviews* 17 (2): 203–215.

McDonough, J., J. McMonagle, T. Copeland, L.D. Zoeffel, and T.M. Shih. 1999. *Comparative evaluation of benzodiazepines for control of soman-induced seizures. Archives of Toxicology* 73 (8-9): 473–478.

McDonough, J., L.D. Zoeffel, J. McMonagle, T. Copeland, C.D. Smith, and T.M. Shih. 2000. *Anticonvulsant treatment of nerve agent seizures: Anticholinergics versus diazepam in soman-intoxicated guinea pigs. Epilepsy Research* 38: 1–14.

McEvoy, G.K., ed. 2012. *Drug information 2012*. Vol. 1299-1303, 2903–2904. Bethesda: American Society of Health-System Pharmacists.

Meaney, C.J., H. Sareh, B.D. Hayes, and J.P. Gonzales. 2013. *Intravenous lipid emulsion in the management of amlodipine overdose. Hospital Pharmacy* 48: 848–854.

Medical Countermeasures Database. 2017. HI-6, last updated 10 Nov 2017.

Menes K, Tintinalli J, Plaster L. 2017. *How one Las Vegas ED saved hundreds of lives after the worst mass shooting in U.S. history*. 3 Nov 2017. http://epmonthly.com/article/not-heroes-wear-capes-one-las-vegas-ed-saved-hundreds-lives-worst-mass-shooting-u-s-history/. Accessed 12 Jan 2018.

Meridian Medical Technologies. 2008. Product label: Pralidoxime Chloride injection [Meridian Medical Technologies, Inc.] Last revised: June 2008. https://dailymed.nlm.nih.gov/dailymed/drugInfo.cfm?setid=a16ac225-bb32-4b27-bb66-9e152cf065eb.

Miller V. 2004. *The health effects of project shad chemical agent: VX nerve agent*. The National Academies. Silver Spring. Contract No. IOM-2794-04-00.

Miller, M., M. Marty, A. Arcus, J. Brown, D. Morry, and M. Sandy. 2016. *Differences between children and adults: Implications for risk Assessment at California EPA*. *International Journal of Toxicology* 21: 403–418.

Mion, G., J.P. Tourtier, F. Petitjeans, F. Dorandeu, G. Lallement, et al. 2003. *Neuroprotective and antiepileptic activities of ketamine in nerve agent poisoning*. *Anesthesiology* 98 (6): 1517.

Mir, S.A., and R. Rasool. 2014. *Reversal of cardiovascular toxicity in severe organophosphate poisoning with 20% Intralipid emulsion therapy: Case report and review of literature*. *Asia Pacific Journal of Medical Toxicology* 3: 169–172.

Moeller B, Espelien B, Weber W, Kuehl P, Doyle-Eisele M, Garner CE et al. (2018): *The pharmacokinetics of ketamine following intramuscular injection to F344 rats,* Drug Testing and Analysis. 2018 Jan 1.

Moshiri, M., E. Darchini-Maragheh, and M. Balali-Mood. 2012. *Advances in toxicology and medical treatment of chemical warfare nerve agents*. *DARU Journal of Pharmaceutical Sciences* 20 (1): 81. https://doi.org/10.1186/2008-2231-20-81.

Ohbu, S., et al. 1997. *Sarin poisoning on Tokyo subway*. *Southern Medical Journal* 90 (6): 587–593.

Okayama, H., T. Aikawa, M. Okayama, H. Sasaki, S. Mue, and T. Takishima. 1987. *Bronchodilating effect of intravenous magnesium sulfate in bronchial asthma*. *JAMA* 257: 1076–1078.

Okumura, S., T. Okumura, S. Ishimatsu, K. Miura, H. Maekawa, and T. Naito. 2005. *Clinical review: Tokyo - protecting the health care worker during a chemical mass casualty event: An important issue of continuing relevance*. *Critical Care* 9 (4): 397–400.

OPCW. 2001. The sarin gas Attack in Japan and the related forensic investigation. OPCW News. 1 June 2001. https://www.opcw.org/media-centre/news/2001/06/sarin-gas-attack-japan-and-related-forensic-investigation. Accessed 13 Dec 2018.

———. 2018. Issues report on technical assistance requested by the United Kingdom. 12 Apr 2018. https://www.opcw.org/media-centre/news/2018/04/opcw-issues-report-technical-assistance-requested-united-kingdom. Accessed 1 Feb 2019.

Ozcan, M.S., and G. Weinberg. 2014. *Intravenous lipid emulsion for the treatment of drug toxicity*. *Journal of Intensive Care Medicine* 29: 59–70.

Pajoumand, A., S. Shadnia, A. Rezaie, M. Abdi, and M. Abdollahi. 2004. *Benefits of magnesium sulfate in the management of acute human poisoning by organophosphorus insecticides*. *Human & Experimental Toxicology* 23 (12): 565–569.

Peronne, J., F. Henretig, M. Sims, M. Beers, and M.A. Grippi. 2003. *A role for ipratropium in chemical terrorism preparedness*. *Academic Emergency Medicine* 10: 290.

Petroianu, G.A., M.Y. Hasan, S.M. Nurulain, N. Nagelkerke, J. Kassa, and K. Kuča. 2007a. *New K-oximes (K-27 and K-48) in comparison with obidoxime (LuH-6), HI-6, trimedoxime (TMB-4), and pralidoxime (2-PAM): Survival in tats exposed IP to the organophosphate paraoxon*. *Toxicology Mechanisms and Methods* 17 (7): 401–408.

Petroianu, G.A., S.M. Nurulain, N. Nagelkerke, M. Shafiullah, J. Kassa, and K. Kuca. 2007b. *Five oximes (K-27, K-48, obidoxime, HI-6 and trimedoxime) in comparison with pralidoxime: Survival in rats exposed to methyl-paraoxon*. *Journal of Applied Toxicology* 27 (5): 453–457.

PHE (2018): Stockpile products. Public Health Emergency. Last updated 17 Dec 2018. https://www.phe.gov/about/sns/Pages/products.aspx. Accessed 21 Jan 2019.

Prall, Y.G., K.K. Gambhir, and F.R. Ampy. 1998. *Acetylcholinesterase: an enzymatic marker of human red blood cell aging*. *Life Sciences* 63: 177–184.

Proakis, A.G., and G.B. Harris. 1978. *Comparative penetration of glycopyrrolate and atropine across the blood-brain and placental barriers in anaesthetised dogs*. *Anaesthesiology* 48: 339–344.

Raipal, S., G. Mittal, R. Sachdeva, M. Chhillar, R. Ali, S.S. Agrawal, R. Kashyap, and A. Bhatnagar. 2009. *Development of atropine sulphate nasal drops and its pharmacokinetic safety evaluation in healthy human volunteers*. *Environmental Toxicology and Pharmacology* 27: 206–211.

Raipal, S., R. Ali, A. Bhatnagar, S.K. Bhandari, and G. Mittal. 2010, 28. *Clinical and bioavailability studies of sublingually administered atropine sulphate. The American Journal of Emergency Medicine*: 143–150.

RamaRao, G., and B.K. Bhattacharya. 2012. *Multiple signal transduction pathways alterations during nerve agent toxicity. Toxicology Letters* 208 (1): 16–22.

RamaRao, G., P. Afley, J. Acharya, and B. Bhattacharya. 2014. *Efficacy of antidotes (midazolam, atropine and HI-6) on nerve agent induced molecular and neuropathological changes. BMC Neuroscience* 15 (1): 47. https://doi.org/10.1186/1471-2202-15-47.

Raveh, L., E. Grauer, J. Grunwald, E. Cohen, and Y. Ashani. 1997. *The stoichiometry of protection against soman and VX toxicity in monkeys pretreated with human butyrylcholinesterase. Toxicology and Applied Pharmacology* 145 (1): 43–53.

Reddy, S.D., and D.S. Reddy. 2015. *Midazolam as an anticonvulsant antidote for organophosphate intoxication--A pharmacotherapeutic appraisal. Epilepsia* 56 (6): 813–821.

Rice, H., T.M. Mann, S.J. Armstrong, M.E. Price, A.C. Green, and J.E.H. Tattersall. 2016. *The potential role of bioscavenger in the medical management of nerve-agent poisoned casualties. Chemico-Biological Interactions* 259: 175–181.

Robenshtok, Eyal, Shay Luria, Zeev Tashma, and Ariel Hourvitz. 2002. *Adverse reaction to atropine and the treatment of organophosphate intoxication. The Israel Medical Association Journal* 4: 535–539.

Rosman, Y., A. Eisenkraft, N. Milk, A. Shiyovich, N. Ophir, S. Shrot, et al. 2014. *Lessons learned from the Syrian sarin attack: Evaluation of a clinical syndrome through social media. Annals of Internal Medicine* 160: 644–648.

Rowe, B.H., M.L. Edmonds, C.H. Spooner, and C.A. Camargo. 2001. *Evidence-based treatments for acute asthma. Respiratory Care* 46 (12): 1380–1390.

Ryniak, S., P. Harbut, W. Gozdzik, J. Sokolowski, P. Paciorek, and J. Halas. 2011. *Whole blood transfusion in the treatment of an acute organophosphorus poisoning – A case report. Medical Science Monitor* 17 (9): CS109–CS111.

Sacan, O., P. White, B. Tufanogullari, and Klein K. Sugammadex. 2007. *Reversal of rocuronium-induced neuromuscular blockade: A comparison with neostigmine–glycopyrrolate and edrophonium–atropine. Anesthesia & Analgesia* 104 (3): 569–574.

Saxena, A., N.B. Hastings, W. Sun, P.A. Dabisch, S.W. Hulet, E.M. Jakubowski, R.J. Mioduszewski, and B.P. Doctor. 2015. *Prophylaxis with human serum butyrylcholinesterase protects Göttingen minipigs exposed to a lethal high-dose of sarin vapor. Chemico-Biological Interactions* 238: 161–169.

Schlager, J.W., T.W. Dolzine, J.R. Stewart, G.L. Wannarka, and M.L. Shih. 1991. *Operational evaluation of three commercial configurations of atropine/HI-6 wet/dry autoinjectors. Pharmaceutical Research* 8 (9): 1191–1194.

Schwartz MD, Sutter ME, Eisnor D, Kirk MA.. 2018. *Contingency medical countermeasures for mass nerve-agent exposure: Use of pharmaceutical alternatives to community stockpiled antidotes,* Disaster Med Public Health Prep. 2018 Oct 15:1–8.

Solana, R., C. Gennings, D. Anderson, W. Carter, R. Carchman, and L. Harris. 1989. *Comparing the response surfaces of two cholinolytics when used in combination with physostigmine as a pretreatment against organophosphate challenge. Drug and Chemical Toxicology* 12 (3-4): 197–219.

Soukup, O., J. Krůšek, M. Kaniaková, U.K. Kumar, M. Oz, D. Jun, J. Fusek, K. Kuča, and G. Tobin. 2011. *Oxime reactivators and their in vivo and in vitro effects on nicotinic receptors. Physiological Research* 60 (4): 679–686.

Stolbach, A., V. Bebarta, M. Beuhler, S. Carstairs, L. Nelson, M. Wahl, P. Wax, and C. McKay. 2018. *ACMT position statement: Alternative or contingency countermeasures for acetylcholinesterase inhibiting agents. Journal of Medical Toxicology* 14 (3): 261–263.

Sungurtekin, H., E. Gürses, and C. Balci. 2008. *Evaluation of several clinical scoring tools in organophosphate poisoned patients. Clinical Toxicology* 44: 121–126. https://doi.org/10.1080/15563650500514350.

Tang, S.Y.H., and J.T.S. Chan. 2002. *A review article on nerve agents. Hong Kong Journal Emergency Medicine* 9: 83–89.

Thiermann, H., S. Seidl, and P. Eyer. 1996. *HI 6 dimethanesulfonate has better dissolution properties than HI 6 dichloride for application in dry/wet autoinjectors. International Journal of Pharmaceutics* 137 (2): 167–176.

Thiermann, H., U. Mast, R. Klimmek, P. Eyer, A. Hibler, R. Pfab, N. Felgenhauer, and T. Zilker. 1997. *Cholinesterase status, pharmacokinetics and laboratory findings during obidoxime therapy in organophosphate poisoned patients. Human & Experimental Toxicology* 16 (8): 473–480.

Towne, A.R., J.M. Pellock, D. Ko, and R.J. DeLorenzo. 1994. *Determinants of mortality in status epilepticus. Epilepsia* 35 (1): 27–34.

Tracey, J.A., and H. Gallaghar. 1990. *Use of glycopyrrolate and atropine in acute organophosphorus poisoning. Human & Experimental Toxicology* 9 (2): 99–100.

UK Government. 2018. Novichok nerve agent use in Salisbury: UK government response, March to April 2018. Last updated 18 Apr 2018. https://www.gov.uk/government/news/novichok-nerve-agent-use-in-salisbury-uk-government-response. Accessed 1 Feb 2019.

United States Senate. 1995. *Global proliferation of weapons of mass destruction: hearings before the Permanent Subcommittee on Investigations of the Committee on Governmental Affairs, United States Senate, One Hundred Fourth Congress, first session*, 24 Chemical Weapons Disarmament in Russia: Problems and Prospects. https://archive.org/stream/globalproliferat01unit/globalproliferat01unit_djvu.txt.

US Army. 2005. Potential military chemical/biological agents and compounds. FeM 3-11.9. US Army Training and Doctrine Command, Fort Monroe, VA. January 2005. Ch. 2 Chemical Warfare agents and their properties.

Uysal, M., and S. Karaman. 2018. *In vivo effects of intravenous lipid emulsion on lung tissue in an experimental model of acute malathion intoxication. Toxicology and Industrial Health* 34 (2): 110–118.

Van Der Schans, M.J., A. Fidder, D. Van Oeveren, A.G. Hulst, and D. Noort. 2008. *Verification of exposure to cholinesterase inhibitors: generic detection of OPCW Schedule 1 nerve agent adducts to human butyrylcholinesterase. Journal of Analytical Toxicology* 32 (1): 125–130.

Vaserhelyi, G., and L. Foldi. 2007. *History of Russia's chemical weapons. AARMS* 6 (1): 135–146.

Vijayakumar, H.N., K. Ramya, D.R. Duggappa, K.V. Gowda, K. Sudheesh, S.S. Nethra, and R.R. Rao. 2016. *Effect of melatonin on duration of delirium in organophosphorus compound poisoning patients: A double-blind randomised placebo controlled trial. Indian Journal of Anaesthesia* 60 (11): 814.

Vijayakumar, H.N., S. Kannan, C. Tejasvi, D.R. Duggappa, K.M. Veeranna Gowda, and S.S. Nethra. 2017. *Study of effect of magnesium sulphate in management of acute organophosphorous pesticide poisoning. Anesthesia, Essays and Researches* 11 (1): 192–196.

Von Bredow, J., K. Corcoran, G. Maitland, A. Kaminskis, N. Adams, and J. Wade. 1991. *Efficacy evaluation of physostigmine and anticholinergic adjuncts as a pretreatment for nerve agent intoxication. Fundamental and Applied Toxicology* 17 (4): 782–789.

Whitmore, C., A.R. Cook, T. Mann, M.E. Price, E. Emery, N. Roughley, D. Flint, S. Stubbs, S.J. Armstrong, H. Rice, and J.E.H. Tattersall. 2018. *The efficacy of HI-6 DMS in a sustained infusion against percutaneous VX poisoning in the guinea-pig. Toxicology Letters* 293: 207–215.

Worek, F., R. Widmann, O. Knopff, and L. Szinicz. 1998. *Reactivating potency of obidoxime, pralidoxime, HI 6 and HLö 7 in human erythrocyte acetylcholinesterase inhibited by highly toxic organophosphorus compounds. Archives of Toxicology* 72 (4): 237–243.

Xue, S.Z., X.J. Ding, and Y. Ding. 1985. *Clinical observation and comparison of the effectiveness of several oxime cholinesterase reactivators. Scandinavian Journal of Work, Environment & Health* 11 (suppl 4): 46–48.

Zivkovic, A.R., J. Bender, T. Brenner, S. Hofer, and K. Schmidt. 2016. *Reduced butyrylcholinesterase activity is an early indicator of trauma-induced acute systemic inflammatory response. Journal of Inflammation Research* 9: 221–230.

Diagnosing the Cause of Disease: Interactive Teaching Approaches

Alastair Hay

1 Introduction

Diseases are an ever present threat for all of us and it is not just humans that are at risk, but the animal and plant kingdom too. Although most diseases are host-specific there are many that cross the species barrier and an illness in cattle, for example, may become a direct human problem too, and not just because of a lack of beef! With plant diseases we are looking at impacts on food supply or maybe devastating changes to the flora around us.

Dealing with natural disease outbreaks is hard enough, but when there is the possibility that the cause was deliberate, complications multiply significantly. No longer is it just national medical or veterinary services involved, or your local ministry of agriculture, but the whole machinery of government becomes party to the issue, and other governments too. Diseases, deliberately initiated, rapidly become an international concern and not just because of the complications involved in containing outbreaks, but because of the urgent need to identify the perpetrator (s). And establishing that the outbreak was malicious requires a high degree of proof as accusations might even lead to hostilities within, and between countries.

2 Interactive Teaching Resource

The accompanying slides (displayed later in the chapter in Annex 1) were part of an European Union project (IFS/2016/376-230) to prepare an interactive teaching resource on diagnosing the cause of a disease, but focussing on how to differentiate a natural outbreak from one caused deliberately. The resource was prepared by two

A. Hay (✉)
LICAAM Institute, School of Medicine, University of Leeds, Leeds LS2 9JT, UK
e-mail: A.W.M.Hay@leeds.ac.uk

M. Martellini, R. Trapp (eds.), *21st Century Prometheus*,
https://doi.org/10.1007/978-3-030-28285-1_14

of us, Tatyana Novossiolova and me (Alastair Hay {AH}) and has been used very successfully with students studying for a masters degree in epidemiology at the University of Leeds (UK). On the first occasion two of us ran the exercise over a 2 hour period and on the second it was AH on his own. There is sufficient content for the exercise to easily stretch into 3 or more hours; it all depends on the emphasis of those doing the teaching, where they choose to focus attention, and how much they engage with the participants during discussions.

This chapter is written in such a way that anyone wishing to carry out the same exercise can simply copy the material. There is a set of some 17 slides, a background sheet which provides context and a handout which can be given out at the end of the session. Also provided are a set of references which provide background details on the biological organism chosen for this exercise, *Coxiella burnetii*, and descriptions of disease outbreaks in animals and humans where *C.burnetti*, a zoonosis, was involved. People develop what is known as Q fever from *C. burnetii.*

2.1 The Approach

To encourage active discussion participants are divided into groups. Ideally, groups should be no larger than six or seven participants as this is a number in which most students feel comfortable speaking up and arguing a case.

Learning goals and specific objectives are explained in the second and third slides. The goals are broad and the expectation is that after the course participants will understand that diagnosing the cause of a disease is a complex process. More specific objectives note that participants will be able to describe some of the complexity involved and be able to identify some of the international organisations or legal instruments that are involved following a major disease outbreak.

Slides 4 and 5 set out how the exercise will be run and explain that it involves discussion of a case and that it will be an evolving scenario.

Before handing out the background sheet it is helpful to run through the questions which are in slide 6 which raises some basic questions about disease and procedures which may have to be considered when dealing with a disease outbreak. It is important to keep reminding participants to always be thinking about these issues.

Slide 6 can be a useful starting point for discussions and groups can be left to consider the issues for 10 or 15 min. Have them choose a spokesperson to report back to the whole class after they have had their group discussion. Encourage use of laptops or mobile phones as participants would resort to these under normal circumstances for information retrieval. Epidemiology students invariably come from a range of backgrounds; some may be clinicians, others have a biology or chemistry first degree, whereas a few may not have this science training but be statisticians or mathematicians. So it is important that all have an understanding about what a disease is and how it might occur.

As a teacher I move around groups and interact with students, and this often involves me raising questions and providing answers to clarify issues which arise in the group discussions. I find that this interaction with each small group helps to maintain the informality of the process and breaks down any barrier between me and participants and, as a result, encourages the more reticent in the group to voice an opinion. When you judge the time right have each group report back to all the participants and involve all in the discussion to clarify any points.

Now hand out the background note about Citytown which describes its geography, population, economy and health service. Give participants a few minutes to read the sheet and then introduce slide 7 which recounts the first cases of an unknown illness involving 3 people initially, with 11 more falling ill over the following 3 days and some developing an atypical pneumonia. Encourage the groups to discuss what the outbreak suggests and to clarify what atypical pneumonia is (defined as an infection of the lung by a bacteria resulting in less severe symptoms than typical pneumonia and not usually requiring hospitalisation; patients with atypical pneumonia are not usually very ill or as fatigued as those with typical pneumonia. Hence, some refer to atypical pneumonia as 'walking pneumonia').

Remind the groups that a festival is to be held within weeks and that there are high hopes that it will attract tourists to the town whose spending will be a significant boost to the local economy. Raise the question as to whether, or not, the festival should go ahead. Do they consider it safe for tourists to visit? Do they have any idea how the disease is spread? Could it be contagious? What will it mean for the town if the festival is cancelled? Follow the procedure you used after discussion about disease in slide 6 and when all groups have had their say introduce slide 8.

The situation has worsened. There are now 40 cases of atypical pneumonia with most in one location. Blood samples have been sent away to a reference laboratory for testing. However, the source of the infection is still not known. The same questions still apply. Have the groups discuss the developments, but with a deteriorating picture answers are even more urgent. My approach is to ask some general questions but not to give too much away. I want participants to work out approaches themselves with little guidance from me, unless really needed. So, you want groups to recognise that patterns need to be established. So patients will need to be questioned about their habits, work, contacts, where they have been, etc. A uniform questionnaire which can be administered needs to be used.

And what about medical personnel providing treatment? How should they continue to protect themselves? The scenario says nothing about any being affected, but you may wish to complicate the picture by having several doctors amongst the ill. Judge whether or not to do this on the basis of the group discussions and questions you are asked when you interact with the small groups. After you have circulated around all get a sense of where participants are at and decide when groups report back. Again, this is only a personal preference, but I tend to ask a different person from each small group to report back after each discussion session. This approach keeps all involved and prevents the more confident dominating conversations.

Groups should now have a clearer idea about what to do and so now introduce slide 9. This slide is simply for information. It describes how diseases are categorised in the UK according to hazard. The hazard groupings are based on whether the disease causes illness in humans, how it is spread and whether or not there are effective treatments. Hazard group 1 is the least dangerous with group 4 being the most concerning as agents classified in it cause serious illness, spread easily in the community and have no effective treatments.

Move on now to slide 10. You are back with the evolving scenario. But now cases have more than doubled since the last revue. There are now 85 cases of atypical pneumonia. The media is reporting a mysterious illness and the festival is but 2 weeks away. Is the situation out of control? Who else needs to know about the outbreak?

This is the stage at which it helps to show slides 11 and 12. The slides refer to the agencies that have an interest in a natural disease outbreak and resources that can be referred to for those setting up laboratory procedures or using existing facilities. The resources relate largely to biosafety and biosafety procedures whereas the agencies referred to are concerned with human disease and the risk of trans- boundary spread. But if there are animals or plants infected, additional agencies have an interest, and if there is a risk of disease spread from animal to human more need to be involved and additional warnings will need to be issued.

If your audience is largely students, as mine were, it is sufficient to simply introduce the need for interagency involvement and point out one or two agencies and perhaps refer to some of the resources. It is important for the participants to know that agencies outside the country may need to be informed. Where the case is being considered by a more experienced audience then the roles of agencies should be an integral part of the discussion in groups.

Having introduced the agencies, refer back to slide 10, and stage III of the scenario. What should happen now? No laboratory results are yet available. But with the media reporting on the outbreak public anxiety is increasing. Decisions are needed on the festival too. Much is at stake. If the festival goes ahead and many tourists descend on the town might they not be infected as well? The risk of this is high as the cause of the outbreak is still unknown. Many tourists becoming ill would be a public relations disaster. Is it better to forgo the much anticipated boost to the economy and delay, or are community expectations too high and cancellation just not feasible?

At this stage of the discussions there should be firm views about the issues. Probe these. Also ask for ideas about the outbreak and its cause. Some may have asked for the location of most of the cases and where these are in relation to the animal farm. Some may ask about any animal illnesses. Encourage all enquiries. There will be anxious calls to the reference laboratory to hurry it up. Judge the nature of the discussions and choose a time to introduce stage IV of the scenario (slide 13).

The cause of the disease outbreak is now known. The laboratory has typed the organism. The illnesses have been caused by *Coxiella burnetii* and the animal farm is the likely source. Have participants consider how they think the disease was spread and what control measures are required to stop the spread. With access to laptops etc the groups should be able to locate articles on *C. burnetti* and discover that the organism is robust, that it is spread by infected animals through their milk, urine and faeces and these fluids will contaminate anything on the ground particularly soil, hay and straw. Dust from these sources is usually how the organism is spread with person-to-person transmission rare, but documented. Rigorous disinfection is required. Culling of animals frequently is deemed the only way to contain the spread with carcases either burnt or buried in lime.

Having come up with a procedure to control spread of the organism, encourage the groups to revisit some of the original questions. Should the festival go ahead, be re-instated, or delayed? How will they know that the control measures are working? How long will they have to wait for the evidence? Will there be any opposition to culling? This is almost certain as slaughtering animals, without the carcases useful as food, directly affects people's livelihood. However, as *C. burnetii* infections in animals causes no symptoms, but is present in the uterus and mammary glands as a site of chronic infection, and associated with abortions in sheep and goats, and infertility in cattle, the long term benefit of a cull is clear.

Now move on to stage V. Groups can now compare what they suggested as the way in which the organism was spread with what is on the screen. Along the way you may wish to discuss what symptoms people experience when infected and how acute infections may become chronic for about 5% of those infected.

In these discussions it should become clear that, although not a fatal disease, *C. burnetii* is a highly infectious organism, with an infectious dose requiring few bacteria and hence the interest in the organism, which causes Q fever, as a candidate for an incapacitating biological weapon by the United States. The US weapons programme was cancelled decades ago.

As a radical group is rumoured to be interested in getting blood samples taken from infected individuals you should now canvas views as to whether groups consider the outbreak a natural occurrence, or a deliberate outbreak. In their searches the groups will have discovered that *C. burnetii* is endemic in many parts of the world in a wide range of wild animals (including birds) and that ticks (and other arthropods) may well be an important route for transmission in these populations, and from them to domestic animals. Tick control and good hygiene can reduce the incidence of infection in domestic animals.

Why do the groups come to the conclusion they have reached? Is it because the source can be traced to the animal farm that it seems more probable that the outbreak is a natural occurrence? Might it have been introduced on the farm deliberately to disguise the fact it was deliberate? Possibly. But this would introduce a good deal of uncertainty and a long lead time before people became ill. Given the fact that there is now only one radical group active and that they seem to know little about how the disease is spread in that they are after blood samples, the evidence (and presumably other intelligence held by authorities) points to a natural outbreak.

But have the groups consider what evidence they would require to suggest that the outbreak was indeed deliberate and if others might have an interest if it was not a natural occurrence. This is a good point for introduction of slide 15 and a discussion of the treaties and procedures in place for deliberate outbreaks. Here again it is important to consider the expertise of the audience. With students it is probably just sufficient to point out some of measures. But a more experienced audience would benefit from discussing some of mechanisms in more detail and perhaps groups could be assigned one or two treaties or measures to look up and report back to broaden the conversation.

After groups have debated the issue introduce slides 16 and 17. These two slides contain information assembled from some of the references at the end and it is worth going through the lists to make sure participants are clear about what is being referred to. It will become clear that whilst there will be pointers suggesting something may be deliberate, definitive evidence is hard to come by. If, of course, there is clear evidence indicating deliberate release, or spread of the organism causing disease (by a person (s)), or aircraft, say, or DNA (or RNA in the case of viruses) signatures suggesting an engineered organism, or strain only used in laboratories, this would indicate deliberate intent. The criteria and scoring systems used to differentiate natural from deliberate outbreaks of disease are improving, but so far none exist which are foolproof. And, inevitably, a great deal of detective work linking events would be required to make a convincing case that an outbreak was deliberate; some, even, may be reluctant to hand over evidence which was incriminating.

Sadly, there are enough real life situations where countries have declined to provide vital evidence and the cause of a disease outbreak remains unknown. In other circumstances the evidence may just not exist and the conclusion will again be 'Source Unknown'.

The references cited under unusual disease outbreaks provide additional material which can be used to show that painstaking research by scientists may often refute the original claim about the type of outbreak made by an accuser.

2.2 *Multiple Uses of Chemicals*

A different teaching approach is one which uses material contained in the online resource Multiple Uses of Chemicals https://multiple.kcvs.ca/. Developed initially by a working group under the auspices of the International Union of Pure and Applied Chemistry (IUPAC) and the Organisation for the Prohibition of Chemical Weapons (OPCW) this resource has since been modified by one of the original authors (Peter Mahaffy) and his students in the Department of Chemistry at Kings Center for Visualisation in Science at Kings University in the Canadian city of Edmonton. The modified resource is interactive, with numerous case studies on subjects ranging from illegal drugs to chemical warfare agents. The cases describe a range of materials and how they are made (with no actual recipes!) but, more importantly, the consequences of their use. With this online resource, users are encouraged to answer questions at various stages with answers available at a click. In some instances there are multiple choice questions, the aim overall being to encourage readers to think about issues and consequences. The aim is to promote discussions about the responsibility of scientists, codes-of-conduct, control of chemicals and very specifically, the 1997 Chemical Weapons Convention.

The 3 slides with the starter 'What is legal? (displayed in Annex 2) are about Crystal Meth and abstract ideas from one of the case studies in Multiple Uses of Chemicals. The case study is about how the pharmaceutical product pseudoephidrine can be converted into the street drug Crystal Meth. Pseudoephidrine, a stimulant, is present in many cough syrups to help relieve congestion, which it does by shrinking swollen nasal mucous membranes.

Teachers using the resource will have many ways in which they introduce the material to students and the following are suggestions on how to make the material even more interactive.

One, more traditional, approach is to divide participants (students) into a number of smaller groups of about 6 or 7 individuals and have half the groups consider under what circumstances they would consider making Crystal Meth and the others why it should not be made. Get the groups to imagine that they are chemistry students with some skills, although not extensive. It is also worth pointing out that the conversion from pseudoephidrine to the street drug is extensively referred to on the Internet.

At this stage I would ask groups not to do any searches on phones, or laptops, but just to consider the issues as they see them. For some, doing the conversion in the laboratory, will simply be about science and perhaps confirming what they have read on the internet. For others, the idea of making an illegal drug will be anathema. And there will be many other views. When the discussion begins to flag have each group report back to the whole room and continue this discussion until all views have been aired.

Now encourage the groups to use the Internet to find out more about Crystal Meth, and its health effects in particular, and to resume their conversations in groups. Allow sufficient time for this. Then invite groups, in turn, to report back. Have any changed their mind following what they have read? Why? Have any differentiated between simply making a chemical in the laboratory, and keeping it there under their control, from what happens if it has wider distribution? Views may again differ.

After this discussion elicit views from the audience about how they view chemists making illegal drugs. Point out that many drugs can be purchased on the Internet and that chemists are available who will make products to order. Can anything be done to control this? What should the training of young chemists entail? Should their access to chemicals be restricted? Should they have a code-of-conduct? How will you enforce this? Or should it simply be a case of chemists being aware, being careful, and restricting access to laboratories to those who only have a right to be there? This debate ought to raise many valuable take-home messages about good practice.

Another approach which I use on the same theme is illustrated in the third slide. This has one header moving from error to misconduct to criminal on the horizontal axis and a vertical scale moving from the non-intentional to intentional. I tend to not make too much of these headers at the beginning of the exercise, but return to them at the end. I also introduce each of the topics 'undeclared conflict of interest', 'curiosity', etc., one-at-a-time, and discuss each before introducing the next.

I have used these slides in large lecture theatres (with up to 450 participants) to have a good discussion and I use an approach to challenge the audience.

The first topic is about conflict of interests. The audience can be told that all presumably understand what this means. Check for nods of heads. Act as though you are taking them into your confidence.

Now remind them about the slide which showed that pseudoephidrine could be converted to Crystal Meth. Make the audience aware that this information is not new and is on the Internet. Ask them to imagine that they are young chemists with sufficient skills to try to make Crystal Meth. Remind them that science is about curiosity. Now ask them if any would be curious enough to see if what was reported on the Internet was true? They all have access to a laboratory. Would they want to see if they could make the drug?

In a large setting many are reluctant to speak up, but if you encourage people to speak to their neighbour about your question, the noise in the auditorium will rise significantly, and you will become aware of many conversations. Allow a minute or two for views to be exchanged and then ask again if any would see if they could make Crystal Meth? Having had a chance to exchange views, perhaps find that some are like-minded, people are now more likely to feel less anxious and more ready to speak up. When they do there may be some who say that seeing if pseudoephidrene can be converted to the street drug is just chemistry (which it is) and they would do it. Others may have an absolute prohibition on trying the exercise. Ask people why they have reached their particular view and use these exchanges to involve the whole audience.

Now ask them if, having decided to make the drug, they would tell anyone about it or keep the information to themselves? Would they be proud of their achievement? Boastful even? All the time I try to make the interaction fun. So think of interjections that will help to do this and when you have conversations with the audience keep these playful. Under no circumstances embarrass those who offer an opinion otherwise the whole conversation, and the interaction you are after, will end. Again, when views are proffered, find out why people decide on particular actions.

You can now remind the audience that science requires reproducible findings. Perhaps the yield, the first time Crystal Meth was made, was poor. Would the pretend young chemists want to repeat the experiment to try to get a better yield? Here you can point out that low yields are not what chemists are after. Play it up! If interaction is flagging (but hopefully not at this juncture) encourage discussions with the person sitting next to them. Again solicit views and how they were arrived at.

The next step is more of a major leap. I sometimes use the excuse of someone having a party and Crystal Meth helping it along. Earlier, I may have asked the name of someone who made the drug and told others that they had done so. Using their name I now ask them if they will now sell me some for my party. They will probably, and hopefully, refuse, citing the fact that the chemical is no longer in their control and they do not know how I will use it. I may attempt some blackmail and point out that their laboratory supervisor had no idea that they had used the lab for making Crystal Meth in the first place. If this doesn't work I will try some other ruse.

What I am after is the audience remaining resolute and refusing to release a drug which is harmful. I want them to recognise that there are legitimate chemistry questions that can be answered by making the drug and that as long as the drug stays in their possession they are in control. Selling any, leads to untold problems, which you can explore with the audience.

If the audience is resolute at this stage then the last issue 'Large scale manufacture' becomes moot. It may evoke laughter when shown. But if a few choose to sell some at the earlier stage you can explore this next stage as well.

In the wrap-up you can draw the audience's attention to the horizontal and vertical headers and what they represent. And at this stage it is worth just reminding all about what is legitimate for chemists to do and what crosses the line into a problematic zone. Perhaps encourage the audience to suggest other topics where chemists might focus their attention, but where there may also be a risk of something moving beyond their control.

This teaching approach was inspired by an article by Mylenna and Simonsen in The Lancet on ethics for scientists and clinicians; their use of headers led me to develop the topic-by-topic approach discussed above.

Annex 1: Diagnosing the Cause of Disease

Rolling Scenario: Disease Outbreak

Background Information

Citytown is a small municipal town in Roseland, a developing country, that until recently has been experiencing political and economic turmoil. Whilst the situation nationwide has become relatively stable, infrastructure in the country remains poor with maintenance and refurbishment dependent on external aid. Roseland is situated in a temperate climate zone.

Prior to the turmoil, the quality of healthcare was of a high standard with many endemic diseases brought under control following the establishment of a nationwide sanitary-epidemiological system. In the recent unrest, however, many facilities have been closed and the remaining ones, including the one in Citytown, suffer from inadequate technical and human resources. A national reference laboratory is operating with a reduced capacity in the capital, Capitalcity, situated some 300 km from Citytown.

Citytown is a predominantly urban area with a population of c. 30 000 inhabitants. The town has renowned architectural features and a centre with many historical monuments and sites of continuing interest to archaeologists which attract tourists from all over the world. The local economy is primarily tourism-based, and has been badly affected by the turmoil with a significant reduction in tourists. A large seasonal festival is approaching which is expected to help revive the economy and generate long-awaited revenue for the local population.

In order to tackle the problem of unemployment in neighbouring villages, a modern farm was set up on the outskirts of Citytown for breeding cattle, goats, and sheep. The farm is Citytown's main source of milk and dairy products.

The overall political situation in Roseland is generally stable. Most of the radical groups active during the turmoil have now been disarmed, with many of their members arrested. A small para-military faction is active, however, and involved in sporadic attacks on both law enforcement services and civilians.

Slide Deck: Differentiating the Cause of Disease

Diagnosing the Cause of a Disease: Distinguishing between a Natural and Deliberate Disease Outbreak

Alastair Hay
University of Leeds, UK

Learning Goals

To develop an understanding that:

- establishing the cause of a disease outbreak is a complex process.
- responding effectively to a disease outbreak (regardless of its cause) requires multi-sectorial coordination and communication.

Learning Objectives

At the end of the session participants are able to:

- Explain how to set about distinguishing between a natural and deliberate disease outbreak;
- List criteria to help interpret the cause of a disease outbreak;
- Cite some of the different international regulations and mechanisms for dealing with biological organisms which affect human, other animal and plant health.

Lesson Plan

- Introduction and explanation of the exercise
- Class divided into groups
- Discussion of a case, review of evidence
- Interpretation of evidence
- Q&A, Debrief

The Scenario

- Each group will be provided with Background information.
- The scenario features several stages.
- Stage 1 will be introduced and each group will assess the situation, any action required and by whom. Identify a spokesperson to report back to the whole class.
- Subsequent stages will be introduced, the same format followed, but with a different spokesperson after each stage. Following this there will be a broader discussion.

Disease Classification

- What is a disease outbreak?
- What could be the possible causes of a disease outbreak?
- During a disease outbreak are there any precautions that might/should be considered for those :

i) Examining patients
ii) Collecting biological samples
iii) Processing samples in the laboratory

Stage I

- On 14 June, three patients independently from one another, seek medical assistance at the local hospital in Citytown; all exhibit flu-like symptoms. Since their condition is serious, but stable, they are not hospitalised and medical treatment at home is prescribed. Over the following 3 days 11 more people with similar symptoms are registered. On 17 June, an elderly man from a local village is admitted to the hospital in a critical condition with what appears to be an atypical pneumonia. On the following day, two of the three patients who sought medical assistance on 14 June are taken into hospital with an atypical pneumonia.

Stage II

- By 24 June, the number of people with atypical pneumonia at the Citytown hospital has grown to 40. Test samples of hospitalised patients are sent to the national reference laboratory for diagnostics. As new cases continue to emerge, local doctors reveal a pattern of disease distribution: it appears that all registered cases in Citytown are clustered in a single neighbourhood, a middle-/upper-middle class estate. The source of the infection remains unknown.

Categorisation of Biological Agents - definitions of hazard groups for clinical laboratories

- **hazard group 1:** A biological agent unlikely to cause human disease.
- **hazard group 2:** A biological agent that can cause human disease and may be a hazard to employees; it is unlikely to spread to the community and there is usually effective prophylaxis or effective treatment available (Eg Salmonella).
- **hazard group 3:** A biological agent that can cause severe human disease and presents a serious hazard to employees; it may present a risk of spreading to the community, but there is usually effective prophylaxis or treatment available (Eg Plague).
- **hazard group 4:** A biological agent that causes severe human disease and is a serious hazard to employees; it is likely to spread to the community and there is usually no effective prophylaxis or treatment (E.g. Ebola virus categories).

http://www.hse.gov.uk/pubns/clinical-laboratories.pdf

Stage III

- By 28 June, another 45 new cases are registered. The media reports a 'mysterious outbreak' in the Citytown area that is occurring two weeks before the start of the festival.

Global Outbreak Alert Response Network (GOARN)

WHO International Health Regulations

FAO Biosecurity Toolkit

FAO Emergency Prevention System for Transboundary Animal and Plant Pests and Diseases (EMPRES)

Strategic Health Operations Centre (SHOC)

Naturally Occurring Disease Outbreak

OIE and FAO Global Framework for Progressive Control of Transboundary Animal Diseases (GF-TADs)

Public Health Emergency Operations Network (EOCNET)

OIE Terrestrial Animal Health Code

World Animal Health Information System (WAHIS)

FAO, OIE and WHO Global Early Warning System for Major Animal Diseases including Zoonosis (GLEWS)

OIE Aquatic Animal Health Code

WHO Laboratory Biosafety Manual

Biorisk Management: Laboratory Biosecurity Guidance

Responsible Life Science Research for Global Health Security: A Guidance Document

Accidental Disease Outbreak

CEN Workshop Agreement on Laboratory Biorisk Management 15793

CEN Workshop Agreement on Biosafety Professional Competence 16335

Measures for Addressing Naturally Occurring Disease Outbreaks

IFBA Professional Certification Programme in Biosafety and Biosecurity

ISO 15189:2012 Medical Laboratory: Requirements for Quality and Competence

UN Model Regulations on the Transport of Dangerous Goods

Stage IV

- The national reference laboratory confirms the presence of *Coxiella burnetii* in the test samples taken from hospitalised patients. *C. burnetii* is the causative agent of Q fever, a zoonotic disease that can cause acute or chronic disease in humans.
- Whilst considered endemic in Roseland and neighbouring countries, Q fever has not been reported in the area of Citytown for over fifty years.
- The affected neighbourhood is the closest to the farm found on the outskirts of the Citytown. It is also revealed that several of the patients in the hospital work on the farm.

Stage V

- Inspections on the farm discover sick goats and sheep. Safety lapses including breaches in waste management, inappropriate staff training in use of personal protective equipment (PPE) provided, and lack of information on potential health hazards are cited as risk factors facilitating the spread of the disease. Dry weather and wind enabled the dissemination of contaminated aerosol from the farm to the affected urban area.
- As the source of the disease is established, local police receive a report that members of the remaining radical faction may attempt to acquire blood samples taken from Q fever patients that are housed in the hospital in Citytown.

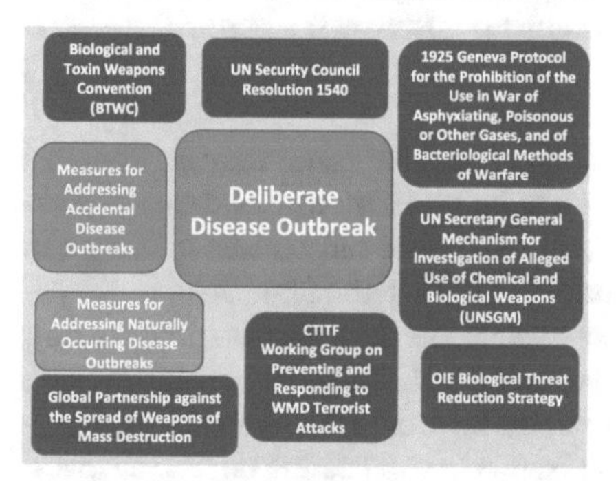

Epidemiological Clues

A highly unusual event with large numbers of casualties

Higher morbidity or mortality than expected

Uncommon disease

Point-source outbreak

Multiple epidemics

Lower attack rates in protected individuals

Dead animals

Reverse spread

Unusual disease manifestation

Downwind plume pattern

Direct evidence

Criteria for Determining the Type of Outbreak

Non-Conclusive	*Conclusive*
A biological risk is present	*Identification of the agent as a biological weapon – DNA or RNA sequences may show this*
A biological threat exists	*Proof that the agent was released deliberately as a biological weapon*
Special characteristics of the biological agent	
High concentration of the biological agent in the environment	
Characteristics of where the biological agent was found	
Specifics about how the biological agent was spread	
The intensity and dynamics of the epidemic	
Timing of any epidemic	
Unusual rapid spread of the epidemic	
Only specific populations affected	
Unusual clinical symptoms	

Thank you !!!

Post-Course Hand-out

Diagnosing the Cause of a Disease: Distinguishing Between a Natural and Deliberate Disease Outbreak

Most diseases are the result of a natural outbreak and the cause may be traceable after diligent investigation. Infective agents can be bacteria, viruses, fungi or rickettsia (microorganisms that resemble bacteria, reproduce only inside a living cell and are transmitted by insect bite to vertebrate hosts, including humans).

The nature of the infection, and the agent responsible, have implications for how the infection is investigated and the outbreak contained. Biological organisms vary considerably in their potency, their stability in the environment, how they are transmitted and their mode of action. All of these factors have a bearing on how an investigation into the cause of an outbreak is conducted. For example, with viral infections like *hepatitis B*, where it is contact with bodily fluids, like blood, that is responsible for the infection being transmitted, simple precautions such as wearing gloves and an apron, avoiding needle stick injuries, and safe disposal of any blood-contaminated material are sufficient to ensure protection for those collecting samples to identify the cause.

But for other agents that may be more contagious and transmitted by aerosol like flu viruses, or the agent in the case under discussion in this training, *Coxiella burnetti*, where not only is it aerosol transmission that is the issue, but the host for the infection is an animal (other than human), usually domestic like sheep or cattle, finding the cause and advising on precautions to take for medical personnel can be difficult. Transmission of *C. burnetii* is often on farmyard dust (from hay or straw from barns) and person-to-person infection is extremely rare, so the disease is not contagious.

C. burnetii is a very infectious organism with a very small quantity of the bacterium required for an infectious dose. The disease occurs in two stages: an acute stage that presents with headaches, chills, and respiratory symptoms, and an insidious chronic stage where mortality is high without treatment. Most infections clear up spontaneously, and treatment with antibiotics tetracycline or doxycycline will reduce the symptomatic duration and reduce the likelihood of chronic infection. A combination of other antibiotics erythromycin and rifampicin is highly effective in curing the disease, and vaccination with several specific vaccines are effective for prevention.

Until biological samples are collected from infected patients identifying the causative agent will be difficult. Advice to tending physicians should be to consider the agent as contagious until identity is established, and suitable measures taken to ensure no contact with bodily fluids (including aerosols) or contaminated surfaces. Laboratory staff should also be advised to take the necessary precautions to avoid skin contact and avoid creating aerosols when separating, centrifuging, or pipetting samples.

In this case the organism was identified in the laboratory, relevant organisations were informed about the outbreak, patients treated and the outbreak contained. The cause was traced to infected animals and the result deemed a natural outbreak.

Crtieria for considering a disease to be a possible deliberate outbreak together with environmental clues to consider have been drawn up and provided as additional handouts. None of the criteria or clues are definitive on their own for saying the outbreak was caused deliberately, with the exception of evidence that an organism had been either deliberately engineered (or prepared) for use as a biological weapon. Evidence of that may be difficult to obtain, but DNA (or RNA in the case of viruses) analysis will invariably identify abnormal strains, and ones which may be particularly virulent which, in turn, will help to identify how the outbreak occurred.

There have been instances in the past where disease outbreaks have been caused deliberately but not identified by investigating authorities, or where accusations have been made that people were deliberately infected, but after rigorous analysis of the evidence it was apparent that the illnesses had nothing to do with the apparent biological weapon. In another instance even when evidence of a specific agent being handled in a military microbiological facility was identified as the cause of over 60 deaths, the government concerned still insisted that the deaths were attributable to infection due to consumption of infected animal carcases.

Moral: *Identifying a disease outbreak as being caused deliberately may be less difficult than all agreeing that it is.*

Annex 2: Multiple Uses of Chemistry

What is Legal?

What is legal?

Alastair Hay
University of Leeds, UK

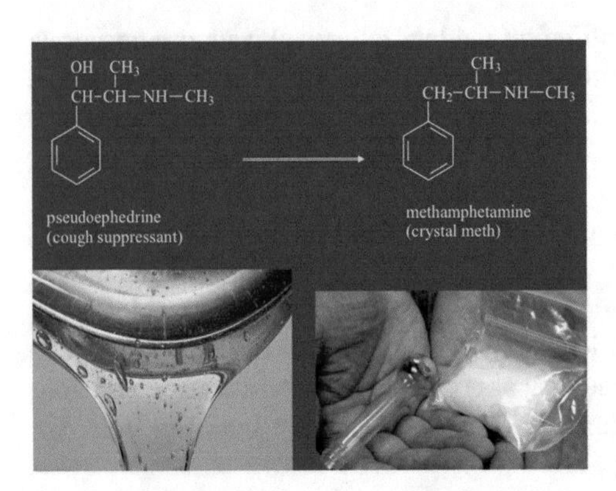

Error → Misconduct → Criminal

Non-intentional

Undeclared conflict of interest
Curiosity
Can chemical be made?
Suppress knowledge
Make it again
Sell some
Large scale manuf.

Intentional

With apologies to M. Nylenna & S. Simansen
Lancet 2006; 367: 1882-4

References

Q fever and C. Burnetii

Anderson, Alicia, et al. 2013. Diagnosis and management of Q Fever – United States, 2013: Recommendations from CDC and the Q fever working group. *Morbidity and Mortality Weekly Report* 62 (RR03): 1–23.
Kamenov, Galin, and Maida Tiholova. 2004. Q fever outbreak in Botevgrad, Bulgaria: May–June 2004. *Eurosurveillance* 8: 35.
Oyston, P.C.F., and C. Davies. 2011. Q fever: The neglected biothreat agent. *Journal of Medical Microbiology* 60: 9–21.
Panaiotov, Stefan, et al. 2009. An outbreak of Q fever in Bulgaria. *Annali dell'Istituto Superiore di Sanità* 45 (1): 83–86.

Differentiating Between Natural and Deliberate Disease Outbreaks

Carus, W.S. 2001. *Bioterrorism and Biocrimes - The Illicit Use of Biological Agents Since 1900.* (February 2001 revision), 50–58. Defense Technical Information Center, Washington, D.C.
Chen, X., A. Chungtai, and C. Raini MacIntyre. 2017. A systematic review of risk analysis tools for differentiating unnatural from natural epidemics. *Military Medicine* 182 (11/12): e1827–e1835.
Grunow, R., and E.J. Finke. 2002. A procedure for differentiating between the intentional release of biological warfare agents and natural outbreaks of disease. https://doi.org/10.1046/j.1469-0691.2002.00524.x.
Radosavljevic, V., and G. Belojevic. 2012. Unusual Epidemic events: A new method of early orientation and differentiation between natural and deliberate epidemics. *Public Health* 126: 77–81.

Unusual Disease Outbreaks

Amerithrax or Anthrax investigation. 2019. https://www.fbi.gov/history/famous-cases/amerithrax-or-anthrax-investigation.
Harris, E.D. 1978. Sverdlovsk and Yellow Rain: Two cases of soviet noncompliance? *International Security* 11 (4): 41–95.
Meselson, M., et al. 1994. The Sverdlovsk Anthrax Outbreak of 1979. *Science* 266 (5188): 1202–1208.
Seeley, Thomas D., Joan W. Nowicke, Matthew Meselson, Jeanne Guillemin, and Pongthep Akratanakul. 1985. Yellow Rain. *Scientific American* 253 (3): 128–137.

Chemistry

Multiple Uses of Chemicals. 2019. https://multiple.kcvs.ca/.
Nylenna, M., and S. Simonsen. 2006. Scientific misconduct: A new approach to prevention. *The Lancet* 367 (1): 1882–1884.

Evaluation Systems for Biological Security Risk Mitigation Training and Education

Giulio Maria Mancini and James Revill

1 Introduction

Over the course of the last decade, engagement, training and education have been widely promoted as ways to address and mitigate biological security risks by building capacity and awareness at different levels, with countries as diverse as US, Denmark, Pakistan Algeria, Morocco, China, Japan and Switzerland exploring biosecurity related education and training (NASEM 2018). The specific rationale for this trend varies, but underlying is a sense that the attitudes and actions of more aware, competent and capable people will lead to lower risks coming from intentional misuse of biological material or technologies.

Another reason why education and training may have become a "popular" biological security risk mitigation measure, with support from both governmental and non-governmental backgrounds (BTWC 1986; IAP 2005; Germany 2005; Russian Federation 2005; India 2005; Rappert 2007; OECD 2007; NSABB 2007; Japan 2008; BTWC 2008; G8 2009; WHO 2010; National Research Council 2010; Australia et al. 2011; BTWC 2012; Netherlands 2015; Deutscher Ethikrat 2014; IAP 2014; NASEM 2018), could be that it would be non-contestable and non-contested security risk mitigation measures when compared with other more

---

Responsibility for the information and views expressed in the chapter lies entirely with the author(s). The content of this chapter does not necessarily reflect the official opinion of the European Union, nor of the Harvard Sussex Program.

G. M. Mancini (✉)
Directorate-General on Migration and Home Affairs of the European Commission, Brussels, Belgium
e-mail: giulio-maria.mancini@ec.europa.eu

J. Revill
Harvard Sussex Program, SPRU and University of Sussex, Brighton, UK

M. Martellini, R. Trapp (eds.), *21st Century Prometheus*,
https://doi.org/10.1007/978-3-030-28285-1_15

intractable measures – from verification to legislating on dual-use redlines – that remain practically or politically difficult.

Despite the general belief that education and training are desirable – and generally positive feedback from experiences of capacity building and engagement (Minehata and Shinomiya 2010; Novossiolova et al. 2013; Dando and Rappert 2005; Mancini and Fasani 2012; Revill et al. 2012) – evaluation of their impacts is often anecdotal or largely based on observations of *ex-post* situations. This is, in part, because capacity building efforts are often not linked with learning outcomes typical of contemporary educational models per se, but a result of the teacher "teaching what the teacher wants" and not necessarily "what the learner needs to learn in order to contribute to risk mitigation"; and, in part, because of the importance of the context in which such activities are being undertaken (NASEM 2018).

Compared to other, possibly more structured, areas where capacity building is applied, rigorous systemic and systematic analyses, and practical examples of predictive and assessment toolboxes for the evaluation of education and training on security risks is at best limited. Or, as was more bluntly noted at a recent meeting in Zagreb on dual use governance, "there is little empirical evidence of what works and what does not with regard to the implementation of dual use governance measures, including the foundational aspects of education… " adding that metrics "need to be further developed, tested, and refined within the context of broader evaluation methodologies for systematically monitoring progress and assessing results" (NASEM 2018).

This paper proposes a methodology for evaluating of the impact of education or training related to biorisks, particularly looking at security-related biorisks. As such, it seeks to evaluate the impact of education and training as a risk mitigation measure, combining and connecting education science with risk management science. The methodology therefore proposes a model to link (expected) outcomes in terms of learning to (expected) outcomes in terms of risk.

From education science, the paper leverages firstly Instructional Systems Design (ISD) approaches (Andrews and Goodson 1980; Edmonds et al. 1994; Gustafson and Branch 2002). In particular it explores the ISD model described by the American Association for Training Development (ASTD) (Hodell 2011), which is one of the most structured and refined applications of what has been informally termed the ADDIE model for ISD. ADDIE comes from the initials of the five main phases of this highly flexible instructional design model: Analysis, Design, Development, Implementation and Evaluation. ADDIE is not a formalized model in itself, nor is it clear who first described the process using this label (Molenda 2003).[1]

[1] The early instructional design model that later developed into the ASTD approach is the Interservice Procedures for Instructional Systems Development (IPISD) designed for the United States Army by Branson et al. (1977). The ADDIE approach includes all the components advanced by other instructional design scholars (Bonner 1982; Gustafson and Branch 2002; McCombs 1986). These components can be grouped under the analysis, design, development, implementation and evaluation categories.

Secondly, from education science the paper also leverages the idea of different levels of effects of education and training, specifically theories of transfer from learning to behaviour and Kirkpatrick's model of the four levels of evaluation. The model has a history of application in training programmes, most commonly for business organizations (Bates 2004) and project management (Steensma and Groeneveld 2010; Buganza et al. 2013). However, the model has also been used in other instructional contexts (Arthur Jr et al. 2003) as well as applied in higher education (Praslova 2010). The model helped to highlight the importance of design for evaluating training programmes (Steensma and Groeneveld 2010), and more generally, basing evaluation of education on learning objectives. In general, Kirkpatrick's typology "has served as a good foundation for training evaluation for many decades" (Salas and Cannon-Bowers 2001 p. 487). This model of evaluation is compatible with instructional systems design and the ADDIE model, as it stresses the importance of setting desired outcomes upfront and then reverse engineering the development and design.

Regarding approaches to risk, the paper follows the conceptualization that risks are measures of the combination of the likelihood and possible consequences of a specific adverse event (Kaplan and Garrick 1981; Kaplan 1997; Garrick 2008; Kates and Kasperson 1983; Bradbury 1989; Hadden 1984; Stern and Fineberg 1996). Risk assessment, therefore, entails estimating the relative likelihood and consequences of a risk (Caskey and Sevilla-Reys 2015; Caskey et al. 2010; Depoy et al. 2005). However it should be noted that in the process of risk assessment, the more rational phase of risk characterisation must be combined with the more subjective process of risk evaluation that acknowledge uncertainty and attend to ambiguity (Mancini and Revill 2017; Apostolakis 2004; Renn 2006; Astuto Gribble et al. 2015).

The paper discusses elements to design evaluation, starting from the design of reaction, learning, behaviour and result objectives. It then presents examples of both learning objectives and tools that can be useful in evaluating behavioural and risk impacts. Subsequently, design and evaluation tools are presented that can improve the effectiveness of education and training related to biological security risks mitigation.

2 Designing Evaluation

Instructional systems design posits that instructional design has to start from the analysis of the situation and the design of objectives and other instructional features. It then moves to the development of the instructional materials and environment, the implementation of the instruction and the measurement of the evaluation.[2] The model of four levels of evaluation is applied to the key steps of objectives

[2] The system is a cycle and as with other management systems approaches, it requires feedback and continuous improvement.

design and evaluation. It posits that capacity building effects act on four levels: reaction, learning, behaviour and result. Achieving success at each level is necessary in order to proceed to the next one; objectives are designed at each level and the process works backwards to the design phase and measured onwards at the evaluation phase of the ADDIE instructional design cycle. In our case, biological security risk mitigation objectives are designed at the results level (level 4), and other objectives are designed on the lower levels to test if and how education and training could impact on risks.

The objectives are defined as lowering biological security risks, something achieved by lowering their likelihood and/or consequences. Level 3 (behaviour) objectives have to be identified as those behaviours, which, if retained and/or generalized, have the potential to lower the likelihood and/or consequence of risks. Level 2 (learning) objectives represent the learning, which, if achieved and transposed in regular practice, have the potential to become behaviour. Level 1 (reaction) objectives correspond to the satisfaction needed to allow learning.

Evaluating performance directly is often complicated or even impossible. It may require, for example, continuously scrutinizing learning on all possible knowledge, attitudes and skills, or a pervasive monitoring of behaviours. Representing, communicating and comparing these measures would also be challenging. As such where possible indicators should be used and instruction designers should plan on what indicator(s) would give a meaningful indication of the extent of actual learning, behaviour, and results. The ideal format for learning objectives is particularly useful as it also represents a format for level 2 indicators. The second key component of an evaluation design is metrics, or how the indicator will be measured. Finally, the evaluation design has to plan how, where and/or when to collect the data for the metrics to measure the indicators.[3]

Based on our and others' experiences of education and training of life scientists or students of life sciences on biological security risks (Mancini and Revill 2017; Whitby and Dando 2010; Revill et al. 2012; Grainger and Turegeldiyeva 2015), we propose a matrix with indicators and metrics on the four levels of evaluation as an example of possible evaluation tools for education as a potential deliberate disease risks mitigation measure. Learning, behaviour and results indicators have been grouped in four thematic areas whose mastery may be relevant in influencing likelihood and/or consequences of biological security risks. The (bio)risk management area would include capacities on understanding the nature and components of risks, assessing risk and taking risk-informed management decisions. The laboratory biosafety area includes capacities on preventing accidental and unintentional harm in the laboratory.[4] The third area includes capacities on laboratory biosecurity.

[3] For each metric the designer can also plan a threshold that will make the educational action "successful" for that indicator, or just plan on observing results.

[4] Such as laboratory acquired infection; contamination; pathogen release in the environment. Capacity in this area, besides contributing to mitigating biological safety risks, would also mitigate the possibility that a biosafety risk evolves into a biosecurity risk (for example because a non-state threat without access to assets, gains access because of inadvertent release)

The fourth area includes capacities related to biosecurity and dual-use management broader than the laboratory approach.

3 Evaluation of Reaction

The first level of evaluation, reaction, "measures how those who participate in the program react to it" (Kirkpatrick and Kirkpatrick 2006 p. 21).[5] Measuring at this level does not depend strictly on the subject matter, and proposed strategies are the same for all four thematic areas. However, it is important that evaluation of reactions is designed, even if methods are not highly structured (Kirkpatrick 1979), and can include data points as simple as students' participation, their use of class materials, etc. Our designed indicator for reaction is if and, if so, how students are satisfied, interested, and/or engaged by the experience. Metrics of satisfaction can include attendance of classes and feedback from students[6]; but also, satisfaction can be gauged by levels of active participation of students and the extent to which they interact with the instructor and raise questions. Data collection methods for those metrics may include attendance sheets; satisfaction surveys or "smile sheets"; observations by the instructor; number of questions raised in class or on an e-learning forum; information collected to a group activity; follow-up communications with students.

4 Evaluation of Learning

Evaluation at the second level, learning, looks at whether the student achieved the learning objectives linked to the activity. The indicators of learning are the learning objectives, categorized according to the cognitive domain of knowledge, attitudes and skills; and to the levels of learning (know, understand, apply, analyse and evaluate). Level two indicators included in the matrix are proposed examples which obviously do not cover all possible learning indicators. However, they are a meaningful collection of indicators that could be reverse engineered from level 4 and level 3 objectives.

A key question to consider in instructional design is, based on the learner population(s) identified, what do we want them to learn that will contribute to

---

[5] Level one evaluation is important as it provides insights for instructors in how students feel about the program as well as establishing a facilitating environment in which student's opinions are recognised as valuable (Kirkpatrick and Kirkpatrick 2007).

[6] The metrics can measure, using for example data from satisfaction surveys, the average appreciation from students and/or the share of students who appreciated. The choice largely depends if instructor value evaluating the enthusiasm ("level of satisfaction") or that all students in the class have a minimal satisfaction ("spread of satisfaction"), or both.

mitigating biological security risks? Instructional systems design models (Hodell 2011) provide guidance on elements that should be included in well-written learning objectives: audience, behaviour, condition, and the area and level of study. Firstly, the learning objective should give a clear definition of who is the "target student", indicating the specific program they are pursuing, at what level in the higher education *cursus* they are, and any other information that helps defining them.[7] Secondly, objectives should define the desired learning from the student. This is where specific levels and domains of learning should be identified. The concept of levels of learning derives from the taxonomy first proposed by Bloom (1956). The taxonomy is frequently used in the design of scientific and medical education (Patel et al. 2009), and it describes how learning occurs in subsequent stages from basic assimilation to more advanced elaboration. The original taxonomy used six categories of "cognitive domains": knowledge, comprehension, application, analysis, synthesis and evaluation. Mastery of one domain was a prerequisite for the student to succeed in the next one. Krathwohl (2002) proposed a revised version of the taxonomy with the following cognitive domains: remember, understand, apply, analyse, evaluate and create.[8] He furthermore proposed singling out "knowledge" as a cognitive domain and breaking it up in three dimensions: factual, conceptual and procedural. We use a similar approach adopting "cognitive domains" to describe three dimensions of theoretical knowledge ("know"), attitude ("feel") and practical skills ("do").[9] The action verbs suggested learning taxonomies (Krathwohl 2002)[10] are of great help to design clear learning objectives. A theoretical, relatively low-level learning objective may be described as "list", "memorize", etc. A theoretical but higher-level learning objective could be described using "critically assess", "judge", or "forecast". Practical learning objectives could be associated with action verbs as "perform", "use", "complete", "build" etc. In any case, a specific verb is preferable than the generic "learn". The achievement (or lack) of the desired learning should be observable and measureable. The learning objective should also state any condition that is needed for the objective to be met.[11] Finally,

---

[7] The audience part of the learning objective should be written with the typical individual student in mind, rather than the whole class, to later make evaluation easier.

[8] Krathwohl (2002) explains, firstly, that it is better to use verbs instead of nouns to describe learning actions, and to do it consistently; that the original knowledge domain is better described by the remember category to avoid confusion with the introduced different plans of knowledge; and that synthesis would change place with evaluation and be renamed "create".

[9] Learning may be different both in terms of complexity (some learning cannot be achieved if others are not completed before) and of cognitive domain. For example, assessing risks and benefits of a situation is a relatively advanced capacity that cannot be mastered if basic concepts are not known. Being able to practically perform a task is a different cognitive domain ("do", "skill") to theoretical reflection ("know", "knowledge"). The two layers are connected but independent: a practical skill may be higher learning level than a theoretical one, but also the other way around.

[10] See also Adelman (2009 pp. 68–70) on the "centrality of the verb" to explain learning objectives in different cognitive domains and levels of a qualification framework.

[11] This may include educational prerequisites, specific learning media, necessary laboratory equipment, etc.

the learning objective should state the degree to which the behaviour has to be met, possibly outlining indicators and acceptable thresholds of success.[12]

Thinking of learning objectives in terms of levels and cognitive domains of learning (see Table 1) is useful in designing learning objectives that are relevant, meaningful, context -appropriate and achievable for the students. The process can also help develop and implement appropriate educational materials in the subsequent phases of the instructional cycle and design suitable evaluative methods.

For example, a learning objective such as "at the end of the course the student must state the definitions of biosafety, biosecurity and dual-use" could be categorized as a "knowledge; know" learning objective according to levels of learning and cognitive domains. Conversely, a learning objective where the student is required to "complete a biosafety and biosecurity risk assessment of a research project, including biological hazards and risk scenarios identification, risk characterization and risk evaluation" could be classified as a "knowledge; evaluate" objective. Finally, a learning objective such as "at the end of the module, given a simple scenario, the student is able to find the correct Personal Protective Equipment (PPE), don and doff[13] it in the correct order in less than 5 min" is a "skill; apply" type of learning objective.

Learning objectives have also to be based on what the learner needs and is able to achieve, and what is possible in the instructional context – information that should be available from the previous analysis phase of the instructional cycle.

Experience of one of the authors in using this structured approach to the definition and categorization of learning objectives within the design phase of the instructional systems design cycle, suggests that instructional designers are more likely to design learning objectives that are more relevant to the target student population. For example, that learning objectives designed for advanced students are of higher levels of learning than those for other students; practical and/or "executer" learning objectives (for example of the apply cognitive domain) are matched to certain populations, and theoretical and/or "managerial" learning objectives (such as of the understand or evaluate cognitive domains) to other populations; or that the analysis of the instructional context informs instructional designers on the feasibility of attaining a learning objective in a certain level or cognitive domain with feasible teaching formats given class sizes or available infrastructure. Learning objectives depend on the local context, interests and priorities and there is no one size fits all approach. However, we can draw some generic considerations. It makes sense to assign to some student populations learning objectives of the "know" or "understand" levels that focus on learning pre-assessed risks. However, other student

---

[12] The concept of learning objectives is already established in training and education systems in most countries. Different systems, and even different instructional institutions within the same system, may define and refer to learning objectives in different ways and using different categories. This is important to consider, to make sure learning objectives are designed in the easiest way for inclusion in the target instructional system.

[13] Contractions of "do on" and "do off", respectively, these verbs indicate the procedures to dress in and take off specific or technical clothing such as PPE.

Table 1 Taxonomy of levels and domains of learning objectives

Levels of Learning \ Cognitive Domains				Cognitive domains		
				Knowledge (Know)	Attitude (Feel)	Skill (Do)
				Theoretical	Emotional	Practical
Level of Learning		Examples of action verbs		↔		
Evaluate	Make value judgement; introduce innovations	Judge, assess, defend, predict, infer, argument	Higher, more complex, abstract, more critical thinking	*Knowledge; Evaluate*	*Attitude; Evaluate*	*Skill; Evaluate*
Analyse	Being able to further abstract, understand why, and transfer learning to others, different situations	Examine, compare, experiment	↕	*Knowledge; Analyse*	*Attitude; Analyse*	*Skill; Analyse*
Apply	Transfer what understood to other similar situations	Use, perform, measure		*Knowledge; Apply*	*Attitude; Apply*	*Skill; Apply*
Understand	Process the meaning; being able to re-state	Explain, describe, recognize		*Knowledge; Understand*	*Attitude; Understand*	*Skill; Understand*
Know	Remember material in the same form	Define, list, memorize	Lower, simpler, factual, less critical thinking	*Knowledge; Know*	*Attitude; Know*	*Skill; Know*

populations should be assigned learning objectives advancing the risk-based approach rather than conventional approaches relying on prescription and "memorization of risk categories and levels"; these students will be expected to "apply", "analyse" and "evaluate". Besides being capable of "critical thinking that includes a thorough risk assessment" (Grainger and Turegeldiyeva 2015), if students learn the principles of risk assessment and management, they are more likely to apply this critical skill in a range of cases, not just on biological security risks and biological risks management.

Education should include designed learning objectives on understanding the factors contributing to risk; how risk is always relative and why this is the case; and the importance of assessment before, and of performance after, as well as on risk mitigation. Generally, instruction on biological security risks in higher education will design more generic learning objectives than training on deliberate disease risks for experienced professionals.[14] The design learning objectives for a higher education context would likely include the responsibilities of scientists or public health professionals in front of the community and society; responsibilities to act to prevent unsafe or un-secure situations; and responsibilities to oversee the potential for misuse to cause harm, including managing the risks connected with dual-use.

5 Evaluation of Behaviour

Evaluation at level 3 looks at specific behaviours that would be expected if the learning on level 2 is not only achieved but also translated to normal practice and sustained over time.[15] Those specific behaviours are designed based on what behaviours would be hypothetically be needed to influence risk factors, such as the "desirable behaviours" of assessing risks, developing and implementing risk mitigation plans; maintain control over hazardous material; respond to incidents;

[14] This may mean that learning objectives for university students will likely be lower in the taxonomy of learning levels; or they could be high level (up to critically "evaluate") but looking at wider cases than the specific scenarios used for in-service training courses.

[15] The third level of evaluation tries to answer questions such as "did the training stick? How much of the training transferred from delivery to the workplace?" (Hodell 2011 p. 74). Another definition of level three could be "achieving behavioral generalization (i.e. applying the skill outside of the training simulation)" (Salas and Cannon-Bowers 2001 p. 485). The transfer from level two to level three is sometimes overlooked in teaching, however "we only care about student performance in school because we believe that it predicts what students will remember and do when they are somewhere else at some other time … if we want transfer, we need to teach in ways that actually enhance the probabilities of transfer. The purpose of formal education is transfer"; and it has an impact on teaching design as "teaching for retention during a single academic term to prepare students for an assessment that will be given to them in the same context in which the learning occurs is very different from teaching for long-term retention and transfer" (Halpern and Hakel 2003 p. 38).

or dissuade knowledgeable personnel from the theft, loss or misuse of pathogens (Young et al. 2014). The passage from level 2 to level 3 evaluation is crucial, but not without challenges. Even in the context of formal educational or training programmes where instructors generally have extensive experiences and instruments to evaluate level 1 and level 2 objectives, they may lack opportunities to evaluate level 3 objectives. Learning is a necessary condition for behaviour, as "if little or no learning has taken place, little or no change in behaviour can be expected" (Kirkpatrick and Kirkpatrick 2006 p. 60). However, educators can often relate with examples of cases when achieved level 2 indicators do not translate in level 3 ones: such as students who pass courses' final tests, demonstrate abilities in a mentored session, but months later do not remember the same notions, or do not keep the learned practice in their laboratory or research behaviour.[16] Learning is a necessary but insufficient condition for behaviour, so level 3 needs to be evaluated specifically.[17]

6 Evaluation of Results

Evaluation at level 4 in general looks at the broader results of engagement, training or education. Here we specifically discuss factors influencing likelihood and/or consequences of biological security risks. These factors should be influenced by a generalized, sustained and formalized adoption of behaviours from level 3. Indicators at this level will be connected to risk likelihood or consequence factors of hazards, threats or situation for biological security risks. Examples of indicators from relevant studies looking at different level 4 results include increased range of service delivery, improvements of clinical care, being involved in teaching, training, committees, international collaborations, co-author peer reviewed articles or books (Anderson et al. 2014); herd productivity and fertility (Knight-Jones and Rushton 2013); or economic value added by the training (Buganza et al. 2013). Specifically for the context of higher education, it has been suggested that level four should evaluate how those educated by higher education institutions would contribute to society, proposing indicators such as qualifications, socialization, subjectification (or how students have developed independent thinking) (Praslova 2010), students' character development and ethical readiness for their roles in society (Boyer and Hechinger 1981; Sax 2004).

Correspondence between indicators in the four levels is not 1:1, and they diminish as levels get higher. Multiple learning objectives concur in the nurturing or

[16] Reasons for learning not translating to behaviour could vary and include for example different equipment; complacency; or poorly designed learning objectives. In any case, translation from learning to behaviour is not granted at all, hence why level 3 should be measured.

[17] It should also be noted that environmental factors may have a role in moderating the translation from learning to behaviour, such as commitments of hierarchy, opportunity, support from peers, and work context (Salas and Cannon-Bowers 2001; Bates 2004; Kontoghiorghes 2004).

modification of a behaviour, and multiple behaviours may influence a risk factor. In case of specific or particularly important competences, one learning objective may correspond to one behaviour and one level 4 goal.

7 Indicators, Metrics and Data

In the thematic area of risk management, examples of knowledge learning indicators (to measure level 2 objectives) we propose including the capacity to define risk; distinguish between biosafety and biosecurity; and prepare a risk assessment. An attitude indicator is that the student feels confident in using a risk management approach, based on assessment, mitigation and performance. Level 3 indicators[18] include that students prepare actual risk assessments. In the thematic area of laboratory biosafety, theoretical learning indicators include the hierarchy of mitigation measures or types of PPE; practical learning indicators include donning and doffing PPE and washing hands. Behaviour indicators include that students continue to appropriately select risk mitigation measures, wash hands, or do not eat and drink in the laboratory. Proposed examples of level 4 indicators look at laboratory or hospital acquired infections, environmental releases, and cross-contaminated samples. In the thematic area of laboratory biosecurity, examples of level 2 indicators include knowing different measures within the pillars of physical security, personnel management, material control and accountability, transport security and information security; while practical indicators include being able to apply those measures, such as securely labelling packages or protecting information. Examples of level 3 indicators include the extent to which participants changed their behaviours and applied learning to their usual work (Buganza et al. 2013), such as when physical security is considered in laboratory design or commissioning projects; or that the chain of custody of assets is maintained. Result indicators may include breaches, thefts, and losses of assets. Finally, in the thematic area of biosecurity and dual-use management, proposed examples of level 2 indicators include knowing experiments of concern and options to manage potential misuse; an understanding of the responsibilities deriving from dual-use issues; or being able to document decisions on dual-use as appropriate.

For all thematic areas, we designed examples of learning objectives (and level 2 indicators) based both on our direct experiences but also on the input of instructors who had been involved for a number of years in promoting education on biological security risks.[19] Learning objectives constructed in the knowledge and attitude cognitive domains could range from the simplest "know" level to the more advanced "analyse" or even "evaluate" levels of learning. Simple "know" objectives may include, for example, that students as a minimum become "aware" of security risks,

---

[18] Level three indicators are also sometimes referred to as "transfer criteria" (Alliger et al. 1997).

[19] This is particularly true for the dual-use and broader biosecurity thematic area.

including the implications of dual use. Proposed examples of learning objectives set at more advanced learning levels may be to "analyse" risk scenarios or "evaluate" a case with security implications, where the student could demonstrate capacity via preparing presentations, essays or self-assessments.

Proposed behaviour indicators in this thematic area are if students contribute to the development of oversight or other management and decision-making systems on dual-use; or are able to enhance reporting practices on managing dual-use material, equipment and information.

Metrics can be more or less formal. At level 2,[20] quite informal tools may include observation, the opinions of facilitators after interaction with students, and exercises completed during the curriculum. In some cases, evaluation can be completed with self-assessment tools.[21] In some cases, however, an informal and/or self-assessed evaluation may not be enough; this will depend on the "degree" component designed into the learning objectives. Some training may need a rigorous, defined and formal evaluation method and a corresponding threshold or "passmark". In the higher education context, formal evaluation is usually a requirement for curricula, sometimes requiring formal pre- and post-tests (Arthur Jr et al. 2003). Appropriate metrics should also be planned with the selected cognitive domain in mind. For learning indicators, proposed metrics could use benchmark or reference definitions and minimal performance requirements. For knowledge-related objectives, and for some attitude-related ones, possible data tools include tests (Kirkpatrick and Kirkpatrick 2007) in the form of multiple choice questions, logical sequences, statement corrections, quizzes; peer evaluation among students; matching exercises; class discussions; exercises based on scenarios or case-studies; essay writing; and presentations by students (Praslova 2010). Data tools for practical learning objectives include simulation, drills, role-playing (Kirkpatrick and Kirkpatrick 2007), or communication exercises. Metrics for level three indicators can include ratings from supervisors or job outputs. Specifically for higher education, it has been suggested that metrics for indicators should include "evidence of student use of knowledge and skills learned in previously taken classes in their following class work, including in research projects or creative productions, in application of learning during internship, and in other behaviours outside the context in which the initial learning occurred" (Praslova 2010 p. 221). Typical data tools for level 3 evaluation include surveys, interviews (with students or educators), visits, reports (including from alumni's work supervisors), checklists, focus groups, Behavioural Observation Data (BOD) (Kirkpatrick and Kirkpatrick 2007). Also for level three, it is suggested that indicators could be measured before and after the training (Steensma and Groeneveld 2010). While this is not always possible, it seems desirable that level three indicators are measured at some point after the training (Table 2).[22]

[20] Level two indicators are also sometimes referred to as "learning criteria" (Praslova 2010).

[21] Which may be an important option especially for adult learners (Kidd 1974; Knowles 1984; Kirkpatrick and Kirkpatrick 2007).

[22] Kirkpatrick and Kirkpatrick (2007) suggest that level three evaluation could be immediately after training courses, with instructors observing behaviour and correcting it, if necessary, as immediate reinforcement. We believe this would actually still be a level two evaluation. With level three, we should really focus on if the learning objective (knowledge, attitude, skill) sustains as situations change and time passes.

Table 2 Four-levels evaluation matrix and example of possible indicators, metrics and data

	L1 Reaction			L2 Learning			L3 Behaviour			L4 Result		
	Indicators	Metrics examples	Data	Indicators (Domain: Level)	Metrics examples	Data	Indicators	Metrics examples	Data	Indicators	Metrics examples	Data
Risk management	The student is satisfied, interested and engaged	≥4,5/5 average evaluation "≥90% participants evaluating positively"	Post-instruction survey	Knowledge: Know: define risk	The student provides a definition of risk that includes: adverse event; likelihood; consequence	Multiple choice questions; self-assessments; case study; statement correction; essay writing; class discussion; peer review; quizzes	Continuing to look at: possible risk scenarios (including what changed); risk likelihood and consequences	One month/one year/three years after training trainees applied what learned	Surveys; interviews; visits (planned or surprise); reports from supervisors; peer review exercises	Available safety and security risk mitigation measures	%	Annual risk assessment; incident reports; audits
		Appreciations, suggestions, and/or asking for additional trainings	Learning activities; follow-up communications; letters; post-instruction survey	Knowledge: Know: define safety and security	The student provides definitions that at least include accidental vs intentional (and/or hazard vs threat)		Consideration of security risks increased; communication lines with law enforcement (where relevant) established			Vulnerability. opportunities available to threats	Qualitative	
		% attendance	Turnout at the event; attendance sheets	Knowledge: Analyse/ Evaluate: prepare a risk assessment	The student demonstrates risk assessment (characterization and evaluation) in context, given a scenario		Actual risk assessment prepared for own context			Reviewed risk assessment	Yes/No	
				Attitude: Understand: feel confident about using the risk management approach	The student describes a risk management process starting from hazard identification within risk assessment, then mentions mitigation, and evaluation and review.							

(continued)

Table 2 (continued)

	L1 Reaction			L2 Learning			L3 Behaviour			L4 Result		
	Indicators	Metrics examples	Data	Indicators (Domain: Level)	Metrics examples	Data	Indicators	Metrics examples	Data	Indicators	Metrics examples	Data
Laboratory biosafety	The student is satisfied, interested and engaged	≥4,5/5 average evaluation	Post-instruction survey	Knowledge: Know: hierarchy of laboratory safety controls	The student orders risk control measures from the most to the least effective.	Multiple choice questions; self-assessments; case study; statement correction; essay writing; class discussion; peer review; quizzes	Applying the highest appropriate mitigation measure; select mitigation measures according to route of exposure	As soon as possible when student starts lab/field work; For higher education: 2 years after graduation	Behavioural Observation Data; SOPs developed by students; follow-up surveys; peer review exercises	Reduced laboratory/ hospital acquired infections; reduced cross-contamination of samples; reduced cases of environmental release.	Year-over-year number	Incident reports; audits; publicly available information; annual documentation
		≥90% participants evaluating positively										
		Appreciations, suggestions, and/or asking for additional trainings	Learning activities; follow-up communications; letters; post-instruction survey	Knowledge: Apply: mitigation measures are based on risk assessment	The student selects mitigation measures after assessment; explains appropriateness of measure with risk.		Proper labelling of samples					
				Knowledge: Understand: advantages and disadvantages of laboratory safety controls	The student provides information such as: hazard elimination is effective but not always possible; engineering controls are expensive, need maintenance; administrative controls are based on authority; practices and procedures on human factors; PPE only protect		Separation between samples and food/ drink in fridges					

		% attendance	Turnout at the event; attendance sheets	Knowledge: Understand: recognize safe and unsafe work practices	The student distinguishes good and bad laboratory work practices		No eating, drinking, smoking in the lab					
				Skill: Understand: washing hands correctly	During simulation, the student washes hands leaving no residues	Simulations; practical exercises; role-play exercise; poster design; write an SOP; peer review	Washing hands properly and when required					
				Skill: Understand: don and doff PPE	The student can don and doff in the correct order and at the first try: coverall, gloves, foot protection, respiratory mask							
				Skill: Apply: classify and segregate biological waste	The student matches wastes with categories (solid, liquid, sharps, pathological)							
				Skill: Evaluate: evaluate appropriateness of operational, facility and management risk mitigation measures	For each provided example, the student explains how the associated biological risk could be reduced		Laboratory biosafety considerations are integrated in work presented at scientific conferences	Occurrences, quantity and quality of consideration	Surveys; interviews; events' programmes			

(continued)

Table 2 (continued)

	L1 Reaction			L2 Learning			L3 Behaviour			L4 Result		
	Indicators	Metrics examples	Data	Indicators (Domain: Level)	Metrics examples	Data	Indicators	Metrics examples	Data	Indicators	Metrics examples	Data
Laboratory biosecurity	The student is satisfied, interested and engaged	≥4,5/5 average evaluation	Post-instruction survey	Knowledge: Understand: importance and rationale of laboratory biosecurity	The student is aware that biological security (deliberate disease) risks exist.	Multiple choice questions; self-assessments; case study; statement correction; essay writing; class discussion; peer review; quizzes	Biosecurity is considered in decision processes like lab design, procurement, etc.			Breaches/thefts reduced; loss of access (keys, locks, etc) mitigated; staff more confident and feel more secure	Year-over-year N	Incident reports; audits; publicly available information; annual documentation
		≥90% participants evaluating positively										
		Appreciations, suggestions, and/or asking for additional trainings	Learning activities; follow-up communications; letters; post-instruction survey	Knowledge: Analyse/ Evaluate: complete a security risk assessment	The student is able to identify, characterize specific security risks and related threats, assets, and vulnerabilities, and evaluate risks based on personal and institutional preferences		Documented biosecurity risk assessment process regularly carried out and revised	Occurrences, quantity and quality.	Surveys; interviews; events' programmes			
							Laboratory biosecurity considerations integrated in work presented at conferences					
				Knowledge: Know: methods for establishing physical security	The student lists at least three examples of methods for physical security		Different/new physical security methods considered in physical security decisions; physical security included in new design/ commissioning		Behavioural Observation Data (BOD); follow-up surveys; peer review exercises			
		% attendance	Turnout at the event; attendance sheets	Skill: Apply: use information protection methods	The student demonstrates use of unique user IDs, passwords and encryption as appropriate.		Protect personal password(s); do not exchange users; change passwords regularly					

Dual-use and broader biosecurity	The student is satisfied, interested and engaged	≥4,5/5 average evaluation	Post-instruction survey	Knowledge: Know: the prohibition norm on causing deliberate disease	The student mentions the prohibition on from legal or ethical, national or international, sources.	Multiple choice questions	The students looks for more information		Behavioural Observation Data (BOD); follow-up surveys; peer review exercises	A well-established "norm" of not doing, not helping, and preventing, the misuse of life/health sciences to cause harm.	Cases	Follow-up surveys
		≥90% participants evaluating positively										
		Appreciations, suggestions, and/or asking for additional trainings	Learning activities; follow-up communications; letters; post-instruction survey	Knowledge: Understand: identify potential dual-use and broader biosecurity issues	The student provides a definition that at least includes that peaceful research, material, equipment or information have the potential to be misapplied to cause harm; and/or explains the criteria to recognize dual-use or potential misuse issues	Multiple choice questions; essay writing;review of research/work proposals and reports	Questions to instructors or supervisors on existing considerations in programs and systems					
				Knowledge: Know: actions to take when faced with a potential misuse ethical dilemma	The student explains the institutional procedure to report an identified dual-use or potential misuse issue.		Questions on dual-use potential of material, equipment or information					
		% attendance	Turnout at the event; attendance sheets	Knowledge: Apply/Analyse: options to manage dual-use and broader biosecurity issues			Demands and proposals for an oversight system					

(continued)

Table 2 (continued)

	L1 Reaction			L2 Learning			L3 Behaviour			L4 Result		
	Indicators	Metrics examples	Data	Indicators (Domain: Level)	Metrics examples	Data	Indicators	Metrics examples	Data	Indicators	Metrics examples	Data
				Attitude: Understand/ Apply: be aware of the responsibilities on dual-use and broader biosecurity and feel ethically accountable	The student discusses oversight and decision-making options on dual-use and broader biosecurity.		Documented decisions on dual-use issues; the student participates to the establishment of a Code of Conduct					
				Knowledge: Analyse/ Evaluate: Document and justify decisions as appropriate			Integration of considerations on dual-use in work presented at conferences	Occurrences, quantity and quality of consideration	Surveys; interviews; events' programmes			

Measuring behaviour can be challenging at times, not least because structured evaluation tools for behaviour are lacking or because it's difficult to observe learners after they completed a specific course. Other challenges may include observing behaviours that implement higher levels of learning, and making sure that behavioural transfer is not prevented by non-training-related problems.[23]

Metrics for level 4 can be qualitative or quantitative – depending on the indicator and how it has been designed – and useful data tools include samples of academic accomplishments, awards (Praslova 2010); incident reports, publicly available information, alumni records, internal and external audits, surveys, and interviews.

8 Conclusions

This paper discussed ideas for possible evaluation systems for assessing engagement, education and training on biological security risk mitigation. The paper first introduced the concepts leveraged from the science of learning of instructional systems design and of levels of impact of learning. We then introduced proposed evaluation strategies for measuring impact of training and education aimed at reducing biological security risks. These strategies included examples of designed indicators, metrics and data sources for the reaction, learning, behaviour and results levels. We discussed guidelines from instructional design on defining learning objectives, as well as different examples of their characterization, detailing the taxonomy of learning objectives based on cognitive domain and level of learning and presenting possible learning indicators in four thematic areas. We also proposed corresponding indicators for behaviour and results evaluation.

Basing on the proposed indicators and metrics, improved competencies identified in four thematic areas, built with education and training and transferred into behaviour, have the potential to reduce biological security risks in specific risk scenarios via impacting factors that primarily influence risk likelihood.

Capacities built through training and education would have the potential to mitigate the likelihood of biological security risks by reducing the chance of future and young scientists becoming involved in state biological weapons programs; reduce the likelihood of themselves becoming insider threats (and make them better able to identify and counter insider threats); and mitigate the opportunity for outsider threats by improving scientists' awareness and capacities to protect their science and to identify and mitigate misuse risks.

For example, education and training could improve capacity in laboratory risk management measures, thus impacting on hazard factors and reducing biological security risk likelihood. They could also instill the ethical norm of prohibition of

[23] Such as a work environment that is not receptive to the awareness and capacities that students acquired or logistical barriers to achieving the expected behaviour

misuse, thus mitigating the likelihood of biological security risks by reducing the likelihood of future and young scientists becoming involved in state programs of developing biological weapons. Furthermore, they could reduce the likelihood of themselves becoming insider threats, and make them better able to identify and counter insider threats; and mitigate the opportunity for outsider threats to misuse by improving scientists' awareness and capacities to protect their science and to identify and mitigate biological security risks. Capacities built would also have the potential to mitigate biological security risks by reducing risk consequences, largely in the laboratory safety and laboratory security thematic areas.

The paper may be of value to those in the biological (and chemical) security community seeking to enhance work on metrics and measures in projects wherein results have proven difficult to determine in a systematic manner and provides a theoretical framework through which projects and programs might be better evaluated.

References

Alliger, G.M., S.I. Tannenbaum, W. Bennett, H. Traver, and A. Shotland. 1997. A meta-analysis of the relations among training criteria. *Personnel Psychology* 50 (2): 341–358.

Anderson, F.W., S.A. Obed, E.L. Boothman, and H. Opare-Ado. 2014. The public health impact of training physicians to become obstetricians and gynecologists in Ghana. *American Journal of Public Health* 104 (S1): S159–S165.

Andrews, D.H., and L.A. Goodson. 1980. A comparative analysis of models of instructional design. *Journal of Instructional Development* 3 (4): 2–16.

Apostolakis, G.E. 2004. How useful is quantitative risk assessment? *Risk Analysis* 24 (3): 515–520.

Arthur, W., Jr., W. Bennett Jr., P.S. Edens, and S.T. Bell. 2003. Effectiveness of training in organizations: A meta-analysis of design and evaluation features. *Journal of Applied Psychology* 88 (2): 234.

Astuto Gribble, L., E. Sangalang Tria, and L. Wallis. 2015. The AMP model. In *Laboratory biorisk management: Biosafety and biosecurity*, ed. R.M. Salerno and J.M. Gaudioso, 31–44. Boca Raton/London/New York: CRC Press, Taylor & Francis Group.

Australia, Canada, Japan, New Zealand, Republic of Korea (on behalf of the 'JACKSNNZ'), Switzerland, Kenya, Sweden and Ukraine. 2011. *Possible approaches to education and awareness raising among life scientists. Working Paper. BWC/CONF.VII/WP.20*. Geneva: Biological and Toxin Weapons Convention.

Bates, R. 2004. A critical analysis of evaluation practice: The Kirkpatrick model and the principle of beneficence. *Evaluation and Program Planning* 27 (3): 341–347.

Bloom, B.S. 1956. *Taxonomy of educational objectives: The classification of educational goals*. Philadelphia: McKay.

Bonner, J. 1982. Systematic lesson design for adult learners. *Journal of Instructional Development* 6 (1): 34–42.

Boyer, E.L., and F.M. Hechinger. 1981. *Higher learning in the Nation's service. A Carnegie Foundation Essay*. ERIC, Washington, DC.

Bradbury, J.A. 1989. The policy implications of differing concepts of risk. *Science, Technology & Human Values* 14 (4): 380–399.

BTWC. 1986. *Second review conference of the parties to the convention on the prohibition of the development, production and stockpiling of bacteriological (biological) and Toxin Weapons and on their Destruction. Final Document. Part II. Final Declaration. BWC/CONF.II/13/II.* Geneva: Biological and Toxin Weapons Convention.

———. 2008. *Meeting of states parties to the convention on the prohibition of the development, production, and stockpiling of bacteriological (biological) and toxin weapons and on their destruction. Report. BWC/MSP/2008/5*. Geneva: Biological and Toxin Weapons Convention.

———. 2012. *Meeting of states parties to the convention on the prohibition of the development, production, and stockpiling of bacteriological (biological) and Toxin Weapons and on their Destruction. Report. BWC/MSP/2012/5*. Geneva: Biological and Toxin Weapons Convention.

Buganza, T., M. Kalchschmidt, E. Bartezzaghi, and D. Amabile. 2013. Measuring the impact of a major project management educational program: The PMP case in Finmeccanica. *International Journal of Project Management* 31 (2): 285–298.

Caskey, S., and E.E. Sevilla-Reys. 2015. Risk assessment. In *Laboratory biorisk management: Biosafety and biosecurity*, ed. R.M. Salerno and J.M. Gaudioso, 45–64. Boca Raton/London/New York: CRC Press, Taylor & Francis Group.

Caskey, S., J. Gaudioso, R.M. Salerno, S. Wagener, M. Shigematsu, G. Risi, J. Kozlovac, and V. Halkjaer-Knudsen. 2010. *Biosafety risk assessment methodology. SAND2010-6487*. Albuquerque: Sandia National Laboratories.

Dando, M.R., and B. Rappert. 2005. Codes of conduct for the life sciences: Some insights from UK Academia. *Bradford briefing papers on strenghtening the biological weapons convention, Second Series*. 16(Second Series).

Depoy, J., J. Phelan, P. Sholander, B. Smith, G.B. Varnado, and G. Wyss. 2005. Risk assessment for physical and cyber attacks on critical infrastructures. In *MILCOM 2005-2005 IEEE Military Communications Conference*, 1961–1969. IEEE.

Deutscher Ethikrat. 2014. *Biosecurity — Freedom and responsibility of research*. Berlin: German Ethics Council. Available from: http://www.ethikrat.org/files/opinion-biosecurity.pdf.

Edmonds, G.S., R.C. Branch, and P. Mukherjee. 1994. A conceptual framework for comparing instructional design models. *Educational Technology Research and Development* 42 (4): 55–72.

G8. 2009. *L'Aquila Statement on Non Proliferation ANNEX B: Recommendations for a coordinated approach in the field of Global WMD Knowledge proliferation and scientists engagement*. L'Aquila: G8 Global Partnership. [Accessed 18 August 2016]. Available from: http://www.g8italia.it/static/G8_Allegato/2._LAquila_Statent_on_Non_proliferation.pdf.

Garrick, B.J. 2008. *Quantifying and controlling catastrophic risks*. Waltham: Academic Press.

Germany. 2005. *Codes of conduct and their application in the life sciences at universities. Working Paper. BWC/MSP/2005/MX/WP.12*. Geneva: Biological and Toxin Weapons Convention.

Grainger, L., and D. Turegeldiyeva. 2015. Biorisk management training. In *Laboratory biorisk management: Biosafety and biosecurity*, ed. R.M. Salerno and J.M. Gaudioso, 101–124. Boca Raton/London/New York: CRC Press, Taylor & Francis Group.

Gustafson, K.L., and R.M. Branch. 2002. *Survey of instructional development models*. 4th ed. Syracuse: ERIC Clearinghouse on Information & Technology.

Hadden, S.G. 1984. Introduction: Risk policy in American institutions. In *Risk analysis, institutions, and public policy*, ed. S.G. Hadden, 3–17. Port Washington: Associated Faculty Press.

Halpern, D.F., and M.D. Hakel. 2003. Applying the science of learning to the university and beyond: Teaching for long-term retention and transfer. *Change: The Magazine of Higher Learning* 35 (4): 36–41.

Hodell, C. 2011. *ISD from the ground up: A no-nonsense approach to instructional design*. 3rd ed. Alexandria: American Society for Training & Development.

IAP. 2005. *IAP statement on biosecurity*. Trieste: InterAcademy Panel. [Accessed 18 Aug 2016]. Available from: http://www.interacademies.net/File.aspx?id=5401.

———. 2014. *IAP statement on realising global potential in synthetic biology: Scientific opportunities and good governance*. Trieste: IAP – the global network of science academies.

India. 2005. *Indian initiatives on codes of conduct for scientists. Working Paper. BWC/MSP/2005/MX/WP.23*. Geneva: Biological and Toxin Weapons Convention.

Japan. 2008. *Oversight, education, awareness raising, and codes of conduct for preventing the Misuse of Bio-Science and Bio-Technology. Working Paper. BWC/MSP/2008/MX/WP.21*. Geneva: Biological and Toxin Weapons Convention.

Kaplan, S. 1997. The words of risk analysis. *Risk Analysis* 17 (4): 407–417.

Kaplan, S., and B.J. Garrick. 1981. On the quantitative definition of risk. *Risk Analysis* 1 (1): 11–27.

Kates, R.W., and J.X. Kasperson. 1983. Comparative risk analysis of technological hazards (a review). *Proceedings of the National Academy of Sciences* 80 (22):7027–7038.

Kidd, J.R. 1974. *How adults learn*. Cambridge: Cambridge Book Co.

Kirkpatrick, D.L. 1979. Techniques for evaluating training programs. *Training and Development*: 178–192.

Kirkpatrick, D.L., and J. Kirkpatrick. 2006. *Evaluating training programs*. 3rd ed. San Francisco: Berrett-Koehler Publishers.

———. 2007. *Implementing the four levels: A practical guide for effective evaluation of training programs*. San Francisco: Berrett-Koehler Publishers.

Knight-Jones, T.J.D., and J. Rushton. 2013. The economic impacts of foot and mouth disease – What are they, how big are they and where do they occur? *Preventive Veterinary Medicine* 112 (3–4): 161–173.

Knowles, M.S. 1984. *The adult learner: A neglected species*. 3rd ed. Houston: Gulf Publishing.

Kontoghiorghes, C. 2004. Reconceptualizing the learning transfer conceptual framework: Empirical validation of a new systemic model. *International Journal of Training and Development* 8 (3): 210–221.

Krathwohl, D.R. 2002. A revision of Bloom's taxonomy: An overview. *Theory Into Practice* 41 (4): 212–218.

Mancini, G.M., and A. Fasani. 2012. Experiences in promoting awareness on biosecurity and dual use issues in European Universities. In *Yearbook of biosecurity education 2012*, ed. J.F. Sture, 76–86. Bradford: University of Bradford.

Mancini, G.M., and J. Revill. 2017. "We"re Doomed!' a critical assessment of risk framing around chemical and biological weapons in the twenty-first century. In *Cyber and chemical, biological, radiological, nuclear, explosives challenges: Threats and counter efforts*, ed. M. Martellini and A. Malizia, 311–325. Cham: Springer International Publishing.

McCombs, B.L. 1986. ERIC/ECTJ Annual Review Paper: The Instructional Systems Development (ISD) Model: A review of those factors critical to its successful implementation. *Educational Communication and Technology*, pp.67–81.

Minehata, M., and N. Shinomiya. 2010. Chapter 5: Japan: Obstacles, lessons and future. In *Education and ethics in the life sciences*, ed. B. Rappert. Canberra: ANU E Press.

Molenda, M. 2003. In search of the elusive ADDIE model. www.ispi.org. 3(May/June 2003), pp. 34–36.

NASEM (National Academies of Sciences, Engineering, and Medicine). 2018. *Governance of dual use research in the life sciences: Advancing global consensus on research Oversight: Proceedings of a workshop*. Washington, D.C.: The National Academies Press. https://doi.org/10.17226/25154.

National Research Council. 2010. *Challenges and opportunities for education about dual use issues in the life sciences*. Washington, D.C.: National Academies Press.

Netherlands. 2015. *Statement on science and Technology, 12-8-2015. Ms. Ayse Aydin, Senior policy adviser, ministry of foreign Affairs of the Kingdom of the Netherlands. 2015 Meeting of Experts of the States Parties of the biological and Toxin Weapons Convention*. Geneva: Biological and Toxin Weapons Convention.

Novossiolova, T., G.M. Mancini, and M.R. Dando. 2013. Effective and sustainable biosecurity education for those in the life sciences: The benefits of active learning. *Bradford Briefing Papers on Strenghtening the Biological Weapons Convention, Third Series*. 7(Third Series).

NSABB. 2007. *Proposed framework for the oversight of dual use life science research: Strategies for minimizing the potential misuse of research information*. Available from: http://osp.od.nih.gov/sites/default/files/biosecurity_PDF_Framework%20for%20transmittal%200807_Sept07.pdf.

OECD. 2007. *OECD Best practice guidelines for biological resource centres*. Paris: Organization for Economic Cooperation and Development Publishing.

Patel, V., N. Yoskowitz, and J. Arocha. 2009. Towards effective evaluation and reform in medical education: A cognitive and learning sciences perspective. *Advances in Health Sciences Education* 14 (5): 791–812.

Praslova, L. 2010. Adaptation of Kirkpatrick's four level model of training criteria to assessment of learning outcomes and program evaluation in Higher Education. *Educational Assessment, Evaluation and Accountability* 22 (3): 215–225.

Rappert, B. 2007. Education for the life sciences: Choices and challenges. In *A web of prevention: "Biological Weapons, life sciences and the governance of research"*, 51–66. London: Routledge.

Renn, O. 2006. *Risk governance. Towards an integrative approach. White Paper No. 1*. Geneva: International Risk Governance Council.

Revill, J., M. Candia Carnevali, A. Fosberg, Z.K. Shinwari, J. Rath, and G.M. Mancini. 2012. Lessons learned from implementing education on dual-use in Austria, Italy, Pakistan and Sweden. *Medicine, Conflict, and Survival* 28 (1): 31–44.

Russian Federation. 2005. *Some reflections on the Ethic Norms and codes of conduct for scientists majoring in Biosciences. Working Paper. BWC/MSP/2005/MX/WP.18*. Geneva: Biological and Toxin Weapons Convention.

Salas, E., and J.A. Cannon-Bowers. 2001. The science of training: A decade of progress. *Annual Review of Psychology* 52 (1): 471–499.

Sax, L.J. 2004. Citizenship development and the American college student. *New Directions for Institutional Research* 2004 (122): 65–80.

Steensma, H., and K. Groeneveld. 2010. Evaluating a training using the 'four levels model'. *Journal of Workplace Learning* 22 (5): 319–331.

Stern, P.C., and H.V. Fineberg. and others1996. *Understanding risk: Informing decisions in a democratic society*. Washington, D.C.: National Academies Press.

Whitby, S., and M.R. Dando. 2010. Biosecurity awareness-raising and education for life scientists: What should be done now? In *Education and ethics in the life sciences*. Canberra: ANU E Press.

WHO. 2010. *Responsible life sciences research for global health security*. Geneva: World Health Organization. Available from: http://www.who.int/csr/resources/publications/HSE_GAR_BDP_2010_2/en/index.html.

Young, S., H.H. Willis, M. Moore, J.G. Engstrom, and National Defense Research Institute, Acquisition and Technology Policy Center and Cooperative Threat Reduction Program of the U.S. Department of Defense. 2014. *Measuring Cooperative Biological Engagement Program (CBEP) performance: Capacities, capabilities, and sustainability enablers for biorisk management and biosurveillance*. Washington, D.C.: National Academies Press.

Microbial Forensics: Detection and Characterization in the Twenty-first Century

K. Lane Warmbrod, Michael Montague, and Nancy D. Connell

1 Introduction

The twenty-first century has been designated the Age of Biology (Glover 2012). Indeed, the global economic value of the biobased economy is estimated to be in the range of USD 388 billion (Issa et al. 2019; Ugalmugale and Swain 2016). The powerful economic drivers of biotechnology lead to remarkable innovation and creativity, accompanied by rapid advances in science and technology. Scientific techniques are becoming more clever and faster. The use of microbial agents as weapons in war, terrorism or crime needs immediate investigation of the source of the material. There are multiple angles from which to approach the task of tracing the origin of use of microbial agents in a deliberate bioweapons event. Traditional forensic science methods and techniques used in criminal investigations (hairs, fibers, fingerprints, handwriting, etc.) or psychological or behavioral sciences (linguistics, psychological profiles, etc,) are used in most forensic investigations. For unlawful acts using microbial agents, bioforensic analysis of microbial evidence must answer the key questions of who, what, where, how, and why. Microbial forensics has been defined as "a scientific discipline dedicated to analyzing evidence from a bioterrorism act, biocrime, or inadvertent microorganism/toxin release for attribution purposes" (Budowle et al. 2003). The field of microbial forensics was explored in depth at an international conference in 2014 held by The National Research Council of the US-National Academies in cooperation with the Croatian Academy of Sciences and Arts, the U.K. Royal Society and the International Union of Microbiological Societies (National Research Council 2014)

"The goal of the microbial forensics process is to use microbial analyses and other evidence to fix a questioned source to a position on a continuum that ranges from "could *not* have

K. L. Warmbrod · M. Montague · N. D. Connell (✉)
Center for Health Security, Johns Hopkins Bloomberg School of Public Health, Baltimore, MD, USA
e-mail: NancyConnell@jhu.edu

M. Martellini, R. Trapp (eds.), *21st Century Prometheus*,
https://doi.org/10.1007/978-3-030-28285-1_16

originated from" to "*consistent* with having originated from" to "absolutely *did* originate from" a known source. Again, identification is simpler than attribution. Exclusion, association, and attribution are dependent on several key factors, with more value and weight given to attribution derived when more possible sources can be eliminated. Uncertainty and confidence must be stated, either qualitatively or quantitatively."

The questions that microbial forensics (National Research Council 2014) seek to answer are:

1. "*What is the threat agent?* Usually establishing this has not been difficult, although it may not occur in an optimal time frame.
2. *Is it probative or relevant?* Establishing certainty here is more difficult. Scientists may be working with trace quantities, for example, or analysis may require an understanding of the sample background to understand the source.
3. *Can it be linked to a source?* Establishing this [link] demands understanding the power of methods used to discriminate and characterize with acceptable confidence limits.
4. *What are the meaning and weight of the conclusion?*"

The field of microbial forensics was relatively new in 2001 when it was put the test in the analysis of the anthrax attacks in the United States. A discussion of the U.S. government's microbial forensic investigation into the release of the anthrax spores through the U.S mail service will serve to illustrate the range of scientific approaches to solving such a crime; other sections of this chapter will survey forensic methodologies related to recent advances in genetic sequencing and other technologies.

2 Anthrax Attacks in the United States

In the fall of 2001, a series of letters bearing spores of *Bacillus anthracis* was released in the U.S. mail, leading to 22 documented infections, of which five were fatal (Amerithrax or Anthrax Investigation n.d.), although these numbers have been disputed (Cymet and Kerkvliet 2004). The subsequent investigation carried out by the US Federal Bureau of Investigation (FBI) lasted 9 years, culminating the FBI's formal closing of the case in February of 2010. The perpetrator was not apprehended; a discussion of specific attribution is outside the scope of this forum. However, the details of scientific arm of the investigation are laid out in detail in a number of academic publications (Griffith et al. 2014; Rasko et al. 2011), books (Decker 2018; Guillemin 2011) and from the FBI itself (The United States Department of Justice 2010) as well as in a consensus study carried out by the US-National Academies (National Research Council 2011). The attack material – anthrax spores - from the letters was subjected to extensive physical and chemical analysis:

- spore preparation and purification
- surrogate preparation and purification
- size and granularity of letter material

- silicon and other elements
- silicon in spore coat
- elemental analysis
- bacterial growth conditions
- media component analysis
- volatile organic compounds
- radiocarbon dating
- stable isotope analysis
- water samples
- envelope measurements.

An extensive network of government, academic and national laboratories as well as private industry worked on these analyses. The key conclusions of the scientific arm of the FBI investigation are summarized below and serve as a comprehensive description of the state of the art of forensic analysis at the time.

In nature, *Bacillus anthracis* bacteria infect animals such as horses and cows, and by replicating to large numbers in the blood, they kill the host. After the animal succumbs, the bacteria continue to replicate by feeding on the decaying remains. Once there are no nutrients left, the bacteria convert into long-lasting spores, remaining viable for years to centuries, with extremely tough coating that confers resistance to heat and desiccation. Once the spores come in contact with nutrients again (as when they infect another host) they begin to grow, and the cycle begins anew. Anthrax spores are thus ideal weapons, as they are highly stable and can infect via air transmission (Goel 2015).

The spores in the material of the 2001 US mail attacks were extremely pure. As part of the investigation, experimental reconstitution of the spores was performed using different growth conditions to try to determine the conditions under which the bacteria were grown and then induced to form spores. The densities of the different attack materials were evaluated to estimate the volume required to create the final spore preparation. Attempts were made to recreate "surrogate" samples, and the size and granularity of the samples were carefully measured. These kinds of analyses were used to offer some indication of where the preparation might have been performed, for example, what kind of laboratory or the size of the vessels needed for growth and preparation (Beecher 2006; National Research Council 2014). Deliberate mailing of anthrax spores might be considered weaponization of spores, whether or not additional treatment of the spores the make them more infectious or better able to "float" in air; historically, offensive biological weapons programs have been described as taking extensive care in how the spores are prepared for dissemination (Riedel 2005). The presence or absence of silicon or bentonite remains a contentious topic, as the physical and chemical analyses have not been considered conclusive by the scientific community (Beecher 2006). In their continued search for signatures, the FBI arranged for extensive elemental analysis using a large suite of chemical approaches, including inductively-coupled plasma-optical emission spectrometry, scanning electron microscopy with energy dispersive X-ray analysis, nanometer secondary ion mass spectrometry, accelerator mass spectrometry, isotope ratio mass spectrometry, gas chromatography-mass spectrometry, liquid chromatography-mass spectrometry, aerodynamic particle sizing and aerodisperser analysis. Radioisotope analysis and carbon dating suggested that at least one of the samples was created between 1998 and 2001. Overall, the NAS report concluded that the data suggested no additives were associated with the attack material and that the extensive chemical analyses, attesting to significant effort on the art of the investigators, resulted in novel data and/or methodology for analysis but no further evidence to determine the source.

Genetic analysis of the attack material was extensive although some of the collection and statistical methods used in the studies were flawed, according the NAS study. In brief, the spores and the patient samples were quickly determined to be harboring the same strain of *B. anthracis*, namely, the experimental "Ames" strain. Interestingly, and as a warning for the future, *Bacillus anthracis* is an unusually genetically stable bacterial species (Keim and Smith 2002); many other so-called Select Agents, known the be potential biological weapons, have unstable genomes, which means comparison between multiple strains of the same species would be much more difficult. The spores of the attack material when germinated (i.e., exposed to nutrients and induced to resume growth) gave rise to colonies among which were a small proportion that exhibited altered morphologies. These "isomorphic strains" were sequenced and associated with specific mutations in their DNA. Expertly validated assays for the presence of these mutations were created and used to assess a 1071-sample "repository" of strains collected from laboratories in the U.S. and other countries. Presence of the mutations in samples from a specific laboratory might indicate that laboratory is the source. Of the repository samples assayed, only one source contained these multiple mutations. This source was provided by a government infectious disease research laboratory. The conclusions of the committee were that the genetic analysis showed genetic similarity between the attack material and the government laboratory, but that other possible explanations were not pursued by the investigation. Clearly, application of high through-put next-generation sequencing to the evidentiary material would have useful and perhaps decisive.

3 Sequencing Techniques

Genetic sequencing techniques are quickly advancing to allow more sequences to be read at greatly reduced cost, enabling the technology to be used more often and applied to a greater diversity of projects (Ziogas et al. 2018). Sequencing technologies are becoming more accurate, enabling more information to be gathered for each sample. Next generation sequencing (NGS), or deep sequencing, refers to a relatively new technology for sequencing genomes. NGS can sequence the entire human genome in a day (Behjati and Tarpey 2013). Because NGS can determine the sequence of many genomes at once and multiple times, the accuracy and sensitivity of sequencing is relatively high and continues to improve as novel methods are developed. NGS technology allows researchers to characterize microbial communities with or without knowing what is in the sample. Shotgun metagenomics is a sequencing method with no selection for which parts of the DNA is sequenced; anything in the sample will be sequenced. The sequencing reads are assembled and annotated using reference genomes to identify what was in the original sample. This method gives the researcher extensive information about the bacterial community. However, the approach also requires complicated and time-consuming analysis and is limited by the availability of known reference genomes. Another method of sequencing, amplicon sequencing, can be used to target a specific gene, organism, or group of organisms. In this method, a specific gene of interest is amplified from every organism in the sample. The amplified products are then sequenced to identify the organisms. This method helps reduce background noise not essential for the experimental inquiry but may exclude helpful information. In addition to these sequencing methods, there are several protocols that can provide other information, such as which genes are active at a given timepoint. NGS enables metagenomic and microbiome studies, discussed below, that previous technologies were not capable of handling. As NGS becomes cheaper and more accessible, the forensic potential for metagenomics and microbiomes increases.

4 Metagenomics in Forensics

Metagenomic analysis of microbial communities enables the identification of microbial species and assessment of some functional capabilities of the microbes within a sample. Metagenomic studies have wide-spread application, from identifying the causative agent of an undiagnosed disease (Brown et al. 2018) to analyzing the ability of bacteria species to degrade polysaccharides (Kougias et al. 2018). For forensic metagenomics, samples are taken from an area or item of interest, such as a footprint and the bottom of a shoe. Once these samples are collected, the genetic material is extracted from the sample and shotgun sequencing is performed. Once sequencing is complete, the individual reads are compiled and analyzed using various bioinformatic approaches. Researchers may then compare the genetic information found in the sample to a reference genome in a database to identify the microbe. The amount of each species can be quantified to assess the population structure from the sample. Additionally, both the genome and transcriptome of the organisms in a given sample may be analyzed in order to assess both the identities of the microbes and the genes expressed by the microbes, allowing investigators to assess the behaviors and functions of the microbes within the sample. Both the diversity and behavior of the organisms within the population may be used to as a microbial fingerprint.

Studies have found patterns in the composition of microbial populations based on geography, resource availability, and interaction with other microbes. External factors influence the evolution and stability of a microbial community, so assessment of the community may be used to understand external factors (Smith et al. 2012; Vieites et al. 2009). The selection pressures imposed on microbial communities by different factors create patterns in populations and individual genomes.

As researchers identify and characterized these patterns that arise in response to specific selection pressures, there is higher potential for the patterns and comparisons to give useful information to investigators. The microbes can tell investigators about the past locations in which the item they were sampled from has previously existed. As the pattern recognition and pool to which compare samples is characterized improves, so will the resolution and specificity can provide.

The composition and behavior of a microbial population for a given location could be similar to the composition and behavior of another microbial community in a different location with similar environmental factors (Edwards et al. 2006). Investigators could take a sample of the microbes on an object and compare that population to known populations. If the sample is similar to a known population, then there is a chance that the item the sample was taken from was at one time in a similar environment to where the known sample was taken. This type of analysis would allow investigators to narrow a field of potential locations. Even within the same geographical area, slight differences between environmental influences within the area may change the microbial community composition and function creating sub-regions within the greater area, allowing for finer definition in the potential map of where an item was previously located (Tringe et al. 2005; Fierer and Jackson 2006; Jansson and Hofmockel 2018). Microbes retain their genomic signatures, allowing for species identification, even as species react and evolve together (Dick et al. 2009). Because identification is possible despite evolution, comparing the genomes of different members of the same species may be used as a time stamp. Measuring the genetic differences between members of the same species to assess the evolutionary differences can become a proxy for time, and potentially provide a window for when two different populations were last in contact. Molecular analysis processes include probability scores, which can be used by investigators and juries to assess the likelihood of claims.

As the field advances, more applications of metagenomics in forensics and surveillance will appear. For example, officials could take samples in areas where they think a weapon of mass destruction has been made, stored, or tested in order to verify their suspicion. Microbes can incorporate distinctive isotopes frequencies into their biomass (Coyotzi et al. 2016). Microbes from the suspected location could be analyzed to determine the length of time since an object shedding these isotopes was in the environment and identify the type of isotopes. This could give investigators conformation that nuclear material was previously located in the area. A similar process could confirm that a specific chemical was previously in a location. In addition to environmental sampling, samples could be taken off suspects of items thought to have come in contact with a weapon to help pinpoint where the weapon had previously been or the type of weapon. Samples taken from shoes may carry distinct soil signatures that match to a specific area, giving investigators a clue about where a suspect has previously been (Khodakova et al. 2014). Metagenomics could provide authorities with critical information for their investigations in each of these situations, especially when combined with information gained from other forensic methods like mass spectrometry.

5 Microbiomes in Forensic Inquiry

On the skin of all creatures there is a vast community of microorganisms that collectively create the microbiome of that individual. The composition of this community is highly impacted and shaped by the individual's environment and lifestyle (Gilbert et al. 2018). Diet, clothing, soaps, medications, and previous locations all affect the microbiome. Because so many different factors contribute to shaping the microbiome, microbiomes between species, people, and even body parts of the same individual are highly diverse (Metcalf et al. 2017; Oh et al. 2014; Ross et al. 2018). Notably, the human microbiome is relatively stable overtime (Costello et al. 2009). While the proportions of different species within the community do change in response to different environmental factors or lifestyle choices, the composition of the community is stable (David et al. 2014; Oh et al. 2016). This stability and diversity allow the microbiome to become a biomarker unique to an individual, creating a microbial fingerprint.

As they go about their day, people shed trace amounts of their microbiomes into the air and onto objects they come in contact with, like a keyboard (Metcalf et al. 2017). In this way, humans shape the indoor microbiome of their homes and workplaces. Metagenomics can match employees to their specific phones or keyboards by comparing the microbiome on employees' hands to the microbiomes found on the office equipment (Fierer et al. 2010; Lax et al. 2014). Matching like this requires authorities to have samples from suspects to compare against. Even if there is no such sample, the unique microbiome signature lifted from an item can give authorities important information. Gender, disease status, lifestyle, home environment, and other characteristics or traits of a suspect can be gleaned from sampling the microbiome left behind. Studies are underway to determine how quickly microbiomes change following a person's initial interaction with the environment, which could provide temporal evidence (Lax et al. 2015). This work will require extensive statistical analysis and computational models which are currently under development.

Privacy is a concern for microbiome forensics. Like DNA, everyone has a unique microbiome. However, trace amounts of the microbiome are shed more often than DNA and there is little a person can do to control this shedding. Unlike DNA, microbiomes can suggest relationships between people who are not family. Microbiomes are similar between members of the same household, such as co-inhabiting couples or roommates, and can be used to identify one another (Kort et al. 2014; Song et al. 2013). There is a great potential for information to be gained from the microbiome of not just one person, but also their contacts. Additionally, microbiomes contain information about an individual's health, information that is typically considered private even for criminals (Gilbert et al. 2018). Privacy must be addressed before metagenomic samples are taken for this purpose, or a repository with metadata containing information about health, diet, and country origin is created.

6 Contamination, Endemism, and Background Noise

Using microbes for forensics has the potential to be a powerful tool for law enforcement and prosecutors. However, there are several challenges that need to be addressed before this becomes a wide-spread and accurate methodology. Accuracy will be especially challenging considering the risk of contamination of samples. Because microbes are nearly ubiquitous in nature, there are many opportunities for contamination of samples. Extreme care will need to be taken to decrease the risk of contamination, including strict, standardized protocols and minimum requirements for tool sterilization. Additionally, comprehensive testing will require several samples. If there is limited biomass available, the accuracy and robustness of any testing will be reduced (Clarke et al. 2017). Currently, there is little evidence concerning the sensitivity and accuracy of trace amounts of microbiomes for forensic applications.

There are some microbes that naturally found only in some geographic areas or types of locations, such as hyperthermophilic archaea in thermal vents (Whitaker 2003) or a bacterial species only found in one part of the world (Dulger et al. 2005). This kind of specificity may enable authorities to track movements by sampling for these endemic pathogens. From a suspect's clothing or an item left behind at a crime scene, investigators can do targeted sequencing looking for one of these endemic microbes. If investigators have a specific region, they believe the item might have previously been found, they could run a series of targeted sequencing protocols to narrow down the location. Additionally, there are some microbes that are only found in association with certain plants or animals, which could give authorities further information about where a suspect or item has previously been. The collective results of such testing could provide a very specific location, especially when these techniques are combined with other tools such as mass spectrometry and machine learning to further narrow the possibilities.

Relying on microbial genetics for forensics is limited by the sample size and sensitivity of detection. Because microbes are ubiquitous in nature, environmental sampling for surveillance or forensics will collect billions of organisms (Daniel 2005). Only a fraction of these organisms will be sequenced from the sample using technology available today. As the field of forensic metagenetics expands, scientists will gain better understanding of which species are most important for their needs. Methods can then be developed to select for these species within a sample before sequencing, enriching the sample to improve the likelihood of researchers getting the information most helpful for their needs. This will decrease help decrease the background readings that will not meaningfully contribute to forensic efforts. Enrichment and selection protocols, as well as sample collection and handling protocols, will need to be standardized to be acceptable evidence in a court.

7 Repositories and Databases

For metagenomics to be a viable option for forensics, there must be a reference database available for sample comparison. Annotated genomes in a database are needed for identifying the microbes in the sample and comparing genome and transcriptome differences. Additionally, authorities will need the ability to compare the sample population to a standard or previous sample to assess how the population composition is different. The database will be a critical resource for microbial forensics and will need to have enough entries for investigators to make meaningful identifications and geolocations for the key microbes.

Several microbial repositories have been created in the last decade. Many are publicly available and specialize in a specific field or sample type, such as soil microbes or human genes (Stenson et al. 2017; Manter et al. 2017). Some of these repositories are coupled with free analysis pipelines so researchers have a standard mechanism for analyzing similar data (Mitchell et al. 2016). These public resources are quickly growing as more researchers add their findings to the databases, allowing others to use the collective work for reference and further projects. Central repositories are especially helpful for decreasing the burden on individual laboratories or organizations, as the creation and upkeep of such systems requires extensive expertise and resources. Databases also have weaknesses: variation among microbes considered to be part of the same species is not readily covered in databases, resulting in ambiguity about consensus sequences. Bioinformatic approaches rely comparing a sample to one or more sequences and determining the amount of variation between these sequences for identification and diversity analysis. When the consensus sequences in the database are not correct due to incorrect sequencing or the consensus sequences has been inappropriately labeled as such, analysis of new samples is biased and potentially incorrect.

There are risks associated with such publicly available databases: anyone can contribute to the collection. Because of their importance for forensics, repository and database could be a target of attack for a malicious actor. Public repositories are especially vulnerable, as anyone could alter the collection. False entries could be intentionally added to the database to misguide officials. Alternatively, the database might be vulnerable to cyber-attacks that shut down the site or erase the information. Information could be knowingly altered to something false without users being aware (Fernandez et al. 2004). For forensics, it will be especially important to verify the trustworthiness and integrity of the data, which could be harder on publicly available resources compared to alternatives with controlled access. Authorities relying on microbial forensics will likely need to develop and curate their own database and repository. This will limit the speed at which new information can be added and prohibit the researchers from accessing the most up to date research, but it will improve security.

High throughput DNA and RNA sequencing analysis requires several statistical tests, including assessing the confidence in the sequence itself. When attempting to match metagenetics from a sample to a particular person or place, there are several comparison steps that will have to measure confidence (Clarke et al. 2017). Novel analysis methods and statistical frameworks must be developed before this technology is ready for deployment by investigative authorities or used in a courtroom.

8 Public Health and Microbial Forensics

The 2014 report "Science Needs for Microbial Forensics" (National Research Council 2014) explored the commonalities between the needs of public and clinical health and those of microbial forensics. Indeed, it is likely that a biological weapons event would begin with an epidemiological investigation. For example, there are epidemiological clues that would suggest a biological or chemical attack, such as unusual disease presentation, cases of common diseases that do not respond to treatment, outbreak that is atypical for population or age group, significant number of cases at the same time ("point source with compressed epidemiologic curve"), etc. As an outbreak unfolds, it is essential to determine as quickly as possible whether the event due to natural causes or deliberate attack, as the nature of the subsequent investigation will shift. In the case of deliberate events, chain of custody must be maintained, and the gathering information must be held to standards that will withstand scrutiny in courts of law.

9 Synthetic DNA Attribution

A genetically engineered region of DNA or RNA, incorporated into an organism or not, may potentially be attributed to its designer based entirely upon design decisions made by that designer. Genetic engineering is no different from any other form of engineering. In engineered constructs (software, architecture, fashion, writing, or a gene), the engineer must make dozens of decisions. Making these decisions is completely unavoidable as they represent solutions to problems. At the same time, the actual solution is arbitrary: any of thousands of equivalent choices would suffice. Below is a brief exploration of some of the design decisions necessary for any genetic engineering project.

- What organism was chosen? Is it a common laboratory organism, or an unknown or uncharacteristic one?
- What gene was chosen?
- Is the gene protein or RNA coding?
- Is the gene inserted on a plasmid or incorporated into the organism's genome?
- How was the gene constructed?
- What regulatory elements surround the gene (promoter, operator, ribosome binding site, terminator, etc.)?

These questions, which are just a select few of the required questions, queries dozens of separate design decisions needed to insert one gene in one organism but is by no means exhaustive. More complex genetic engineering, involving multiple genes or multiple organisms, represents an exponentially larger list of such design decisions. For each new gene or organism, each of these questions will have a separate answer, creating a very complicated decision-making pathway that will create patterns. As complexity of the engineering increases, the number of design decisions also increases exponentially, potentially increasing the ease of attribution.

Different genetic engineers will likely choose different solutions to these same design issues, even when trying to build the same biological function or organism. The basis for this variance incorporates multiple factors including personal style and preference. The most influential factor will be the individual engineer knowledge and training. For example, a person who knows little about a gene's upstream regulatory structure would likely not think to try fine-tuning the ribosome binding site of a gene to control its expression and would instead swap the entire promoter/upstream sequence with alternatives until the correct function is achieved. In the end, either solution would work, but they illustrate certain kinds of knowledge in the designer. Crucially, both patterns of knowledge convey identifying information about their respective designers to people examining the engineered organism after the fact. While no individual design decision will be conclusive, the collective decisions can elucidate a more specific actor profile, creating a designer fingerprint. The unavoidable nature of a design-decision fingerprint creates a bioinformatic capability by which a biological attack using an engineered organism might be attributed to the weapon's designer and builder.

Preliminary work demonstrating that genetic engineering design decisions can produce such a finger-print allowing for attribution has been published (Nielsen and Voigt 2018). In this research, machine learning was employed to analyze a global, non-profit repository of engineered plasmid DNA constructs as a proxy for a database of engineered biological weapons (Addgene: Homepage n.d.). The machine learning algorithm was able to identify patterns in the DNA sequence of submitted plasmids that correlated with and predicted the submitting lab of individual entries in the database.

The decision-making process will enable authorities to narrow the list of potential suspects following the release of an engineered agent but may fail to conclusively identify the exact actor. Complete attribution to the exact responsible parties is not necessary for attribution to be useful in general. More general conclusions, such as determination that development of the agent required a certain level of funding and expertise or resources, is still useful even though the exact identity of biological engineers is not determined. Additionally, attribution conclusions will be combined with other classes of less technical information to build a more complex picture of the parties behind any specific attack. This is similar to investigations of other crimes- many classes of clue contribute to the prosecutor's evidence.

10 Conclusion: Gaps and Challenges

The field of microbial forensics continues to mature as advances in science and technology bring increased precision, expanding databases and reduced costs. Next generation/deep sequencing has revolutionized the molecular characterization of microbes and their communities. These kinds of analyses will increase the utility of microbial forensics into investigations of bioterrorism and biowarfare. Importantly, these new approaches will also provide impetus for the application of methods of microbial genetic analysis to other WMD categories. However, significant challenges remain: a systematic and international effort to categorize microbial background; high confidence protocols for distinguishing quickly between natural and deliberate outbreaks; validated and standardized protocols and methodology for sample analyses; mechanisms for data sharing across borders; and global, shared access to efficient molecular diagnostics.

References

Addgene: Homepage. n.d.. https://www.addgene.org/. Accessed 27 June 2019.

Amerithrax or Anthrax Investigation. n.d. Federal Bureau of Investigation. https://www.fbi.gov/history/famous-cases/amerithrax-or-anthrax-investigation. Accessed 29 June 2019.

Beecher, D. 2006. Forensic application of microbiological culture analysis to identify mail intentionally contaminated with Bacillus Anthracis spores. *Applied and Environmental Microbiology* 72 (8): 5304–5310.

Behjati, S., and P. Tarpey. 2013. What is next generation sequencing? *Archives of Disease in Childhood. Education and Practice Edition* 98 (6): 236–238.

Brown, J., T. Bharucha, and J. Breuer. 2018. Encephalitis diagnosis using metagenomics: Application of next generation sequencing for undiagnosed cases. *Journal of Infection* 76 (3): 225–240.

Budowle, B., S. Schutzer, A. Einseln, L. Kelley, A. Walsh, J. Smith, B. Marrone, J. Robertson, and J. Campos. 2003. Building microbial forensics as a response to bioterrorism. *Science* 301 (5641): 1852–1853.

Clarke, T., A. Gomez, H. Singh, K. Nelson, and L. Brinkac. 2017. Integrating the microbiome as a resource in the forensics toolkit. *Forensic Science International: Genetics* 30: 141–147.

Costello, E., C. Lauber, M. Hamady, N. Fierer, J. Gordon, and R. Knight. 2009. Bacterial community variation in human body habitats across space and time. *Science* 326 (5960): 1694–1697.

Coyotzi, S., J. Pratscher, J. Murrell, and J. Neufeld. 2016. Targeted metagenomics of active microbial populations with stable-isotope probing. *Current Opinion in Biotechnology* 41 (2016): 1–8.

Cymet, C., and G. Kerkvliet. 2004. What is the true number of victims of the postal anthrax attack of 2001? *The Journal of the American Osteopathic Association* 104 (11): 452.

Daniel, R. 2005. The metagenomics of soil. *Nature Reviews Microbiology* 3 (6): 470–478.

David, L., A. Materna, J. Friedman, M. Campos-Baptista, M. Blackburn, A. Perrotta, S. Erdman, and E. Alm. 2014. Host lifestyle affects human microbiota on daily timescales. *Genome Biology* 15 (7): R89.

Decker, S. 2018. *Recounting the anthrax attacks: terror, the task force, and the evolution of forensics in the FBI.* Rowman & Littlefield Publishers, Lanham, MD.

Dick, G., A. Andersson, B. Baker, S. Simmons, B. Thomas, A. Yelton, and J. Banfield. 2009. Community-wide analysis of microbial genome sequence signatures. *Genome Biology* 10 (8): R85.

Dulger, B., E. Ugurlu, C. Aki, T. Suerdem, A. Camdeviren, and G. Tazeler. 2005. Evaluation of antimicrobial activity of some endemic Verbascum., Sideritis., and Stachys. Species from Turkey. *Pharmaceutical Biology* 43 (3): 270–274.

Edwards, R., B. Rodriguez-Brito, L. Wegley, M. Haynes, M. Breitbart, D. Peterson, M. Saar, S. Alexander, E.C. Alexander, and F. Rohwer. 2006. Using pyrosequencing to shed light on deep mine microbial ecology. *BMC Genomics* 7 (1): 57.

Fernandez, E., M. Larrondo Petrie, and T. Sorgente. 2004. Security models for medical and genetic information. In *Proceedings of the IADIS international conference (e-Society 2004)*, 509–516. Avila.

Fierer, N., and R. Jackson. 2006. The diversity and biogeography of soil bacterial communitites. *PNAS* 103 (3): 626–631.

Fierer, N., C. Lauber, N. Zhou, D. McDonald, E. Costello, and R. Knight. 2010. Forensic identification using skin bacterial communities. *PNAS* 107 (14): 6477–6481.

Gilbert, J., M. Blaser, J.G. Caporaso, J. Jansson, S. Lynch, and R. Knight. 2018. Current understanding of the human microbiome. *Nature Medicine* 24 (4): 392–400.

Glover, A. 2012. *The 21st century: The age of biology*. Presented at the OECD Forum on Global Biotechnology, Paris, November 12. https://www.oecd.org/sti/emerging-tech/A%20Glover.pdf. Accessed 26 June 2019.

Goel, A. 2015. Anthrax: A disease of biowarfare and public health importance. *World Journal of Clinical Cases* 3 (1): 20–33.

Griffith, J., D. Blaney, S. Shadomy, M. Lehman, N. Pesik, S. Tostenson, L. Delaney, R. Tiller, A. DeVries, T. Gomez, M. Sullivan, C. Blackmore, D. Stanek, R. Lynfield, and The Anthrax Investigation Team. 2014. Investigation of inhalation anthrax case, United States. *Emerging Infectious Diseases* 20 (2): 280–283.

Guillemin, J. 2011. American anthrax: Fear, crime, and the investigation of the nation's deadliest bioterror attack. *Times Books*.

Issa, I., S. Delbrück, and U. Hamm. 2019. Bioeconomy from experts' perspectives – Results of a global expert survey. *PLoS One* 14 (5): e0215917.

Jansson, J., and K. Hofmockel. 2018. The soil microbiome—From metagenomics to metaphenomics. *Current Opinion in Microbiology* 43 (2018): 162–168.

Keim, P., and K.L. Smith. 2002. Bacillus Anthracis evolution and epidemiology. *Current Topics in Microbiology and Immunology* 271: 21–32.

Khodakova, A., R. Smith, L. Burgoyne, D. Abarno, and A. Linacre. 2014. Random whole metagenomic sequencing for forensic discrimination of soils. *PLoS One* 9 (8): e104996.

Kort, R., M. Caspers, A. van de Graaf, W. van Egmond, B. Keijser, and G. Roeselers. 2014. Shaping the oral microbiota through intimate kissing. *Microbiome* 2 (1): 41.

Kougias, P., S. Campanaro, L. Treu, P. Tsapekos, A. Armani, and I. Angelidaki. 2018. Spatial distribution and diverse metabolic functions of lignocellulose-degrading uncultured bacteria as revealed by genome-centric metagenomics. *Applied and Environmental Microbiology* 84 (18): e01244–e01218.

Lax, S., D.P. Smith, J. Hampton-Marcell, S. Owens, K. Handley, N. Scott, S. Gibbons, P. Larson, B. Shogan, S. Weiss, J. Metcalf, L. Ursell, Y. Vazquez-Baeza, W. Van Rreuren, N. Hasan, M. Gibson, R. Colwell, G. Dantas, R. Knight, and J. Gilbert. 2014. Longitudinal analysis of microbial interaction between humans and the indoor environment. *Science* 345 (6200): 1048–1052.

Lax, S., J. Hampton-Marcell, S. Gibbons, G. Colares, D. Smith, J. Eisen, and J. Gilbert. 2015. Forensic analysis of the microbiome of phones and shoes. *Microbiome* 3: 21–29.

Manter, D., J. Delgado, H. Blackburn, D. Harmel, A. Pérez de León, and C. Honeycutt. 2017. Opinion: Why we need a national living soil repository. *Proceedings of the National Academy of Sciences* 114 (52): 13587–13590.

Metcalf, J., Z. Xu, A. Bouslimani, P. Dorrestein, D. Carter, and R. Knight. 2017. Microbiome tools for forensic science, trends in biotechnology. *Special Issue: Environmental Biotechnology* 35 (9): 814–823.

Mitchell, A., F. Bucchini, G. Cochrane, H. Denise, P. ten Hoopen, M. Fraser, S. Pesseat, S. Potter, M. Scheremetjew, P. Sterk, and R. Finn. 2016. EBI metagenomics in 2016 - an expanding and evolving resource for the analysis and archiving of metagenomic data. *Nucleic Acids Research* 44 (1): D595–D603.

National Research Council. 2011. *Review of the scientific approaches used during the FBI's investigation of the 2001 anthrax letters*. Washington, DC: National Academies.

———. 2014. *Science needs for microbial forensics: Developing initial international research priorities*. Washington, DC: National Academies.

Nielsen, A., and C. Voigt. 2018. Deep learning to predict the lab-of-origin of engineered DNA. *Nature Communications* 9 (1): 3135–3145.

Oh, J., A. Byrd, C. Deming, S. Conlan, H. Kong, and J. Segre. 2014. Biogeography and individuality shape function in the human skin metagenome. *Nature* 514 (7520): 59–64.

Oh, J., A. Byrd, M. Park, H. Kong, and J. Segre. 2016. Temporal stability of the human skin microbiome. *Cell* 165 (4): 854–866.

Rasko, D., P. Worsham, T. Abshire, S. Stanley, J. Bannan, M. Wilson, R. Langham, R. Decker, L. Jiang, T. Read, A. Phillippy, S. Salzberg, M. Pop, M. van Ert, L. Kenefic, P. Keim, C. Fraser-Liggett, and J. Ravel. 2011. Bacillus Anthracis comparative genome analysis in support of the amerithrax investigation. *PNAS* 108 (12): 5027–5032.

Riedel, S. 2005. Anthrax: A continuing concern in the era of bioterrorism. *Proceedings Baylor University Medical Center* 8 (3): 234–243.

Ross, A., K. Müller, J. Weese, and J. Neufeld. 2018. Comprehensive skin microbiome analysis reveals the uniqueness of human skin and evidence for phylosymbiosis within the class mammalia. *PNAS* 115 (25): E5786–E5795.

Smith, R., T. Jeffries, B. Roudnew, A. Fitch, J. Seymour, M. Delpin, K. Newton, M. Brown, and J. Mitchell. 2012. Metagenomic comparison of microbial communities inhabiting confined and unconfined aquifer ecosystems: Aquifer metagenomics. *Environmental Microbiology* 14 (1): 240–253.

Song, S., C. Lauber, E. Costello, C. Lozupone, G. Humphrey, D. Berg-Lyons, J. Caporaso, D. Knights, J. Clemente, S. Nakielny, J. Gordon, N. Fierer, and R. Knight. 2013. Cohabiting family members share microbiota with one another and with their dogs. *ELife* 2 (4): e00458.

Stenson, P., M. Matthew, E.V. Ball, K. Evans, M. Hayden, S. Heywood, M. Hussain, A.D. Phillips, and D. Cooper. 2017. The human gene mutation database: Towards a comprehensive repository of inherited mutation data for medical research, genetic diagnosis and next-generation sequencing studies. *Human Genetics* 136 (6): 665–677.

The United States Department of Justice. 2010. Amerithrax investigative summary.

Tringe, S., C. von Mering, A. Kobayashi, A. Salamov, K. Chen, H. Chang, M. Podar, J. Short, E. Mathur, J. Detter, P. Bork, P. Hugenholtz, and E. Rubin. 2005. Comparative metagenomics of microbial communities. *Science* 308 (5721): 554–557.

Ugalmugale, S., and R. Swain. 2016. *Biotechnology market share size 2018–2024 growth forecast report*, GMI784, Global Market Insights. https://www.gminsights.com/industry-analysis/biotechnology-market. Accessed 15 June 2019.

Vieites, M., M. Guazzaroni, A. Beloqui, P. Golyshin, and M. Ferrer. 2009. Metagenomics approaches in systems microbiology. *FEMS Microbiology Reviews* 33 (1): 236–255.

Whitaker, R. 2003. Geographic barriers isolate endemic populations of hyperthermophilic archaea. *Science* 301 (5635): 976–978.

Ziogas, D., I. Kyrochristos, and D. Roukos. 2018. Next-generation sequencing: From conventional applications to breakthrough genomic analyses and precision oncology. *Expert Review of Medical Devices* 15 (1): 1–3.

Printed in the United States
by Baker & Taylor Publisher Services